ELECTROMAGNETIC WAVES PROPAGATION IN COMPLEX MATTER

Edited by **Ahmed A. Kishk**

Electromagnetic Waves Propagation in Complex Matter
Edited by Ahmed A. Kishk

Published by InTech
Janeza Trdine 9, 51000 Rijeka, Croatia

Publishing Process Manager Iva Lipovic
Technical Editor Teodora Smiljanic

First published June, 2011
Printed in Croatia

A free online edition of this book is available at www.intechopen.com
Additional hard copies can be obtained from orders@intechweb.org

Electromagnetic Waves Propagation in Complex Matter, Edited by Ahmed A. Kishk
p. cm.
ISBN 978-953-307-445-0

For EU product safety concerns:
IN TECH d.o.o., Prolaz Marije Krucifikse Kozulić 3, 51000 Rijeka, Croatia,
info@intechopen.com or visit our website at intechopen.com.

Meet the editor

Ahmed A. Kishk is a Professor at Concordia University, Montreal, Canada (since 2011). He is an IEEE Fellow since 1998, Fellow of the Electromagnetic Academy, Fellow of Applied Electromagnetic Society. He is now an Editor of Antennas & Propagation Magazine. He was an Editor-in-Chief of the ACES Journal from 1998 to 2001. He was the chair of Physics and Engineering division of the Mississippi Academy of Science (2001-2002). Dr. Kishk is a recipient of many awards. Dr. Kishk received the 1995 and 2006 outstanding paper awards for ACES Journal. He received the 1997 Outstanding Engineering Educator Award from Memphis section of the IEEE. He received the Outstanding Engineering Faculty Member of the Year in 1998 and 2009, Faculty research award for outstanding performance in research in 2001 and 2005. He received the Award of Distinguished Technical Communication for the entry of IEEE Antennas and Propagation Magazine, 2001. He received The Valued Contribution Award for outstanding Invited Presentation in ACES. He received the Microwave Theory and Techniques Society Microwave Prize 2004.

Contents

Preface

This book is based on the contributions of several authors in electromagnetic waves propagations. Several issues are considered. The contents of most of the chapters are highlighting non classic presentation of wave propagation and interaction with matters. This book bridges the gap between physics and engineering in these issues. Each chapter keeps the author notation that the reader should be aware of as he reads from chapter to the other. The author's notations are kept in order to eliminate any possible unintentional errors that might lead to confusion. We would like to thank all authors for their excellent contributions.

In chapter 1, the problem of radiation of arbitrarily distributed currents in boundless uniaxial anisotropic media is considered through the method of generalized solutions of the system of Maxwell's equations in an exact form. The solution resolves into two independent solutions. The first corresponds to the isotropic solution for currents directed along the crystal axis, while the second corresponds to the anisotropic solution when the currents are perpendicular to the axis. The independent solutions define the corresponding polarization of electromagnetic waves. The generalized solutions obtained in vector form by the fundamental solutions of the Maxwell's equations are valid for any values of the elements of the permeability tensor, as well as for sources of the electromagnetic waves described by discontinuous and singular functions. The solutions can be also represented with the help of vector potentials by the corresponding fundamental solutions. The problems for tensors of the dielectric and magnetic permeabilities are considered separately. In particular, the solutions for elementary electric and magnetic dipoles have been deduced. Through the use of the expressions for current density of the point magnetic and electric dipoles using delta-function representations, the formulae for the radiated electromagnetic waves, as well as the corresponding radiation patterns, are derived. The obtained solution in the anisotropic case yields the well-known solutions for the isotropic case as a limiting case. The radiation patterns for Hertz radiator and point magnetic dipole are represented. Directivity diagrams of radiation of point magnetic and electric dipoles are constructed at parallel and perpendicular directions of an axis of a crystal.Validity of the solutions has been checked up on balance of energy by integration of energy flow on sphere. The numerical calculation of the solution of Maxwell's equations shows that it satisfies the energy conservation law, i.e. the time average value of energy flux through the surface of a sphere with a point dipole placed at its center remains independent of the radius of the sphere. Numerical calculation shows that its values keep with the high accuracy. The rigorous solving of system of Maxwell equations in an anisotropic media can be used in construction of the integral equations for solving the class of respective boundary problems.

In chapter 2, the consistent physic-mathematical model of propagation of an electromagnetic wave in a heterogeneous medium is constructed using the generalized wave equation and the Dirichlet theorem. Twelve conditions at the interfaces of adjacent media are obtained and justified without using a surface charge and surface current in explicit form. The conditions are fulfilled automatically in each section of the heterogeneous medium and are conjugate, which make it possible to use through-counting schemes for calculations. The effect of concentration of "medium-frequency" waves with a length of the order of hundreds of meters at the fractures and wedges of domains of size 1-3 μm is established. Numerical calculations of the total electromagnetic energy on the wedges of domains are obtained. It is shown that the energy density in the region of wedges is maximum and in some cases may exert an influence on the motion, sinks, and the source of dislocations and vacancies and, in the final run, improve the near-surface layer of glass due to the "micromagnetoplastic" effect. The results of these calculations are of special importance for medicine, in particular, when microwaves are used in the therapy of various diseases. For a small, on the average, permissible level of electromagnetic irradiation, the concentration of electromagnetic energy in internal angular structures of a human body (cells, membranes, neurons, interlacements of vessels, etc) is possible. A consistent physicomathematical model of interaction of nonstationary electric and thermal fields in a layered medium with allowance for mass transfer is constructed. The model is based on the methods of thermodynamics and on the equations of an electromagnetic field and is formulated without explicit separation of the charge carriers and the charge of an electric double layer. The relations for the electric-field strength and the temperature are obtained, which take into account the equality of the total currents and the energy fluxes, to describe the electric and thermal phenomena in layered media where the thickness of the electric double layer is small compared to the dimensions of the object under study. The heating of an electrochemical cell with allowance for the influence of the electric double layer at the metal-electrolyte interface is numerically modeled. The calculation results are in satisfactory agreement with experimental data.

Chapter 3 demonstrates the fabrication process, structure and magnetic properties of metal (alloy) coated cenosphere composites by heterogeneous precipitation thermal reduction method to form metal-coated core-shell structural composites. These composites can be applied for advanced functional materials such as electromagnetic wave absorbing materials.

In chapter 4, a novel approach based on a periodic distribution of planar or quasi-planar chiral particles is proposed for the design of artificial chiral media. The metal particles are replaced by dielectric ones, so that a high contrast between the permittivity of the new dielectric particles and the host medium is achieved. This approach would allow the design of materials with lower losses and more simply scalable in frequency. Both approaches are presented by dealing with the aspects of design and realization of different "basic cells". Numerical analysis in time and frequency-domain using commercial software program are used to treat the propagation. Characterization of the media and their propagation properties are verified experimentally.

Soil is a complex, potentially heterogeneous, lossy, and dispersive material. Propagation and scattering of electromagnetic waves in soil is, hence, more challenging than air or other less complex media. Chapter 5 explains the fundamentals of modeling electromagnetic wave propagation and scattering in soil by solving Maxwell's equations using a finite difference time domain (FDTD) model. The chapter explains how the lossy and dispersive soil medium (in both dry and water-saturated conditions), two different types of transmitting antennae (a monopole and a dipole), and required absorbing boundary conditions can be modeled. A sample problem is simulated to demonstrate the scattering effects of a dielectric anomaly in soil. Thereafter, the details about preparation and conduct of an experimental simulation are discussed. The precautions necessary to perform a repeatable experiment is explained in detail as well. The results of the numerically simulated example is compared and validated against experimental data.

In Chapter 6, Einstein field equations (EFE) are extended to explain electromagnetism by charge distributions in like manner, which should not be confused with the Einstein-Maxwell equations, in which electromagnetic fields energy contents were added onto those as attributed to the presence of matter, to account for gravitational motions. This chapter is substituting the term "electric charge" for energy, and electromagnetism for gravity, i.e., a geometrization of the electromagnetic force. Einstein field equations describe one space-time, but in this chapter two are proposed: one for "particles" and the other for "waves;" to wit, there are two gravitational constants. The gravitational motions in a "combined space-time 4-manifold" are unified. Also, the reader finds that the chapter proves that electromagnetic fields as produced by charges, in analogy with gravitational fields as produced by energies, cause space-time curvatures, not because of the energy contents of the fields but because of the Coulomb potential of the charges. As a result, a special constant of proportionality between an electromagnetic energy-momentum tensor and Einstein tensor are derived.

In chapter 7, a ray tracing approach based on the Jacobi-Hamiltonian theory is used as rays are defined by their characteristic vector and the slowness (inverse of phase velocity) vector along the ray. Both are functions of the integration variable and of the initial conditions (launching point and direction). The characteristic vector satisfies the Hamilton differential equations. The Hamiltonian function describes the wave propagation in the considered medium. Rays trajectories are functions of the unique integration parameter. Hamilton-Jacobi theory guarantees that there is always a domain of representation in which solutions are monodrome functions. Here, the wave fronts are monodrome function of the ray launching angle. Amplitude is computed through paraxial rays. Caustics arise when the ray field folds. These events are carefully accounted so that proper phase shifts can be applied to the field.

In chapter 8, an overview of the whole process to compute electromagnetic field levels based on the high frequency technique modified equivalent current approximation is presented. Moreover, three new fast algorithms are briefly described to solve the visibility problem. Those are used to complete a modified equivalent current approximation. They can also be helpful in other disciplines of engineering.

In chapter 9, the field theory on electrodynamics and derive of the Maxwell equation and the Lorenz force are introduced. The multiconductor transmission-line (MCTL) equations for the TEM mode are developed. Solutions of the MCTL equations for a normal mode without coupling to the common and antenna modes are provided as well as a solution of one antenna system for emission and absorption of radiation. A three-conductor transmission-line system and the symmetrization for the decoupling of the normal mode from the common and antenna modes are discussed.

A fundamental and accurate technique to compute the propagation constant of waves in a lossy rectangular waveguide is proposed in chapter 10. The formulation is based on matching the fields to the constitutive properties of the material at the boundary. The electromagnetic fields are used in conjunction of the concept of surface impedance to derive transcendental equations, whose roots give values for the wavenumbers in the transverse directions of the waveguide axis for different transverse electric and transverse magnetic modes. The new boundary-matching method is validated by comparing the attenuation of the dominant mode with the transmition coefficients measurement, as well as that obtained from the power-loss method. The attenuation curve plotted using the new method matches with the power-loss method at a reasonable range of frequencies above the cutoff. There are, however, two regions where both curves are found to differ significantly. At frequencies below the cutoff, the power-loss method diverges to infinity with a singularity at cutoff frequency. The new method, however, shows that the signal increases to a highly attenuating mode as the frequencies drop below the cutoff. Such result agrees very closely with the measurement result, therefore, verifying the validity of the new method. At frequencies above 100 GHz, the attenuation obtained using the new method increases beyond that predicted by the power-loss method. At frequency above the millimeter wavelengths, the field in a lossy waveguide can no longer be approximated to those of the lossless case. The additional loss predicted by the new boundary-matching method is attributed to the presence of the longitudinal electric field component in hybrid modes.

Chapter 11 is concerned with a numerical problem which can be solved by parallel regular computations performed in points of a rectangular mesh that spans over irregular computational areas. A hierarchical approach to the optimized program macro data flow graph design for execution in parallel systems is presented. The presented Re-deployment with Connectivity-based Distributed Node Clustering (RCDC) algorithm consists of two independent methods for the finite difference time domain (FDTD) data flow graph optimization: the cell re-deployment and the Connectivity-based Distributed Node Clustering (CDC) algorithm. There are several differences between these two methods. The first method is fully centralized and the macro data flow graph is created in three phases: computational area partitioning, merging and re-deployment. The CDC method is decentralized with only local knowledge of the simulation area. In the RCDC algorithm, both methods are merged in order to obtain better parallel simulation speedup (comparable to the speedup obtained in the CDC) and to shorten the execution time of the optimization. It turned out that such a hierarchical combination of the two algorithms has improved partitioning of data flow graphs for the FDTD problem, and additionally, such hierarchical optimization takes significantly less time than the CDC method.

Ahmed Kishk
Professor
Departmentof Electrical and Computer Engineering
Concordia University
Teir 1 Canada Research Chair
Montreal, Quebec, Canada

Part 1

Solutions of Maxwell's Equations in Complex Matter

1

The Generalized Solutions of a System of Maxwell's Equations for the Uniaxial Anisotropic Media

Seil Sautbekov
Eurasian National University
Kazakhstan

1. Introduction

Media with anisotropic properties are widely used in modern radio electronics, an astrophysics, and in plasma physics. Anisotropic materials have found wide application in the microcircuits working on ultrahigh frequencies. Thin films from monocrystals are effectively used as waveguide's systems. At present, artificial anisotropic materials are especially actual for the design of microwave integrated circuits and optical devices. The technology advances are making the production of substrates, dielectric anisotropic films and anisotropic material filling more and more convenient.

Due to the complexity caused by the parameter tensors, the plane wave expansion (Born & Wolf, 1999) is often used in the analysis of anisotropic media. The standard mathematical technique for treating propagation through a homogeneous anisotropic medium is to investigate the independent time-harmonic plane wave solutions of Maxwell's equations. And consequently, the Fourier transform is widely applied (Chen, 1983; Kong, 1986; Ren, 1993; Uzunoglu et al., 1985).

The radiation field of a dipole in a anisotropic medium is considered in greater detail and devised by (Bunkin, 1957; Clemmow, 1963a;b; Kogelnik & Motz, 1963). It is shown (Clemmow, 1963a;b) that each such field is related by a simple scaling procedure to a corresponding vacuum field. The vacuum field is expressed as the superposition of a transverse magnetic field, in which the magnetic vector is everywhere perpendicular to the axis of symmetry of the anisotropic medium, and a coplanar transverse electric field; and different scaling is applied separately to each partial field. But because of the inevitable complication of any such general analysis it also seemed desirable to isolate the simplest non-trivial case.

Using these methods, (Uzunoglu et al., 1985) found the solution of the vector wave equation in cylindrical coordinates for a gyroelectric medium. (Ren, 1993) continued that work for spherical coordinates in a similar procedure and obtained spherical wave functions and dyadic Green's functions in gyroelectric media. The dyadic Green's functions for various kinds of anisotropic media with different structures have been studied by many authors (Barkeshli, 1993; Cottis, 1995; Lee & Kong, 1983; Weiglhofer, 1990; 1993). The problems, however, are mostly analyzed in spectral domain in terms of Fourier transform, due to the difficulty of finding the expansion of the dyadic Green's functions in terms of vector wave functions for anisotropic media. It shows the necessity of better characterizing the anisotropic media and producing more realistic models for the components that use them.

The purpose of this chapter is to obtain explicit expressions for the stationary problem of the field produced by a given distribution of external currents in an infinite homogeneous uniaxial anisotropic medium without using the scaling procedure to a corresponding vacuum field and the dyadic Green's functions.

This chapter is organized as follows: in the following section, section 2, we present the method of generalized function to solve the Maxwell's equations for isotropic media. To demonstrate the method we deduce the general solutions of Maxwell's equations by means of the unique Green's function. Interrelation between the Green's function and electrodynamic potentials and also the Hertz potential is shown. Some forms of the Green's function for lower dimension are presented. In section 3, the method of generalized functions is employed for solving the problem of uniaxial crystals. We introduce there a stationary electromagnetic field ($\boldsymbol{E}$, $\boldsymbol{H}$) of frequency ω and formulate the problem. We use the generalized method of the Fourier transformation for solving the matrix form of Maxwell's equations. Radiation of electric and magnetic dipole is considered and their directivity diagrams are obtained. In section 4, we adduce all points of section 3 only for magnetic media. Finally, in section 5 conclusions and future related research are presented. In this chapter, we can note the following main results. The fundamental solutions of a system of Maxwell's equations for uniaxial anisotropic media are obtained. Due to the fundamental solutions, general exact expression of an electromagnetic field in boundless uniaxial crystal is obtained in the vector type by the method of generalized functions. The results are valid for any values of the elements of the permeability tensor, as well as for sources of the electromagnetic waves described by discontinuous and singular functions. In particular, the solutions for elementary electric and magnetic dipoles have been deduced. Through the use of the expressions for current density of the point magnetic and electric dipoles using delta-function representations, the formulae for the radiated electromagnetic waves, as well as the corresponding radiation patterns, are derived. The obtained solution in the anisotropic case yields the well-known solutions for the isotropic case as a limiting case. Validity of the solutions have been checked up on balance of energy by integration of energy flow on sphere. The numerical calculation of the solution of Maxwell's equations shows that it satisfies the energy conservation law.

By means of the method of generalized functions it is possible to represent the electromagnetic field in the integral form with respect to their values on an arbitrary closed surface. Subsequently analytical solutions of Maxwell's equations obtained in this chapter for unbounded anisotropic media allow to construct the integral equations for solving the class of corresponding boundary problems. The obtained solutions can be easily generalized taking into account magnetic currents. Because electric and magnetic currents are independent, it makes it possible to decompose a solution on separate independent solutions. The solutions can be also represented with the help of vector potentials for the corresponding fundamental solutions.

2. Method of generalized functions

To present the method of generalized functions (Alekseyeva & Sautbekov, 1999; 2000) we shall consider stationary Maxwell's equations for isotropic media

$$\begin{cases} \nabla \times \boldsymbol{H} + i\omega \boldsymbol{D} = \boldsymbol{j}, \\ \nabla \times \boldsymbol{E} - i\omega \boldsymbol{B} = 0. \end{cases} \tag{1}$$

The linear relation between the induction and the intensity of electric field in isotropic dielectric mediums is:

$$\boldsymbol{D} = \varepsilon\varepsilon_0 \boldsymbol{E} \tag{2}$$

and the vector of magnetic induction is given by:

$$\boldsymbol{B} = \mu\mu_0 \boldsymbol{H}, \tag{3}$$

μ, ε are the relative magnetic and dielectric permeability respectively, $\boldsymbol{E}$ and $\boldsymbol{H}$ are the intensity of electric and magnetic fields respectively, and $\boldsymbol{j}$ is vector of current density.
We will present the system (1) in the matrix form:

$$\mathbf{MU} = \mathbf{J}, \tag{4}$$

where

$$\mathbf{M} = \begin{pmatrix} i\omega\varepsilon_0\varepsilon \boldsymbol{I} & \boldsymbol{G}_0 \\ \boldsymbol{G}_0 & -i\omega\mu_0\mu \boldsymbol{I} \end{pmatrix}, \qquad \boldsymbol{G}_0 = \begin{pmatrix} 0 & -\partial_z & \partial_y \\ \partial_z & 0 & -\partial_x \\ -\partial_y & \partial_x & 0 \end{pmatrix}, \tag{5}$$

$$\mathbf{U} = \begin{pmatrix} \boldsymbol{E} \\ \boldsymbol{H} \end{pmatrix}, \quad \mathbf{J} = \begin{pmatrix} \boldsymbol{j} \\ \boldsymbol{0} \end{pmatrix}, \quad \boldsymbol{E} = \begin{pmatrix} E_x \\ E_y \\ E_z \end{pmatrix}, \quad \boldsymbol{H} = \begin{pmatrix} H_x \\ H_y \\ H_z \end{pmatrix}, \quad \boldsymbol{j} = \begin{pmatrix} j_x \\ j_y \\ j_z \end{pmatrix}, \quad \boldsymbol{0} = \begin{pmatrix} 0 \\ 0 \\ 0 \end{pmatrix}, \tag{6}$$

ω is the constant frequency of electromagnetic field, $\mathbf{M}$ is Maxwell's operator, $\boldsymbol{I}$ is a identity matrix 3×3.
A method generalized functions based on the theory of the generalized function of the Fourier transformation is used for solving the matrix equation (4) (Vladimirov, 2002):

$$\tilde{\boldsymbol{E}}(\boldsymbol{k}) = \mathbf{F}[\boldsymbol{E}(\boldsymbol{r})] = \int_{\mathbb{R}^3} \boldsymbol{E}(\boldsymbol{r}) \exp(-i\boldsymbol{k}\boldsymbol{r}) dV \tag{7}$$

where

$$\boldsymbol{E}(\boldsymbol{r}) = \mathbf{F}^{-1}[\tilde{\boldsymbol{E}}(\boldsymbol{k})] = \int_{\mathbb{R}^3} \tilde{\boldsymbol{E}}(\boldsymbol{k}) \exp(i\boldsymbol{k}\boldsymbol{r}) \frac{d^3k}{(2\pi)^3}, \tag{8}$$

$$\boldsymbol{k} = (k_x, k_y, k_z), \quad d^3k = dk_x k_y k_z, \quad dV = dxdydz, \quad \boldsymbol{r} = (x, y, z).$$

By means of direct Fourier transformation we will write down the system of equations (1) or (4) in matrix form:

$$\tilde{\mathbf{M}}\tilde{\mathbf{U}} = \tilde{\mathbf{J}}, \tag{9}$$

$$\tilde{\mathbf{M}} = \begin{pmatrix} i\varepsilon_0\varepsilon\omega \boldsymbol{I} & \tilde{G}_0 \\ \tilde{G}_0 & -i\mu_0\mu\omega \boldsymbol{I} \end{pmatrix}, \qquad \tilde{G}_0 = i \begin{pmatrix} 0 & -k_z & k_y \\ k_z & 0 & -k_x \\ -k_y & k_x & 0 \end{pmatrix}. \tag{10}$$

The solution of the problem is reduced to determination of the system of the linear algebraic equations relative to Fourier-components of the fields, where $\tilde{\mathbf{U}}$ is defined by means of inverse matrix $\tilde{\mathbf{M}}^{-1}$:

$$\tilde{\mathbf{U}} = \tilde{\mathbf{M}}^{-1}\tilde{\mathbf{J}}. \tag{11}$$

By introducing new function according to

$$\tilde{\psi}_0 = \frac{1}{k_0^2 - k^2}, \tag{12}$$

we define the inverse matrix:

$$\tilde{\mathbf{M}}^{-1} = \tilde{\psi}_0 \begin{pmatrix} -(\varepsilon_0\varepsilon)^{-1}\tilde{\mathbf{G}}_1 & -\tilde{\mathbf{G}}_0 \\ -\tilde{\mathbf{G}}_0 & (\mu_0\mu)^{-1}\tilde{\mathbf{G}}_1 \end{pmatrix}, \tag{13}$$

where

$$\tilde{G}_1 = \frac{1}{i\omega} \begin{pmatrix} k_1^2 - k_0^2 & k_1k_2 & k_1k_3 \\ k_1k_2 & k_2^2 - k_0^2 & k_2k_3 \\ k_1k_3 & k_2k_3 & k_3^2 - k_0^2 \end{pmatrix}, \tag{14}$$

$$k_0 \equiv \omega\sqrt{\varepsilon_0\varepsilon\mu_0\mu}, \qquad k^2 = k_x^2 + k_y^2 + k_z^2. \tag{15}$$

By considering the inverse Fourier transformation

$$\mathbf{M}^{-1} = \mathrm{F}^{-1}[\tilde{\mathbf{M}}^{-1}], \quad \mathbf{J} = \mathrm{F}^{-1}[\tilde{\mathbf{J}}], \quad \mathbf{U} = \mathrm{F}^{-1}[\tilde{\mathbf{U}}] \tag{16}$$

and using the property of convolution:

$$\mathrm{F}^{-1}[\tilde{\mathbf{M}}^{-1}\tilde{\mathbf{J}}] = \mathbf{M}^{-1} * \mathbf{J}, \tag{17}$$

where symbol "$*$" denotes the convolution on coordinates x, y, z, it is possible to get the solution of the Maxwell equations (4) as:

$$\mathbf{U} = \mathbf{M}^{-1} * \mathbf{J}, \tag{18}$$

where

$$\mathbf{M}^{-1} = \begin{pmatrix} -(\varepsilon_0\varepsilon)^{-1}\boldsymbol{G}_1 & -\boldsymbol{G}_0 \\ -\boldsymbol{G}_0 & (\mu_0\mu)^{-1}\boldsymbol{G}_1 \end{pmatrix}\psi_0, \quad \boldsymbol{G}_1 = \frac{-1}{i\omega}\begin{pmatrix} \frac{\partial^2}{\partial x^2} + k_0^2 & \frac{\partial^2}{\partial x\partial y} & \frac{\partial^2}{\partial x\partial z} \\ \frac{\partial^2}{\partial x\partial y} & \frac{\partial^2}{\partial y^2} + k_0^2 & \frac{\partial^2}{\partial y\partial z} \\ \frac{\partial^2}{\partial x\partial z} & \frac{\partial^2}{\partial y\partial z} & \frac{\partial^2}{\partial z^2} + k_0^2 \end{pmatrix} \tag{19}$$

or

$$\begin{pmatrix} \boldsymbol{E} \\ \boldsymbol{H} \end{pmatrix} = \begin{pmatrix} (i\varepsilon_0\varepsilon\omega)^{-1}(\nabla\nabla + k_0^2)(\boldsymbol{j} * \psi_0) \\ -\nabla \times (\boldsymbol{j} * \psi_0) \end{pmatrix} = \begin{pmatrix} (\varepsilon_0\varepsilon)^{-1}\nabla\rho * \psi_0 - i\mu_0\mu\omega\psi_0 * \boldsymbol{j} \\ \boldsymbol{j} \times * \nabla\psi_0 \end{pmatrix}, \tag{20}$$

according to the charge conservation low

$$\nabla \boldsymbol{j} - i\omega\rho = 0. \tag{21}$$

Here ρ is charge density, ψ_0 is the Green's function or a fundamental solution of the Helmholtz operator for isotropic medium (Vladimirov, 2002):

$$\psi_0 = \mathrm{F}^{-1}[\tilde{\psi}_0] = -\frac{1}{4\pi}\frac{e^{ik_0 r}}{r}, \qquad r = \sqrt{x^2 + y^2 + z^2} \tag{22}$$

which satisfy the equation:

$$(\triangle + k_0^2)\psi_0 = \delta(\boldsymbol{r}). \tag{23}$$

On the other hand, we can obtain Eqn. (23), after application of the inverse Fourier transformation from Eqn. (12).

Thus, the general solution of the system of stationary Maxwell's equations for a three-dimensional unbounded isotropic media was deduced by means of solely one fundamental solution ψ_0. Hence, the solution preserves the same form concerning the fundamental solution ψ_0 for two-dimensional and one-dimensional problems. The useful forms of the fundamental solutions are adduced below by means of Fourier transformations:

$$\psi_0(x, y, k_z, \omega) = -\frac{i}{4}\mathrm{H}_0^{(1)}\left(\sqrt{k_0^2 - k_z^2}\sqrt{x^2 + y^2}\right) \quad \text{(two-dimensional case)},$$

$$\psi_0(k_x, y, k_z, \omega) = \frac{\exp(ih|y|)}{2ih}, \qquad h = \sqrt{k_0^2 - k_x^2 - k_z^2} - \text{const} \quad \text{(one-dimensional case)}.$$

2.1 Electrodynamic potentials and Hertz vector

It should be noted that all electrodynamic quantities of isotropic media can be expressed by function ψ_0, including the electrodynamic potentials and Hertz vector. By designating the scalar potential φ and vector potential $\boldsymbol{A}$:

$$\varphi = -(\varepsilon_0\varepsilon)^{-1}\psi_0 * \rho, \qquad \boldsymbol{A} = -\mu_0\mu\psi_0 * \boldsymbol{j}, \tag{24}$$

the solution (18) or (20) can be presented in known form by vector potential

$$\boldsymbol{E} = -\nabla\varphi + i\omega\boldsymbol{A}, \qquad \boldsymbol{H} = (\mu_0\mu)^{-1}\nabla \times \boldsymbol{A}. \tag{25}$$

It should be noted that physical sense of Lorentz gauge of potentials consists in the charge conservation law (21), indeed

$$\nabla\boldsymbol{A} - i\omega\varepsilon_0\varepsilon\mu_0\mu\varphi = -\mu_0\mu\psi_0 * (\nabla\boldsymbol{j} - i\omega\rho) = 0. \tag{26}$$

Similarly, by designating Hertz vector

$$\boldsymbol{\Pi} = (i\omega\varepsilon_0\varepsilon)^{-1}\boldsymbol{j} * \psi_0 \tag{27}$$

solution (18) can be written as

$$\begin{aligned} \boldsymbol{E} = (\nabla\nabla + k_0^2)\boldsymbol{\Pi} = \nabla \times \nabla \times \boldsymbol{\Pi} + (i\varepsilon_0\varepsilon\omega)^{-1}\boldsymbol{j}, \\ \boldsymbol{H} = -i\varepsilon_0\varepsilon\omega \bigtriangledown \times\boldsymbol{\Pi}. \end{aligned} \tag{28}$$

It is easy to take notice that the relation between the electrodynamic potentials and Hertz potential is

$$\boldsymbol{A} = -i\omega\varepsilon_0\varepsilon\mu_0\mu\boldsymbol{\Pi}, \qquad \varphi = -\nabla\boldsymbol{\Pi}. \tag{29}$$

3. The generalized solutions of Maxwell equations for the uniaxial crystal

The exact analytical solutions of Maxwell's equations are constructed by means of method of generalized functions in vector form for unlimited uniaxial crystals (Sautbekov et al., 2008). The fundamental solutions of a system of Maxwell's equations for uniaxial crystals are obtained. The solution of the problem was analyzed in Fourier space and closed form analytical solutions were derived in Section (3.2), above. Then, when the current distribution is defined in such a medium, the corresponding radiated electric and magnetic fields can be calculated anywhere in space. In particular, the solutions for elementary electric dipoles have been deduced in Section(3.3), and the radiation patterns for Hertz radiator dipole are represented. The governing equations and radiation pattern in the case of an unbounded isotropic medium were obtained as a special case. Validity of the solutions have been checked up on balance of energy by integration of energy flow on sphere.

3.1 Statement of the problem

The electric and magnetic field strengths satisfy system of stationary the Maxwell's equations (1), which is possible to be presented in matrix form (4), where

$$\mathbf{M} = \begin{pmatrix} -i\omega\varepsilon_0\hat{\varepsilon} & \boldsymbol{G}_0 \\ \boldsymbol{G}_0 & i\mu_0\mu\boldsymbol{I} \end{pmatrix}, \qquad \hat{\varepsilon} = \begin{pmatrix} \varepsilon_1 & 0 & 0 \\ 0 & \varepsilon & 0 \\ 0 & 0 & \varepsilon \end{pmatrix}. \tag{30}$$

The relation between the induction and the intensity of electric field in anisotropic dielectric mediums is:

$$\boldsymbol{D} = \hat{\varepsilon}\varepsilon_0\boldsymbol{E}.$$

If we choose a frame in main axes of dielectric tensor, the constitutive equation will be written as:

$$D_x = \varepsilon_1\varepsilon_0 E_x, \qquad D_y = \varepsilon\varepsilon_0 E_y, \qquad D_z = \varepsilon\varepsilon_0 E_z.$$

The elements of the dielectric permeability tensor $\hat{\varepsilon}$ correspond to a one-axis crystal, moreover the axis of the crystal is directed along axis x. Moreover, it is required to define the intensities of the electromagnetic field $\boldsymbol{E}, \boldsymbol{H}$ in the space of generalized function.

3.2 Problem solution

By means of direct Fourier transformation, we write down the system of equations in matrix form (9). The solution of the problem is reduced to determination of the system of the linear algebraic equations relative to Fourier-components of the fields, where $\tilde{\mathbf{U}}$ is defined by means of inverse matrix $\tilde{\mathbf{M}}^{-1}$. By introducing new functions according to $\tilde{\psi}_0$ (12) and

$$\tilde{\psi}_1 = (k_n^2 - \frac{\varepsilon_1}{\varepsilon}k_x^2 - k_y^2 - k_z^2)^{-1}, \qquad \tilde{\psi}_2 = (\frac{\varepsilon_1}{\varepsilon} - 1)\tilde{\psi}_1\tilde{\psi}_0 \tag{31}$$

the components of the electromagnetic field after transformations in Fourier space can be written as follows:

$$\begin{cases} \tilde{E}_x = (i\varepsilon\varepsilon_0\omega)^{-1}(k_0^2\tilde{j}_x - k_x\boldsymbol{k}\tilde{\boldsymbol{j}})\tilde{\psi}_1, \\ \tilde{E}_y = (i\varepsilon\varepsilon_0\omega)^{-1}(k_0^2\tilde{\psi}_0\tilde{j}_y - k_y\tilde{\psi}_1\boldsymbol{k}\tilde{\boldsymbol{j}} - k_0^2k_y\tilde{\psi}_2\boldsymbol{k}\tilde{\boldsymbol{j}}_\perp), \\ \tilde{E}_z = (i\varepsilon\varepsilon_0\omega)^{-1}(k_0^2\tilde{\psi}_0\tilde{j}_z - k_z\tilde{\psi}_1\boldsymbol{k}\tilde{\boldsymbol{j}} - k_0^2k_z\tilde{\psi}_2\boldsymbol{k}\tilde{\boldsymbol{j}}_\perp), \end{cases} \tag{32}$$

$$\begin{cases} \tilde{H}_x = -i\tilde{\psi}_0(\boldsymbol{k} \times \tilde{\boldsymbol{j}})_x, \\ \tilde{H}_y = -i(k_z j_x \tilde{\psi}_1 - k_x k_z \tilde{\psi}_2 \boldsymbol{k}\tilde{\boldsymbol{j}}_\perp + k_x \tilde{j}_z \tilde{\psi}_0), \\ \tilde{H}_z = -i(k_x \tilde{j}_y \tilde{\psi}_0 + k_y \tilde{j}_x \tilde{\psi}_1 + k_x k_y \tilde{\psi}_2 \boldsymbol{k}\tilde{\boldsymbol{j}}_\perp), \end{cases} \tag{33}$$

where,

$$\tilde{\boldsymbol{j}} = \tilde{\boldsymbol{j}}_\perp + \tilde{\boldsymbol{j}}_0, \qquad \tilde{\boldsymbol{j}}_0 = (\tilde{j}_x, 0, 0), \qquad \tilde{\boldsymbol{j}}_\perp = (0, \tilde{j}_y, \tilde{j}_z), \qquad k_n^2 = k_0^2 \varepsilon_1 / \varepsilon, \tag{34}$$

that k_n is the propagation constant along the axis of the crystal (x-axis), and k_0 is the propagation constant along the y and z axis. It is possible to present the electromagnetic fields in vector type:

$$\tilde{\boldsymbol{E}} = -\frac{i}{\varepsilon_0 \varepsilon \omega} \left(k_0^2 \{ \tilde{\boldsymbol{j}}_\perp \tilde{\psi}_0 + \tilde{\boldsymbol{j}}_0 \tilde{\psi}_1 + (k_x \boldsymbol{k}\tilde{\boldsymbol{j}}_\perp - \boldsymbol{k}(\boldsymbol{k}\tilde{\boldsymbol{j}}_\perp)) \tilde{\psi}_2 \} - \boldsymbol{k}(\boldsymbol{k}\tilde{\boldsymbol{j}}) \tilde{\psi}_1 \right), \tag{35}$$

$$\tilde{\boldsymbol{H}} = -i\boldsymbol{k} \times \left(\tilde{\boldsymbol{j}}_\perp \tilde{\psi}_0 + \tilde{\boldsymbol{j}}_0 \tilde{\psi}_1 + \boldsymbol{k}_x (\boldsymbol{k}\tilde{\boldsymbol{j}}_\perp) \tilde{\psi}_2 \right), \qquad \boldsymbol{k}_x = \mathbf{e}_x \left(\mathbf{e}_x \boldsymbol{k} \right), \tag{36}$$

where $\mathbf{e}_x$ is the unit vector along x-axis. Using the property of convolution and by considering the inverse Fourier transformation it is possible to get the solution of the Maxwell equations in the form of the sum of two independent solutions:

$$\boldsymbol{E} = \boldsymbol{E}_1 + \boldsymbol{E}_2, \qquad \boldsymbol{H} = \boldsymbol{H}_1 + \boldsymbol{H}_2. \tag{37}$$

The first of them, fields $\boldsymbol{E}_1$ and $\boldsymbol{H}_1$, is defined by one Green function ψ_1 and the density of the current $\boldsymbol{j}_0$ along the axis of the crystal:

$$\begin{cases} \boldsymbol{E}_1 = (i\varepsilon_0 \varepsilon \omega)^{-1} (\nabla\nabla + k_0^2)(\psi_1 * \boldsymbol{j}_0), \\ \boldsymbol{H}_1 = - \nabla \times (\psi_1 * \boldsymbol{j}_0), \end{cases} \tag{38}$$

where $r' = \sqrt{x^2 \varepsilon / \varepsilon_1 + y^2 + z^2}$ and

$$\psi_1 = -\frac{1}{4\pi} \sqrt{\frac{\varepsilon}{\varepsilon_1}} \frac{e^{ik_n r'}}{r'}. \tag{39}$$

The second solution, fields $\boldsymbol{E}_2$ and $\boldsymbol{H}_2$, can be written by using the component of the density of the current $\boldsymbol{j}_\perp$ perpendicular to axis x and Green functions ψ_0 , ψ_1 and ψ_2 :

$$\begin{cases} \boldsymbol{E}_2 = (i\varepsilon_0 \varepsilon \omega)^{-1} (k_0^2 (\boldsymbol{j}_\perp * \psi_0 + \nabla_\perp \nabla \boldsymbol{j}_\perp * \psi_2) + \nabla\nabla(\boldsymbol{j}_\perp * \psi_1)), \\ \boldsymbol{H}_2 = -\nabla \times (\boldsymbol{j}_\perp * \psi_0 - \mathbf{e}_x \frac{\partial}{\partial x} \nabla \boldsymbol{j}_\perp * \psi_2), \end{cases} \tag{40}$$

$$\psi_2 = (\frac{\varepsilon_1}{\varepsilon} - 1)\psi_0 * \psi_1, \tag{41}$$

$$\nabla_\perp \equiv \nabla - \mathbf{e}_x \frac{\partial}{\partial x}. \tag{42}$$

We note here that the function ψ_0 (22) is a fundamental solution of the Helmholtz operator for isotropic medium, while ψ_1 (39) corresponds to the functions ψ_0 for the space deformed along the axis of the crystal. Furthermore, the following useful identities are valid:

$$\mathrm{F}^{-1}[\tilde{\psi}_0(k_0^2-k_x^2)^{-1}] = \psi_0 * \mathrm{F}^{-1}[(k_0^2-k_x^2)^{-1}] = \\ -\frac{i}{8\pi k_0}\Big(e^{ik_0x}(\mathrm{Ci}(k_0(r-x))+i\,\mathrm{si}(k_0(r-x)))+e^{-ik_0x}(\mathrm{Ci}(k_0(r+x))+i\,\mathrm{si}(k_0(r+x)))\Big), \tag{43}$$

$$\mathrm{F}^{-1}[\tilde{\psi}_1(k_0^2-k_x^2)^{-1}] = \psi_1 * \mathrm{F}^{-1}[(k_0^2-k_x^2)^{-1}] = -\frac{i}{8\pi k_0}\Big(e^{ik_0x}(\mathrm{Ci}(k_nr'-k_0x)+ \\ i\,\mathrm{si}(k_nr'-k_0x))+e^{-ik_0x}(\mathrm{Ci}(k_nr'+k_0x)+i\,\mathrm{si}(k_nr'+k_0x))\Big), \tag{44}$$

$$\mathrm{F}^{-1}[\tilde{\psi}_2] = \mathrm{F}^{-1}[\tilde{\psi}_0(k_0^2-k_x^2)^{-1}] - \mathrm{F}^{-1}[\tilde{\psi}_1(k_0^2-k_x^2)^{-1}]. \tag{45}$$

Therefore, by also using (44), (45) above we find that the function ψ_2 is given by:

$$\psi_2 = \frac{1}{8\pi k_0 i}\Big(e^{ik_0x}(\mathrm{Ci}(k_0(r-x))+i\,\mathrm{si}(k_0(r-x)))+e^{-ik_0x}(\mathrm{Ci}(k_0(r+x))+i\,\mathrm{si}(k_0(r+x)))- \\ e^{ik_0x}(\mathrm{Ci}(k_nr'-k_0x)+i\,\mathrm{si}(k_nr'-k_0x))-e^{-ik_0x}(\mathrm{Ci}(k_nr'+k_0x)+i\,\mathrm{si}(k_nr'+k_0x))\Big), \tag{46}$$

where integral cosine and integral sine functions are defined by the following formulae:

$$\mathrm{Ci}(z) = \gamma + \ln(z) + \int_0^z \frac{\cos(t)-1}{t}dt, \qquad \mathrm{si}(z) = \int_0^z \frac{\sin(t)}{t}dt - \frac{\pi}{2} \tag{47}$$

and Euler constant $\gamma = 0,5772$.
Solutions (19), (20) and (22) can be also represented with the help of vector potentials $\boldsymbol{A}_0$, $\boldsymbol{A}_1$ and $\boldsymbol{A}_2$ as follows:

$$\begin{cases} \boldsymbol{E} = i\omega\Big(\boldsymbol{A}_0 + \mathbf{e}_x(\mathbf{e}_x\boldsymbol{A}_1) + \nabla_\perp\nabla\boldsymbol{A}_2 + \frac{1}{k_0^2}\nabla\nabla\boldsymbol{A}_1\Big), \\ \boldsymbol{H} = \frac{1}{\mu\mu_0}\nabla\times\Big(\boldsymbol{A}_0 + \mathbf{e}_x(\mathbf{e}_x\boldsymbol{A}_1) - \mathbf{e}_x\frac{\partial}{\partial x}\nabla\boldsymbol{A}_2\Big). \end{cases} \tag{48}$$

The vector potentials $\boldsymbol{A}_0$, $\boldsymbol{A}_1$ and $\boldsymbol{A}_2$ satisfy the following equations:

$$(\triangle + k_0^2)\boldsymbol{A}_0 = -\mu\mu_0\boldsymbol{j}_\perp, \qquad (\triangle' + k_n^2)\boldsymbol{A}_1 = -\mu\mu_0\boldsymbol{j}, \tag{49}$$

$$(\triangle + k_0^2)(\triangle' + k_n^2)\boldsymbol{A}_2 = -\mu\mu_0(\varepsilon_1/\varepsilon - 1)\boldsymbol{j}_\perp, \tag{50}$$

where $\triangle$ is the Laplace operator, the prime in $\triangle$ corresponds a replacement $x \to x\,\varepsilon/\varepsilon_1$.
The solutions of the equations (49), (50) can be written as follows:

$$\boldsymbol{A}_0 = -\mu\mu_0\cdot\boldsymbol{j}_\perp * \psi_0, \qquad \boldsymbol{A}_1 = -\mu\mu_0\cdot\boldsymbol{j} * \psi_1, \qquad \boldsymbol{A}_2 = -\mu\mu_0\cdot\boldsymbol{j}_\perp * \psi_2. \tag{51}$$

3.3 Hertz radiator in one-axis crystals

On the basis of the results obtained above, we shall consider the radiation of the electric Hertzian dipole in unbounded one-axis crystals. The point dipole moment is given by

$$\boldsymbol{p} = \boldsymbol{n} p_e \exp(-i\omega t), \tag{52}$$

where p_e is a constant, $\boldsymbol{n}$ is a unit vector parallel to the direction of the dipole moment, and the current density is defined by means of Dirac delta-function :

$$\boldsymbol{j} = -i\omega \boldsymbol{p}\delta(\boldsymbol{r}), \qquad \boldsymbol{p} = \boldsymbol{p}_0 + \boldsymbol{p}_\perp. \tag{53}$$

The last formula of current density which follows from the expression of charge density for the point dipole is given by:

$$\rho = -(\boldsymbol{p}\nabla)\delta(\boldsymbol{r}) \tag{54}$$

and also the charge conservation law (21).

Furthermore, the expression of the radiated electromagnetic field for electric Hertzian dipole will take the following form, when the direction of the dipole moment $\boldsymbol{p}_0$ is parallel to the axis x of the crystal (Fig. 1):

$$\begin{cases} \boldsymbol{E}_1 = -(\varepsilon\varepsilon_0)^{-1}(\nabla\nabla + k_0^2)(\psi_1 \boldsymbol{p}_0), \\ \boldsymbol{H}_1 = i\omega\nabla \times (\psi_1 \boldsymbol{p}_0). \end{cases} \tag{55}$$

Also, when the direction of the dipole moment is perpendicular to the axis x, we obtain (Fig. 2):

$$\begin{cases} \boldsymbol{E}_2 = -\dfrac{1}{\varepsilon\varepsilon_0}\left(k_0^2(\boldsymbol{p}_\perp\psi_0 + \nabla_\perp\nabla(\boldsymbol{p}_\perp\psi_2)) + \nabla\nabla(\boldsymbol{p}_\perp\psi_1)\right), \\ \boldsymbol{H}_2 = -i\omega\nabla \times \left(\boldsymbol{p}_\perp\psi_0 - \mathbf{e}_x\dfrac{\partial}{\partial x}\nabla(\boldsymbol{p}_\perp\psi_2)\right) \qquad (\boldsymbol{p}_0 \perp \boldsymbol{p}_\perp). \end{cases} \tag{56}$$

Moreover, we note that the independent solutions (38) and (40) define the corresponding polarization of electromagnetic waves. In addition, when ε_1 tends to ε, from (31) it follows that the potential ψ_2 tends to zero and the well-known expressions of electromagnetic field followed from formula (38) are obtained :

$$\boldsymbol{E} = i\omega(k_0^{-2}\nabla\nabla + 1)\boldsymbol{A}, \tag{57}$$

$$\boldsymbol{H} = (\mu\mu_0)^{-1}\nabla \times \boldsymbol{A}, \tag{58}$$

where the known vector potential of electromagnetic field for isotropic mediums is defined from (51) as (24):

$$\boldsymbol{A}(\boldsymbol{r}) = \frac{\mu\mu_0}{4\pi}\int_V \boldsymbol{j}(\boldsymbol{r}')\frac{\exp(ik_0|\boldsymbol{r} - \boldsymbol{r}'|)}{|\boldsymbol{r} - \boldsymbol{r}'|}dV. \tag{59}$$

The obtained generalized solutions of the Maxwell equations are valid for any values of ε_1 and ε, as well as for sources of the electromagnetic waves, described by discontinuous and singular functions.

Below as a specific application radiation from a Hertzian dipole in such a medium was examined and the corresponding radiation patterns were presented.

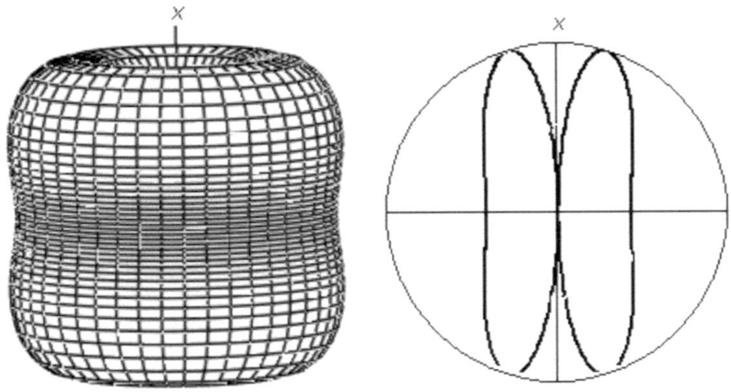

(a) Case $\varepsilon_1 = 9$ (b) View in meridian surface

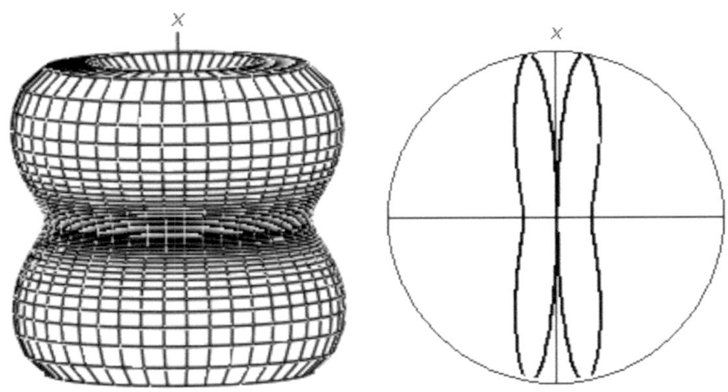

(c) Case $\varepsilon_1 = 25$ (d) View in meridian surface

Fig. 1. Directivity diagrams, the axis of dipole is parallel to axis of a crystal

It should be note that the pattern in Fig. 1 remains invariable, and independent of r. The radiation pattern of the Hertzian dipole in isotropic medium is shown in Fig. 3, which of course possesses rotation symmetry around x-axis.
Furthermore, we note here that the numerical calculation of the above solution of Maxwell equations satisfies the energy conservation law. Poynting vector

$$< \mathbf{\Pi} >= \frac{1}{2}\text{Re}(\boldsymbol{E} \times \bar{\boldsymbol{H}})$$

is necessary to calculate time-averaged energy-flux on spherical surface

$$\Phi = \int_{S_{sph}} < \Pi_r > dS,$$

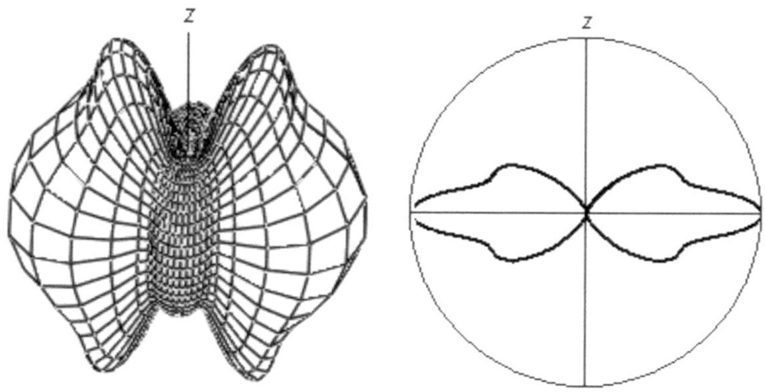

(a) Case at $r = 5$, $\varepsilon_1 = 7$ (b) View in meridian surface

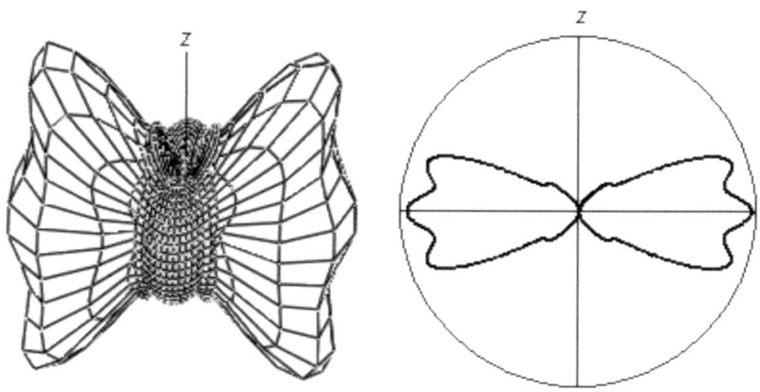

(c) Case at $r = 5$, $\varepsilon_1 = 10$ (d) View in meridian surface

Fig. 2. Directivity diagram, the axis of dipole (z) is perpendicular to axis of a crystal

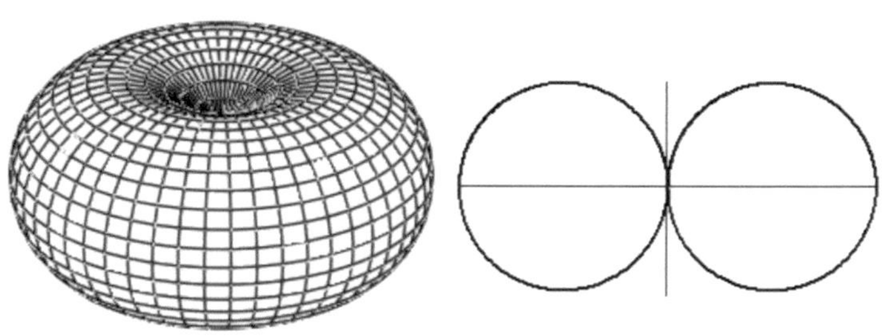

Fig. 3. Directivity diagram of the Hertzian dipole, the isotropic medium ($\varepsilon = \varepsilon_1 = 1$)

which is a constant at various values of radius of sphere, where $\bar{\boldsymbol{H}}$ is complex conjugate function.

3.4 Directivity diagrams of the magnetic moment of a dipole in one-axis crystals

Exact analytical solution of Maxwell's equations for radiation of a point magnetic dipole in uniaxial crystals are obtained. Directivity diagrams of radiation of a point magnetic dipole are constructed at parallel and perpendicular directions of an axis of a crystal.

On the basis of the obtained results (38) and (40), we will consider radiation of point magnetic dipole moment in an uniaxial crystal. Let's define intensity of an electromagnetic field for the concentrated magnetic dipole at a parallel and perpendicular direction to a crystal axis in the anisotropic medium and we will construct diagrams of directivity for both cases.

For a point radiator with the oscillating magnetic dipole moment

$$\boldsymbol{m} = \boldsymbol{n} p_m e^{-i\omega t} \qquad (p_m = const) \tag{60}$$

the electric current density is defined by using Dirac delta-function:

$$j = -(\boldsymbol{m} \times \nabla)\delta(\boldsymbol{r}). \tag{61}$$

Components of a current density (61) are:

$$j_0 = \mathbf{e}_x (m_z \frac{\partial}{\partial y} - m_y \frac{\partial}{\partial z})\delta(\boldsymbol{r}), \tag{62}$$

$$\boldsymbol{j}_\perp = \mathbf{e}_y (m_x \frac{\partial}{\partial z} - m_z \frac{\partial}{\partial x})\delta(\boldsymbol{r}) + \mathbf{e}_z (m_y \frac{\partial}{\partial x} - m_x \frac{\partial}{\partial y})\delta(\boldsymbol{r}). \tag{63}$$

It is possible to express the magnetic dipole moment $\boldsymbol{m}$ in the form of the sum of two components of magnetic moment:

$$\boldsymbol{m} = \boldsymbol{m}_0 + \boldsymbol{m}_\perp, \qquad \boldsymbol{m}_0 = \mathbf{e}_x m_x. \tag{64}$$

Relation between density of an electric current $\boldsymbol{j}_\perp$ and the magnetic dipole moment $\boldsymbol{m}$ in the anisotropic medium is defined from (63), in case $\boldsymbol{m} = \boldsymbol{m}_0$:

$$\boldsymbol{j}_\perp = m_x (\mathbf{e}_y \frac{\partial}{\partial z} - \mathbf{e}_z \frac{\partial}{\partial y})\delta(\boldsymbol{r}), \tag{65}$$

$$\nabla \boldsymbol{j}_\perp = 0. \tag{66}$$

3.4.1 The parallel directed magnetic momentum

Taking into account equality (66), from solutions (38) intensities of the electromagnetic field of the magnetic dipole moment are defined, in the case when the magnetic dipole moment $\boldsymbol{m}$ is directed parallel to the crystal axis x:

$$\begin{cases} \boldsymbol{E} = -i\mu_0 \mu\omega \nabla \times (\psi_0 \boldsymbol{m}_0), \\ \boldsymbol{H} = -\nabla \times \nabla \times (\psi_0 \boldsymbol{m}_0). \end{cases} \tag{67}$$

It is necessary to notice, that expressions (67) correspond to the equations of an electromagnetic field in isotropic medium (Fig. 4). In Fig. 4, directivity diagrams of magnetic dipole moment in the case that the magnetic moment is directed in parallel to the crystal axis are shown. The given directivity diagram coincides with the directivity diagram of a parallel

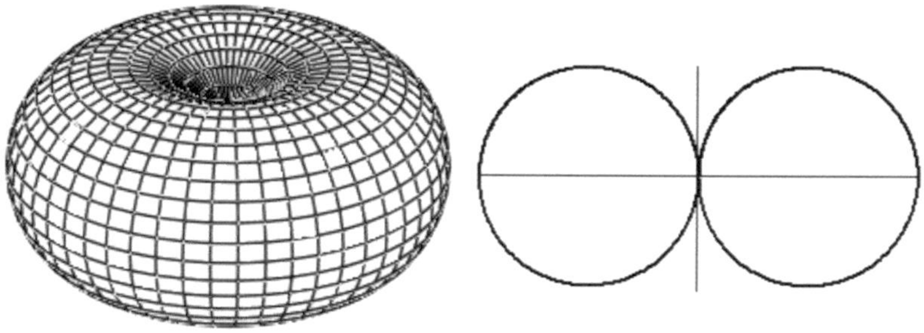

Fig. 4. Directivity diagram of the magnetic dipole, the isotropic medium ($\varepsilon = \varepsilon_1 = 1$)

directed electric dipole in isotropic medium. This diagram looks like a toroid, which axis is parallel to the dipole axis. Cross-sections of the diagram are a contour on a plane passing through an axis of the toroid. It has the shape of number 'eight' ; cross-sections perpendicular to the axis of the toroid represent circles.

3.4.2 The perpendicular directed magnetic momentum

Relation between density of an electric current $\boldsymbol{j}$ and the magnetic dipole momentum $\boldsymbol{m}_\perp$ is defined from expression (61), if it is directed on axis z:

$$j = m_z(\mathbf{e}_x \frac{\partial}{\partial y} - \mathbf{e}_y \frac{\partial}{\partial x})\delta(\boldsymbol{r}). \tag{68}$$

For the point magnetic dipole $\boldsymbol{m}_\perp$ which is perpendicular to crystal axes, by substituting (68) in solutions (40), we define components of field intensity (Fig. 5, Fig. 6):

$$\begin{cases} E_x = -i\mu_0\mu\omega m_z \dfrac{\partial\psi_1}{\partial y}, \\ E_y = i\mu_0\mu\omega m_z \dfrac{\partial}{\partial x}\left(\psi_0 + \dfrac{\partial^2\psi_2}{\partial y^2}\right), \\ E_z = i\mu_0\mu\omega m_z \dfrac{\partial^3\psi_2}{\partial x\partial y\partial z}, \end{cases} \tag{69}$$

$$\begin{cases} H_x = -m_z \dfrac{\partial^2\psi_0}{\partial x\partial z}, \\ H_y = -m_z \dfrac{\partial^2}{\partial y\partial z}\left(\psi_1 + \dfrac{\partial^2\psi_2}{\partial x^2}\right), \\ H_z = m_z\left(\dfrac{\partial^2}{\partial y^2}(\psi_1 + \dfrac{\partial^2\psi_2}{\partial x^2}) + \dfrac{\partial^2\psi_0}{\partial x^2}\right). \end{cases} \tag{70}$$

In Figs. 5 and 6, directivity diagrams of the magnetic moment of a dipole perpendicular a crystal axis at different values of radius are shown, for two (30) values of dielectric permeability ratio, $\varepsilon_1/\varepsilon = 9$ and $\varepsilon_1/\varepsilon = 15$. The magnetic dipole is directed along an axis z (a

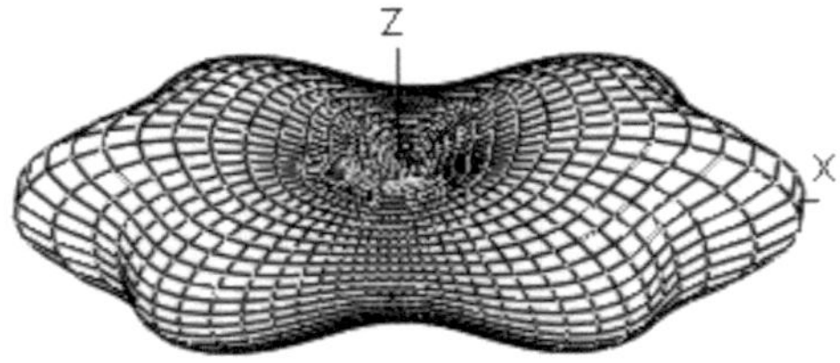

(a) Case $r = 3$

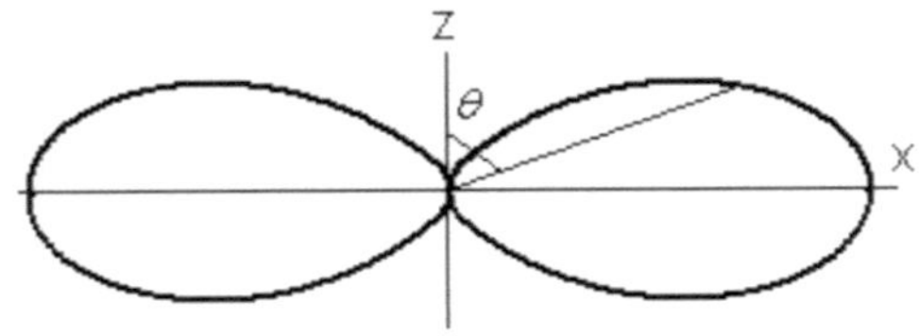

(b) View in meridian surface $r = 3$, $\varphi = \pi/2$

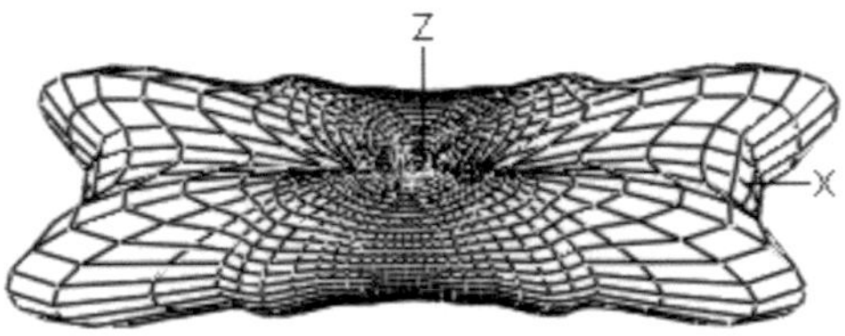

(c) Case at $r = 9$

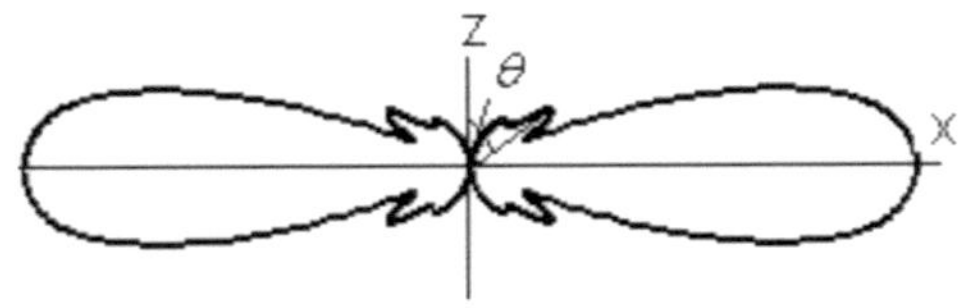

(d) View in meridian surface $r = 9$, $\varphi = \pi/2$

Fig. 5. Directivity diagram, the axis of magnetic dipole (z) is perpendicular to axis of a crystal at $\varepsilon_1/\varepsilon = 9$

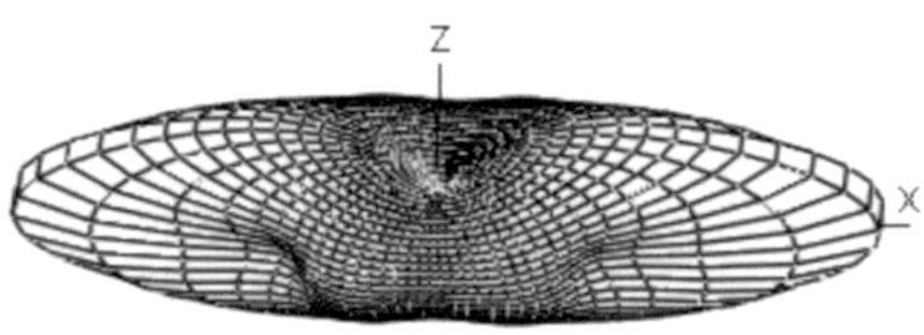

(a) Case $r = 1$

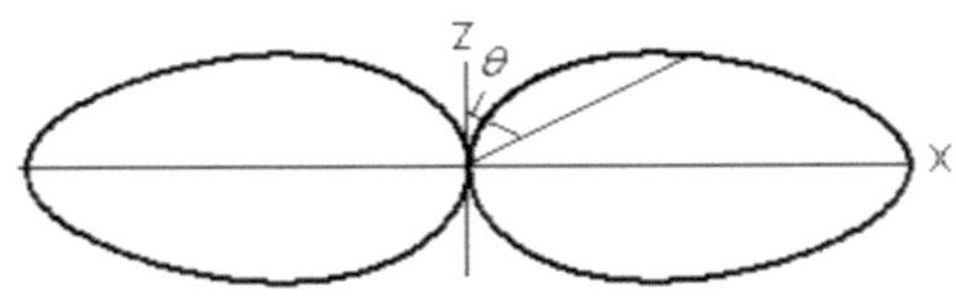

(b) View in meridian surface $r = 1$, $\varphi = \pi/2$

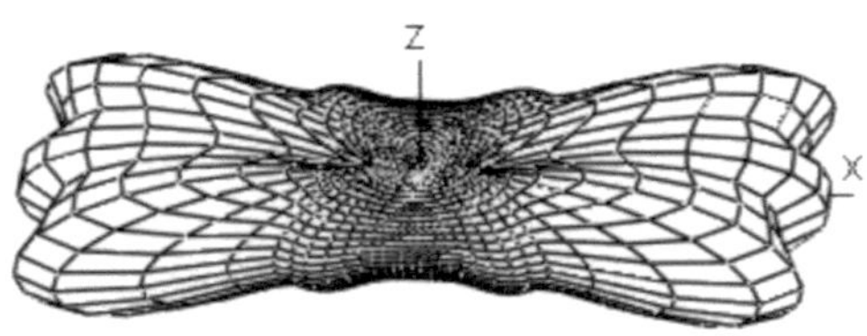

(c) Case at $r = 5$

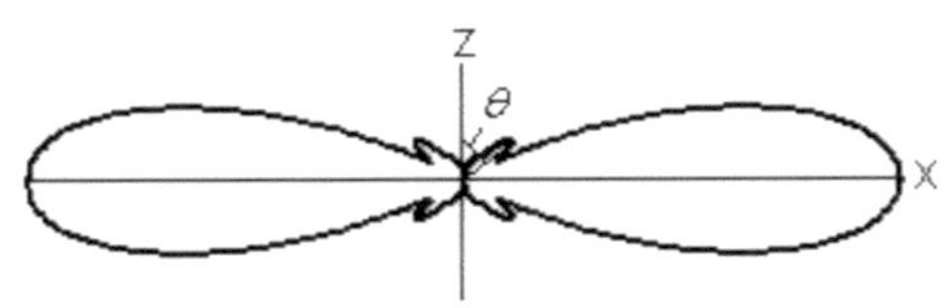

(d) View in meridian surface $r = 5$, $\varphi = \pi/2$

Fig. 6. Directivity diagram, the axis of magnetic dipole (z) is perpendicular to axis of a crystal at $\varepsilon_1/\varepsilon = 15$.

crystal axis - along x). As one can see from Fig. 5, that radiation in a direction of the magnetic moment does not occur, it propagates in a direction along an axis of a crystal.
Validity of solutions has been checked up on performance of the conservation law of energy. Time-averaged energy flux of energy along a surface of sphere for various values of radius was calculated for this purpose. Numerical calculations show that energy flux over the above mentioned spherical surfaces, surrounding the radiating magnetic dipole, remains constant, which means that energy conservation is preserved with high numerical accuracy.

The obtained generalized solutions of the Maxwell equations are valid for any values of ε_1 and ε, and also for any kind of sources of the electromagnetic waves, described by discontinuous and singular functions.

4. Radiation of electric and magnetic dipole antennas in magnetically anisotropic media

The electric and magnetic field intensities satisfy the system of stationary Maxwell's equations (1) which is possible to be presented in matrix form (4), where

$$\mathbf{M} = \begin{pmatrix} -i\omega\varepsilon_0\varepsilon \boldsymbol{I} & \boldsymbol{G}_0 \\ \boldsymbol{G}_0 & i\omega\mu_0\hat{\mu} \end{pmatrix}, \qquad \hat{\mu} = \begin{pmatrix} \mu & 0 & 0 \\ 0 & \mu & 0 \\ 0 & 0 & \mu_1 \end{pmatrix}. \tag{71}$$

M is Maxwell's operator and μ represents the magnetic permeability.
In magnetically anisotropic media the relation between induction and intensity of the magnetic field is:

$$\boldsymbol{B} = \mu_0\hat{\mu}\boldsymbol{H}, \qquad B_x = \mu_0\mu H_x, \qquad B_y = \mu_0\mu H_y, \qquad B_z = \mu_0\mu_1 H_z. \tag{72}$$

The elements of the magnetic permeability tensor $\hat{\mu}$ are chosen so that the axis of anisotropy is directed along axis z. It is required to define the intensities of the electromagnetic field $\boldsymbol{E}, \boldsymbol{H}$ in the space of generalized functions.

4.1 Solution of the problem

By means of direct Fourier transformation we write down the system of the equations in matrix form (9). The solution of this problem is reduced to the solution of the system, where $\tilde{\mathbf{U}}$ is defined by means of inverse matrix $\tilde{\mathbf{M}}^{-1}$.
By introducing new functions according to

$$\tilde{\psi}_1^m = \frac{1}{k_n^2 - k_x^2 - k_y^2 - \frac{\mu_1}{\mu}k_z^2}, \tag{73}$$

$$\tilde{\psi}_2^m = \left(\frac{\mu_1}{\mu} - 1\right)\tilde{\psi}_1^m\tilde{\psi}_0, \tag{74}$$

the components of the electromagnetic field after transformations in spectral domain can be written as follows:

$$\begin{cases} \tilde{E}_x = (i\varepsilon\varepsilon_0\omega)^{-1}\left(k_0^2(\tilde{j}_x\tilde{\psi}_0 + k_y(\boldsymbol{k}\times\tilde{\boldsymbol{j}})_z\tilde{\psi}_2^m) - k_x\boldsymbol{k}\tilde{\boldsymbol{j}}\tilde{\psi}_0\right), \\ \tilde{E}_y = (i\varepsilon\varepsilon_0\omega)^{-1}\left(k_0^2(\tilde{j}_y\tilde{\psi}_0 - k_x(\boldsymbol{k}\times\tilde{\boldsymbol{j}})_z\tilde{\psi}_2^m) - k_y\boldsymbol{k}\tilde{\boldsymbol{j}}\tilde{\psi}_0\right), \\ \tilde{E}_z = (i\varepsilon\varepsilon_0\omega)^{-1}(k_0^2\tilde{j}_z - k_z\tilde{\boldsymbol{j}})\tilde{\psi}_0, \end{cases} \tag{75}$$

$$\begin{cases} \tilde{H}_x = -i\left((\boldsymbol{k}\times\tilde{\boldsymbol{j}})_x\tilde{\psi}_0 + k_xk_z(\boldsymbol{k}\times\tilde{\boldsymbol{j}})_z\tilde{\psi}_2^m\right), \\ \tilde{H}_y = -i\left((\boldsymbol{k}\times\tilde{\boldsymbol{j}})_y\tilde{\psi}_0 + k_yk_z(\boldsymbol{k}\times\tilde{\boldsymbol{j}})_z\tilde{\psi}_2^m\right), \\ \tilde{H}_z = -i(\boldsymbol{k}\times\tilde{\boldsymbol{j}})_z\tilde{\psi}_1^m, \end{cases} \tag{76}$$

$$((\boldsymbol{k}\times\tilde{\boldsymbol{j}})_z \equiv k_x\tilde{j}_y - k_y\tilde{j}_x).$$

It is possible to represent the electromagnetic fields in (75) and (76) in vector form as following

$$\tilde{\boldsymbol{E}} = (i\varepsilon\varepsilon_0\omega)^{-1}\left(k_0^2(\tilde{\boldsymbol{j}}\tilde{\psi}_0 + \boldsymbol{k}\times\mathbf{e}_z(\boldsymbol{k}\times\tilde{\boldsymbol{j}}_\perp)_z\tilde{\psi}_2^m) - \boldsymbol{k}(\boldsymbol{k}\tilde{\boldsymbol{j}})\tilde{\psi}_0\right), \tag{77}$$

$$\tilde{\boldsymbol{H}} = i(\mathbf{e}_z k_z - \boldsymbol{k})k_z(\boldsymbol{k}\times\tilde{\boldsymbol{j}}_\perp)_z\tilde{\psi}_2^m + i\mathbf{e}_z(\boldsymbol{k}\times\tilde{\boldsymbol{j}}_\perp)_z(\tilde{\psi}_0 - \tilde{\psi}_1^m) - i\boldsymbol{k}\times\tilde{\boldsymbol{j}}\tilde{\psi}_0, \tag{78}$$

where

$$\tilde{\boldsymbol{j}} = \tilde{\boldsymbol{j}}_\perp + \tilde{\boldsymbol{j}}_0, \qquad \tilde{j}_0 = (0, 0, \tilde{j}_z), \qquad \tilde{\boldsymbol{j}}_\perp = (\tilde{j}_x, \tilde{j}_y, 0), \qquad k_0^2 = \omega^2\varepsilon_0\varepsilon\mu\mu_0, \qquad k_n^2 = k_0^2\frac{\mu_1}{\mu}.$$

It should be noted that the following useful formulae follow from (12), (73) and (74):

$$\tilde{\psi}_0 - \tilde{\psi}_1^m = (k_0^2 - k_z^2)\tilde{\psi}_2^m, \tag{79}$$

$$\tilde{\psi}_0 - \frac{\mu_1}{\mu}\tilde{\psi}_1^m = (k_x^2 + k_y^2)\tilde{\psi}_2^m. \tag{80}$$

With the help of identity in (79) and (80), the last equation (78) can be represented as

$$\tilde{\boldsymbol{H}} = i(k_0^2\mathbf{e}_z - k_z\boldsymbol{k})(\boldsymbol{k}\times\tilde{\boldsymbol{j}}_\perp)_z\tilde{\psi}_2^m - i(\boldsymbol{k}\times\tilde{\boldsymbol{j}})\tilde{\psi}_0. \tag{81}$$

Using the property of convolution (17) and considering inverse Fourier transformation it is possible to get the solution of the Maxwell equations in form (18).
So, after the inverse Fourier transformations from (77) and (81) we obtain:

$$\boldsymbol{E} = (i\varepsilon\varepsilon_0\omega)^{-1}\left((k_0^2 + \nabla\nabla)\boldsymbol{j}*\psi_0 - k_0^2\nabla\times\left(\mathbf{e}_z(\nabla\times\boldsymbol{j}_\perp)_z\right)*\psi_2^m\right), \tag{82}$$

$$\boldsymbol{H} = (k_0^2\mathbf{e}_z + \frac{\partial}{\partial z}\nabla)(\nabla\times\boldsymbol{j}_\perp)_z * \psi_2^m - \nabla\times\boldsymbol{j}*\psi_0. \tag{83}$$

This solution can be written in the form of the sum of two solutions:

$$\boldsymbol{E} = \boldsymbol{E}_1 + \boldsymbol{E}_2, \qquad \boldsymbol{H} = \boldsymbol{H}_1 + \boldsymbol{H}_2.$$

It should be noted that the first of them is the 'isotropic' solution. It is defined by Green's function ψ_0 and the density of the current $\boldsymbol{j}_0$ directed parallel to axis z (of the anisotropy):

$$\begin{cases} \boldsymbol{E}_1 = (i\varepsilon\varepsilon_0\omega)^{-1}(\nabla\nabla + k_0^2)(\psi_0 * \boldsymbol{j}_0), \\ \boldsymbol{H}_1 = -\nabla\times(\psi_0 * \boldsymbol{j}_0), \end{cases} \tag{84}$$

where ψ_0 is the Green's function (22). The second solution can be written by using the component of the density of the current $\boldsymbol{j}_\perp$ perpendicular to axis z and the Green's functions ψ_0 and ψ_2^m :

$$\begin{cases} \boldsymbol{E}_2 = -\dfrac{i}{\varepsilon\varepsilon_0\omega}\left((k_0^2 + \nabla\nabla)\boldsymbol{j}_\perp * \psi_0 - k_0^2\nabla\times\left(\mathbf{e}_z(\nabla\times\boldsymbol{j}_\perp)_z\right)*\psi_2^m\right), \\ \boldsymbol{H}_2 = (k_0^2\mathbf{e}_z + \dfrac{\partial}{\partial z}\nabla)(\nabla\times\boldsymbol{j}_\perp)_z * \psi_2^m - \nabla\times\boldsymbol{j}_\perp * \psi_0, \end{cases} \tag{85}$$

$$\psi_1^m = \mathrm{F}^{-1}[\tilde{\psi}_1^m] = -\frac{1}{4\pi}\sqrt{\frac{\mu}{\mu_1}}\frac{\exp(ik_n r')}{r'}, \tag{86}$$

$$\psi_2^m = (\frac{\mu}{\mu_1} - 1)\psi_0 * \psi_1^m, \tag{87}$$

$$r' = \sqrt{x^2 + y^2 + \frac{\mu}{\mu_1} z^2}. \tag{88}$$

Furthermore, the following transformation is valid similarly to (45):

$$\mathrm{F}^{-1}[\tilde{\psi}_2^m] = \mathrm{F}^{-1}[\tilde{\psi}_0(k_0^2 - k_z^2)^{-1}] - \mathrm{F}^{-1}[\tilde{\psi}_1^m(k_0^2 - k_z^2)^{-1}].$$

We find that function ψ_2^m , Eqn. (87) is given by:

$$\psi_2^m = -\frac{i}{8\pi k_0}\Big(e^{ik_0 z}(\mathrm{Ci}(k_0(r-z)) + i\mathrm{si}(k_0(r-z))) + e^{-ik_0 z}(\mathrm{Ci}(k_0(r+z)) + i\mathrm{si}(k_0(r+z))) - \\ -e^{ik_0 z}(\mathrm{Ci}(k_n r' - k_0 z) + i\mathrm{si}\,(k_n r' - k_0 z)) - e^{-ik_0 z}(\mathrm{Ci}(k_n r' + k_0 z) + i\mathrm{si}\,(k_n r' + k_0 z))\Big). \tag{89}$$

It should be noted that ψ_2 and ψ_2^m are similar for magnetic and dielectric crystals.

4.2 Radiation patterns of Hertzian radiator in magnetically anisotropic media

On the basis of the results obtained above, we consider now numerical results for the radiation of the electric dipole.

The dipole moment of point electric dipole is given by $\boldsymbol{p}$, Eqn. (52). It corresponds to the current density defined by means of the Dirac delta-function (53).

The expression of the electromagnetic field for electric radiator will take the following form as for isotropic medium, when the direction of the dipole moment is parallel to the axis z ($\boldsymbol{p} = \boldsymbol{p}_0$) (Fig. 7):

$$\begin{cases} \boldsymbol{E}_1 = -(i\varepsilon\varepsilon_0)^{-1}(\nabla\nabla + k_0^2)(\psi_0 \boldsymbol{p}_0), \\ \boldsymbol{H}_1 = i\omega\nabla \times (\psi_0 \boldsymbol{p}_0). \end{cases} \tag{90}$$

Also when the direction of the dipole moment is perpendicular to axis z, we obtain ($\boldsymbol{p} = \boldsymbol{p}_\perp$) (Fig. 8, Fig. 9):

$$\begin{cases} \boldsymbol{E}_2 = (\varepsilon\varepsilon_0)^{-1}(k_0^2\nabla \times (\mathbf{e}_z(\nabla \times \boldsymbol{p}_\perp \psi_2^m)_z) - (k_0^2 + \nabla\nabla)(\boldsymbol{p}_\perp \psi_0)), \\ \boldsymbol{H}_2 = (k_0^2\mathbf{e}_z + \frac{\partial}{\partial z}\nabla)(\nabla \times \boldsymbol{p}_\perp \psi_2^m)_z - \nabla \times (\boldsymbol{p}_\perp \psi_0). \end{cases} \tag{91}$$

4.3 Radiation patterns of point magnetic dipole in magnetically anisotropic media

On the basis of the obtained results, we consider now radiation of point magnetic dipole moment. For a point radiator with the oscillating magnetic dipole moment, similarly to the magnetic dipole case in one-axis crystals, Eqn. (60), the electric current density is defined by using Dirac's delta-function, Eqn. (61). It is possible to express the magnetic dipole moment $\boldsymbol{m}$ in the form of the sum of two components of magnetic moment:

$$\boldsymbol{m} = \boldsymbol{m}_0 + \boldsymbol{m}_\perp, \qquad \boldsymbol{m}_0 = \mathbf{e}_z m_z. \tag{92}$$

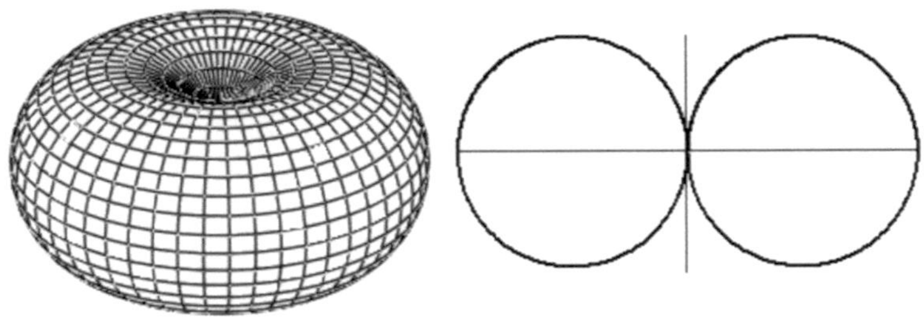

Fig. 7. Directivity diagram. Electric dipole moment is parallel to the axis z ($\boldsymbol{p} = \boldsymbol{p}_0$)

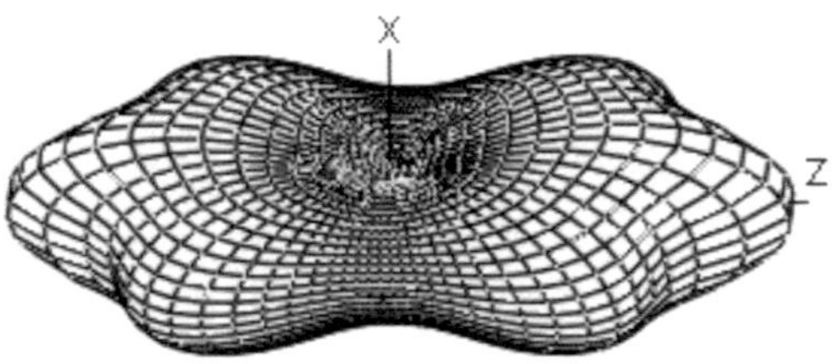

(a) Case $r = 3$

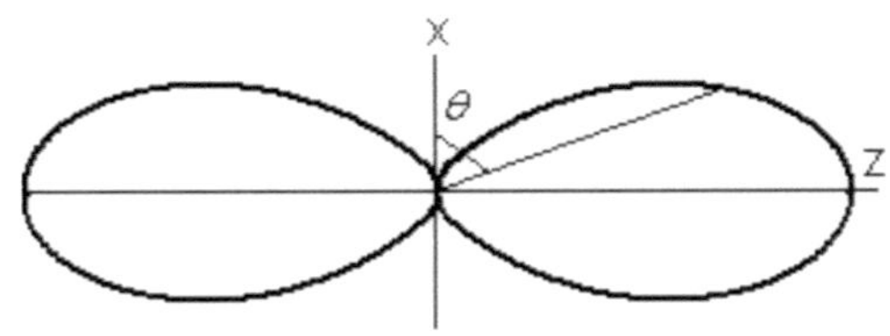

(b) View in meridian surface $r = 3$, $\varphi = \pi/2$

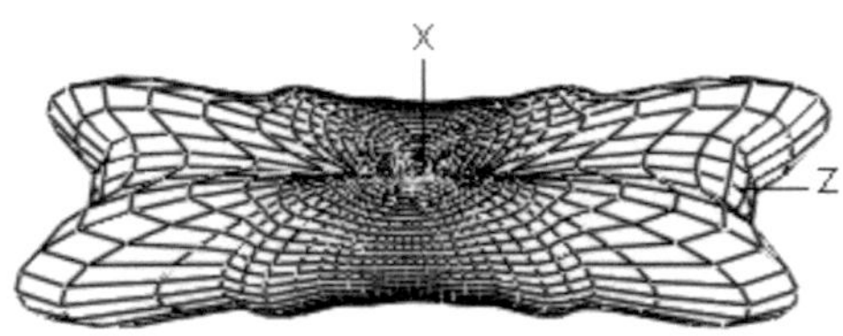

(c) Case $r = 9$

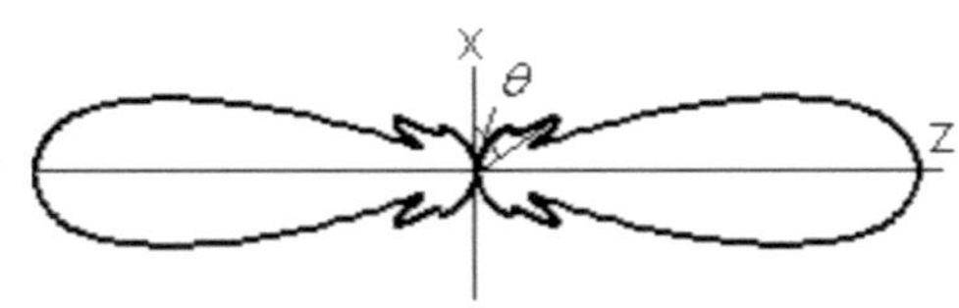

(d) View in meridian surface $r = 9$, $\varphi = \pi/2$

Fig. 8. Directivity diagram. The axis of electric dipole is perpendicular to axis z ($\boldsymbol{p} = \boldsymbol{p}_\perp$), $\mu_1/\mu = 9$

Components of a current density, Eqn. (61), have the following form:

$$j_0 = \mathbf{e}_z(m_y \frac{\partial}{\partial x} - m_x \frac{\partial}{\partial y})\delta(\boldsymbol{r}), \tag{93}$$

and

$$\boldsymbol{j}_\perp = \left(\mathbf{e}_x(m_z \frac{\partial}{\partial y} - m_y \frac{\partial}{\partial z}) + \mathbf{e}_y(m_x \frac{\partial}{\partial z} - m_z \frac{\partial}{\partial x})\right)\delta(\boldsymbol{r}). \tag{94}$$

Case $\boldsymbol{m} = \boldsymbol{m}_0$.

Relation between density of an electric current $\boldsymbol{j}_\perp$ and the magnetic dipole moment is defined from (93) in case when the magnetic dipole moment $\boldsymbol{m}$ is directed lengthwise z:

$$\boldsymbol{j}_\perp = m_z(\mathbf{e}_x \frac{\partial}{\partial y} - \mathbf{e}_y \frac{\partial}{\partial x})\delta(\boldsymbol{r}). \tag{95}$$

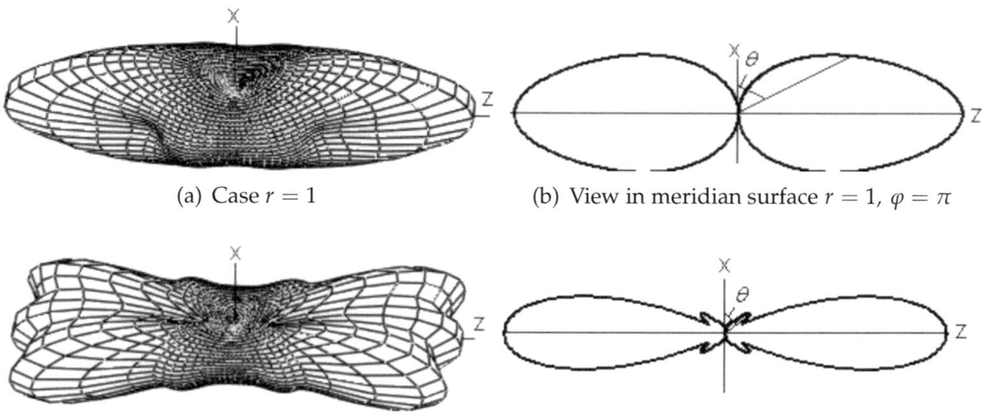

(a) Case $r = 1$ (b) View in meridian surface $r = 1$, $\varphi = \pi$

(c) Case $r = 5$ (d) View in meridian surface $r = 5$, $\varphi = \pi$

Fig. 9. Directivity diagram. The axis of electric dipole is perpendicular to axis z ($\boldsymbol{p} = \boldsymbol{p}_\perp$), $\mu_1/\mu = 15$

It should be noted that the following useful formula hold:

$$\nabla \boldsymbol{j}_\perp = 0, \qquad j_0 = 0. \tag{96}$$

Taking into account (96), intensities of the electromagnetic field by the magnetic dipole moment are defined from the solutions (85) in this case, as following :

$$\begin{cases} \boldsymbol{E} = -i\mu_0\mu_1\omega m_0 \nabla \times (\mathbf{e}_z \psi_1^m), \\ \boldsymbol{H} = \boldsymbol{m}_0 \triangle \psi_1^m + \boldsymbol{m}_0 (\frac{\mu_1}{\mu} - 1) \frac{\partial^2}{\partial z^2} \psi_1^m - m_0 \frac{\mu_1}{\mu} \frac{\partial}{\partial z} \nabla \psi_1^m. \end{cases} \tag{97}$$

We have taken advantage of the next formulae which followed from (79) after inverse Fourier transformation

$$\frac{\mu_1}{\mu} \psi_1^m = \psi_0 + (\frac{\partial^2}{\partial x^2} + \frac{\partial^2}{\partial y^2}) \psi_2^m, \tag{98}$$

$$\psi_0 = \psi_1^m + (\frac{\partial^2}{\partial z^2} + k_0^2) \psi_2^m. \tag{99}$$

Directional diagrams are represented in Fig. 10.

Case $\boldsymbol{m} = \boldsymbol{m}_\perp$.

For the point magnetic dipole moment $\boldsymbol{m}_\perp$ which is perpendicular to axis z, by substituting (61) (or (93) and (94)) in (84) and (85), we define intensities of electromagnetic field as following (Fig. 11, Fig. 12):

$$\begin{cases} \boldsymbol{E} = i\mu_0\mu\omega (\frac{\partial}{\partial z} \{ \boldsymbol{m}_\perp \times \mathbf{e}_z \frac{\mu_1}{\mu} \psi_1^m - \nabla_\perp (\nabla \times \psi_2^m \boldsymbol{m}_\perp)_z \} - \mathbf{e}_z (\nabla \times \psi_0 \boldsymbol{m}_\perp)_z), \\ \boldsymbol{H} = (k_0^2 \mathbf{e}_z + \frac{\partial}{\partial z} \nabla) \frac{\partial}{\partial z} \nabla (\boldsymbol{m}_\perp \psi_2^m) - \nabla \times \nabla \times (\boldsymbol{m}_\perp \psi_0). \end{cases} \tag{100}$$

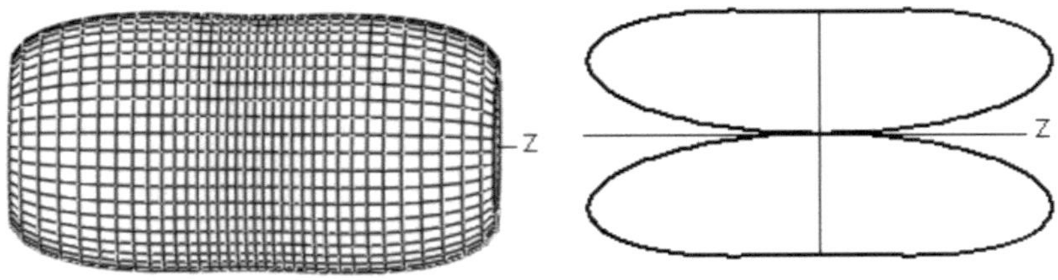

(a) Case $\mu_1/\mu = 7$ (b) View in meridian surface ($\varphi = 0$)

Fig. 10. Directivity diagram. The axis of magnetic dipole is parallel to axis z ($\boldsymbol{m} = \boldsymbol{m}_0$)

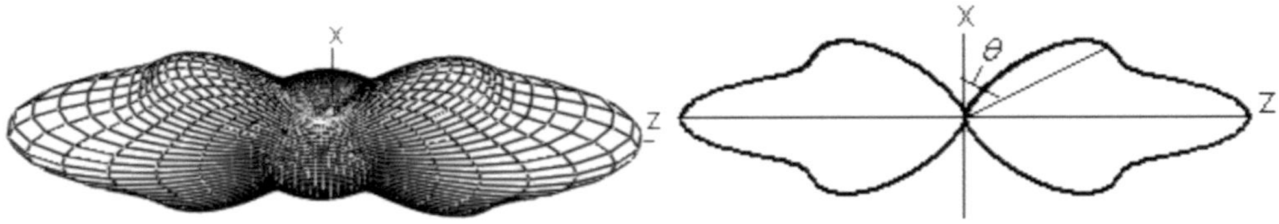

(a) Case $r = 5$ (b) View in meridian surface $r = 5$, $\varphi = \pi/2$

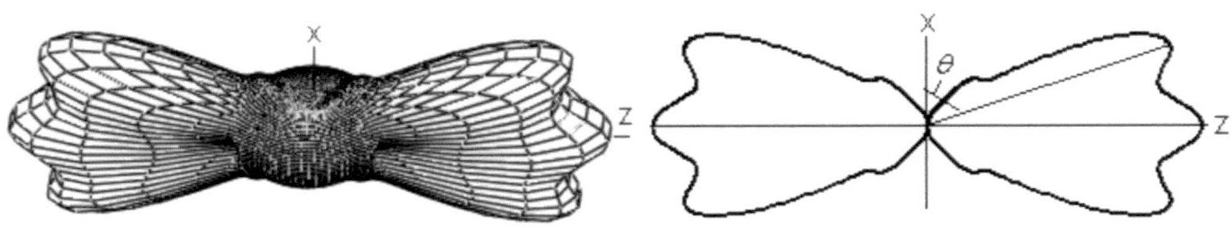

(c) Case $r = 10$ (d) View in meridian surface $r = 10$, $\varphi = \pi/2$

Fig. 11. Directivity diagram. The axis of magnetic dipole is perpendicular to z, $\mu_1/\mu = 7$

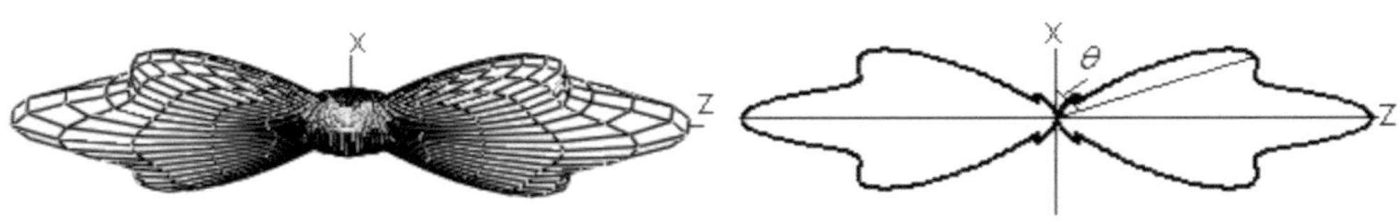

(a) Case $r = 5$ (b) View in meridian surface $r = 6$, $\varphi = \pi/2$

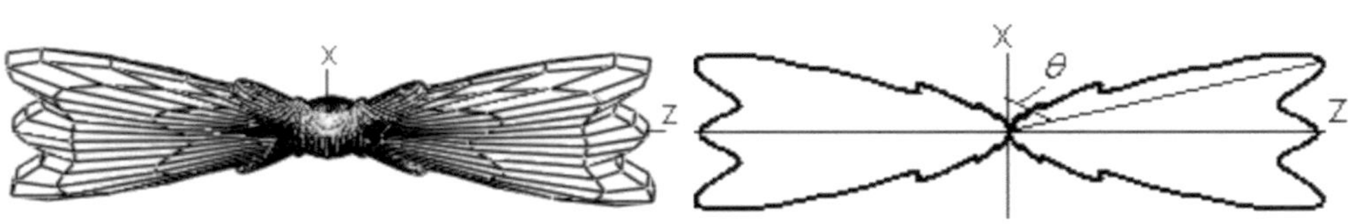

(c) Case $r = 10$ (d) View in meridian surface $r = 10$, $\varphi = \pi/2$

Fig. 12. Directivity diagram. The axis of magnetic dipole is perpendicular to z, $\mu_1/\mu = 15$

The numerical calculation of the solution of Maxwell equations satisfies the energy conservation law. Numerical computation shows that time average value energy flux on a

surface of sphere from a point dipole remains independent from its radius, with very high accuracy. As shown in the electric dipole directional diagrams, medium becomes isotropic for such radiator if its moment is directed along anisotropy axis.
The dipole pattern in isotropic media is shown in Fig. 1 and directional diagram itself possesses the rotation symmetry. However the point magnetic moment does not possess such property.
When μ_1 approaches μ, the potential ψ_2 tends to zero and the well-known expressions of electromagnetic field in form (57), (58) and (59) follow from (84) and (85).
The obtained generalized solutions of the Maxwell equations are valid for any values of μ_1 and μ, also near the sources of the electromagnetic waves, described by discontinuous and singular functions.

5. Conclusion

We were able to solve Maxwell's equations for uniaxial anisotropic medium actually showing that the exact general solution in vector form is given by integral convolutions of the fundamental solutions ψ_0 , ψ_1 and ψ_2 with external arbitrary current density, without using the scaling procedure to a corresponding vacuum field and the dyadic Green's functions. In the particular case of a current density, the radiation field of a dipole is considered in greater detail. That method may be useful in analytical treatment of corresponding boundary value problems.

6. Acknowledgment

I wish to thank Professor Panayiotis Frangos, School of Electrical and Computer Engineering, National Technical University of Athens, Greece, for providing me with the encouragement to pursue research in this field.

7. References

Alekseyeva, L. A. & Sautbekov, S. S. (1999). Fundamental Solutions of Maxwell's Equations. *Diff. Uravnenia*,Vol. 35, No. 1, 125-127.

Alekseyeva, L. A. & Sautbekov, S. S. (2000). Method of Generalized Functions For Solving Stationary Boundary Value Proplems For Maxwell's Equations. *Comput. Math. and Math. Phys.*, Vol. 40, No. 4, 582-593.

Barkeshli, S. (1993). An efficient asymptotic closed-form dyadic Green's function for grounded double-layered anisotropic uniaxial material slabs, *J. Electromagnetic Waves and Applications*, Vol. 7, No. 6, 833-856.

Born, M. & Wolf, E. (1999). *Principles of Optics. Electromagnetic Theory of Propagation, Interference and Dffraction of Light*, 7th ed. Cambridge U. Press, Cambridge.

Bunkin, F.V. (1957). On Radiation in Anisotropic Media. *Sov. Phys. JETP* Vol. 5, No.2, 277-283.

Chen, H. C. (1983). *Theory of Electromagnetic Waves*, McGraw-Hill, New York.

Clemmow, P.C. (Jun. 1963a). The theory of electromagnetic waves in a simple anisotropic medium, *Proc. IEE*, pp. 101-106, Vol. 110, No. 1, Jun 1963.

Clemmow, P.C. (Jun. 1963b). The resolution of a dipole field in transverse electric and transverse magnetic waves, *Proc. IEE*, pp. 107-111, Vol. 110, No. 1, Jun 1963.

Cottis, P. G. & Kondylis, G. D. (Feb. 1995). Properties of the dyadic Green's function for an unbounded anisotropic medium, *IEEE Trans. Ant. Prop.*, Vol. 43, No. 2, Feb. 1995, 154-161.

Kogelnik, H. & Motz, H. (1963). Electromagnetic Radiation from Sources Embedded in an Infinite Anisotropic Medium and the Significance of the Poynting Vector, *Proc. Symp. on Electromagnetic Theory and Antennas*, pp. 477-493, 1963, Pergamon Press, New York.

Kong, J. A. (1986). *Electromagnetic Wave Theory*, The 2nd ed., John Wiley and Sons, New York.

Lee, J. K. & Kong, J. A. (1983). Dyadic Green's functions for layered anisotropic medium. *Electromagnetics*, Vol. 3, 111-130.

Ren, W. (Jan. 1993). Contributions to the Electromagnetic Wave Theory of Bounded Homogeneous Anisotropic Media. *Physical Review E*, Vol. 47, No. 1, 664-673.

Sautbekov, S.; Kanymgazieva, I. & Frangos, P. (2008). The generalized solutions of Maxwell equations for the uniaxial crystal, *Journal of Applied Electromagnetism (JAE)*, Vol. 10, No. 2, 43-55, ISSN 1392-1215.

Uzunoglu, N. K.; Cottis, P. G. & Fikioris, J. G. (Jan. 1985). Excitation of Electromagnetic Waves in a Gyroelectric Cylinder, *IEEE Trans. Ant. Propa.*, Vol. AP-33, No. 1, 90-99.

Vladimirov, V. S. (2002). *Methods of the theory of generalized functions*, Taylor and Francis, ISBN 0-415-27356-0, London.

Weiglhofer, W. S. (Feb. 1990). Dyadic Greens functions for general uniaxial media, *IEE Proc. H.*, Vol. 137, No. 1, 5-10.

Weiglhofer, W. S. (1993). A dyadic Green's function representation in electrically gyrotropic media, *AEÜ*, Vol. 47, No. 3, 125-130.

2

Fundamental Problems of the Electrodynamics of Heterogeneous Media with Boundary Conditions Corresponding to the Total-Current Continuity

N.N. Grinchik[1], O.P. Korogoda[2], M.S. Khomich[2], S.V. Ivanova[3], V.I. Terechov[4] and Yu.N. Grinchik[5]
[1]A.V. Luikov Heat and Mass Transfer Institute of the National Academy of Science, Minsk,
[2] Scientific-Engineering Enterprise "Polimag", Minsk,
[3]National Research Nuclear University "MEPhI", Moscow,
[4]Thermal Physics Institute, Novosibirsk,
[5]Belarusian State University, Minsk,
[1,2,5]Republic of Belarus
[3,4]Russian Federation

1. Introduction

Let us consider the interface *S* between two media having different electrophysical properties. On each of its side the magnetic-field and magnetic-inductance vectors as well as the electric-field and electric-displacement vectors are finite and continuous; however, at the surface *S* they can experience a discontinuity of the first kind. Furthermore, at the interface there arise induced surface charges σ and surface currents *i* (whose vectors lie in the plane tangential to the surface *S*) under the action of an external electric field.

The existence of a surface charge at the interface *S* between the two media having different electrophysical properties is clearly demonstrated by the following example. We will consider the traverse of a direct current through a flat capacitor filled with two dielectric materials having relative permittivities ε_1 and ε_2 and electrical conductivities λ_1 and λ_2. A direct-current voltage *U* is applied to the capacitor plates; the total resistance of the capacitor is *R* (Fig. 1). It is necessary to calculate the surface electric charge induced by the electric current.

From the electric-charge conservation law follows the constancy of flow in a circuit; therefore, the following equation is fulfilled:

$$\lambda_1 E_{n_1} = \lambda_2 E_{n_2} = U / (RS) \tag{1}$$

where E_{n_1} and E_{n_2} are the normal components of the electric-field vector.

At the interface between the dielectrics, the normal components of the electric-inductance vector change spasmodically under the action of the electric field by a value equal to the value of the induced surface charge σ:

$$\varepsilon_0 \varepsilon_1 E_{n_1} - \varepsilon_0 \varepsilon_2 E_{n_2} = \sigma \tag{2}$$

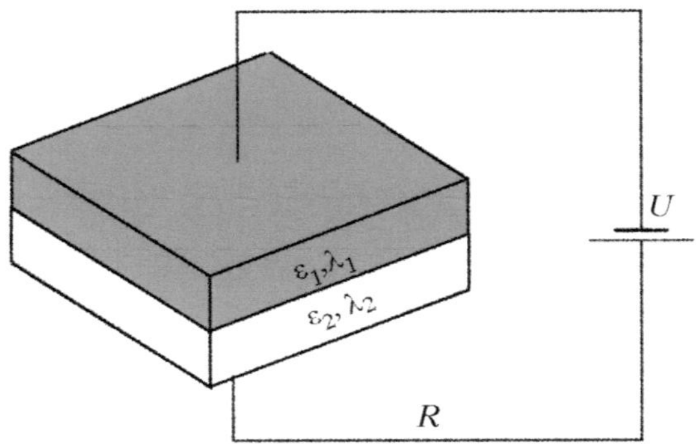

Fig. 1. Dielectric media inside a flat capacitor

Solving the system of Eqs. (see Equations 1 and 2), we obtain the expression for σ

$$\sigma = (U / RS)\varepsilon_0 \left[(\varepsilon_1 / \lambda_1) - (\varepsilon_2 / \lambda_2) \right] \tag{3}$$

It follows from (see Equation 3) that the charge σ is determined by the current and the multiplier accounting for the properties of the medium. If

$$(\varepsilon_1 / \lambda_1) - (\varepsilon_2 / \lambda_2) = 0 \tag{4}$$

a surface charge σ is not formed. What is more, recent trends are toward increased use of micromachines and engines made from plastic materials, where the appearance of a surface charge is undesirable. For oiling of elements of such machines, it is best to use an oil with a permittivity ε_{oil} satisfying the relation

$$\varepsilon_1 < \varepsilon_{oil} < \varepsilon_2 \tag{5}$$

This oil makes it possible to decrease the electrization of the moving machine parts made from dielectric materials. In addition to the charge σ, a contact potential difference arises always independently of the current.

An electric field interacting with a material is investigated with the use of the Maxwell equation (1857)

$$\mathbf{j}_{total} = \nabla \times \mathbf{H}, \nabla \cdot \mathbf{D} = \rho \tag{6}$$

$$-\frac{\partial \mathbf{B}}{\partial t} = \nabla \times \mathbf{E}, \nabla \cdot \mathbf{B} = 0 \tag{7}$$

where $\mathbf{j}_{total} = \lambda \mathbf{E} + \frac{\partial \mathbf{D}}{\partial t}; \mathbf{B} = \mu\mu_0 \mathbf{H}; \mathbf{D} = \varepsilon\varepsilon_0 \mathbf{E}$. In this case, at the interface S the above system of equations is supplemented with the boundary conditions (Monzon, I.; Yonte,T.; Sanchez-Soto, L., 2003; Eremin,Y. & Wriedt,T., 2002)

$$D_{n_1} - D_{n_2} = \sigma \tag{8}$$

$$E_{\tau_1} - E_{\tau_2} = 0 \tag{9}$$

$$B_{n_1} - B_{n_2} = 0 \tag{10}$$

$$\mathbf{H}_{\tau_1} - \mathbf{H}_{\tau_2} = \left[\vec{\mathbf{i}}_\tau \cdot \vec{\mathbf{n}} \right] \tag{11}$$

The indices (subscripts) n and τ denote the normal and tangential components of the vectors to the surface S, and the indices 1 and 2 denote the adjacent media with different electrophysical properties. The index τ denotes any direction tangential to the discontinuity surface. At the same time, a closing relation is absent for the induced surface charge σ, which generates a need for the introduction of an impedance matrix (Wei Hu & Hong Guo, 2002; Danae, D. et al., 2002; Larruquert, J. I., 2001; Koludzija, B. M., 1999; Ehlers, R. A. & Metaxas, A. C., 2003) that is determined experimentally or, in some cases, theoretically from quantum representations (Barta, O.; Pistora, I.; Vesec, I. et al., 2001; Broe, I. & Keller, O., 2002; Keller, 1995; Keller, O., 1995; Keller, O., 1997).

The induced surface charge σ not only characterizes the properties of a surface, but also represents a function of the process, i.e., $\sigma(\mathbf{E}(\partial\mathbf{E}/\partial t, \mathbf{H}(\partial\mathbf{H}/\partial t)))$; therefore, the surface impedances (Wei Hu & Hong Guo, 2002; Danae, D. et al., 2002; Larruquert, J. I., 2001; Koludzija, B. M., 1999; Ehlers, R. A. & Metaxas, A. C., 2003) are true for the conditions under which they are determined. These impedances cannot be used in experiments conducted under other experimental conditions.

The problem of determination of surface charge and surface current on metal-electrolyte boundaries becomes even more complicated in investigating and modeling nonstationary electrochemical processes, e.g., pulse electrolysis, when lumped parameters L, C, and R cannot be used in principle.

We will show that σ can be calculated using the Maxwell phenomenological macroscopic electromagnetic equations and the electric-charge conservation law accounting for the special features of the interface between the adjacent media.

Separate consideration will be given to ion conductors. In constructing a physicomathematical model, we take into account that $\vec{E}$ and $\vec{H}$ are not independent functions; therefore, the wave equation for $\vec{E}$ or $\vec{H}$ is more preferable than the system of equations (see Equations 6 and 7).

2. Electron conductors. New closing relations on the boundaries of adjacent media

2.1 Generalized wave equation for $\vec{E}$ and conditions on the boundaries in the presence of strong discontinuities of the electromagnetic field

2.1.1 Physicomathematical model

We will formulate a physicomathematical model of propagation of an electromagnetic field in a heterogeneous medium. Let us multiply the left and right sides of the equation for the total current (see Equation 6) by $\mu\mu_0$ and differentiate it with respect to time. Acting by the operator rot on the left and right sides of the first equation of Eq. (see Equation 7) on condition that μ=const we obtain

$$\frac{\partial \mathbf{j}_{total}}{\partial t} = \frac{1}{\mu\mu_0}\nabla^2\mathbf{E} - \frac{1}{\mu\mu_0} grad(div\mathbf{E}) \tag{12}$$

In Cartesian coordinates, Eq. (see Equation 12) will take the form

$$\frac{\partial \mathbf{j}_{totalx}}{\partial t} = \frac{1}{\mu\mu_0}\left(\frac{\partial^2 E_x}{\partial x^2} + \frac{\partial^2 E_x}{\partial y^2} + \frac{\partial^2 E_x}{\partial z^2}\right) - \frac{1}{\mu\mu_0}\frac{\partial}{\partial x}\left(\frac{\partial E_x}{\partial x} + \frac{\partial E_y}{\partial y} + \frac{\partial E_z}{\partial z}\right) \tag{13}$$

$$\frac{\partial \mathbf{j}_{totaly}}{\partial t} = \frac{1}{\mu\mu_0}\left(\frac{\partial^2 E_y}{\partial x^2} + \frac{\partial^2 E_y}{\partial y^2} + \frac{\partial^2 E_y}{\partial z^2}\right) - \frac{1}{\mu\mu_0}\frac{\partial}{\partial y}\left(\frac{\partial E_x}{\partial x} + \frac{\partial E_y}{\partial y} + \frac{\partial E_z}{\partial z}\right) \tag{14}$$

$$\frac{\partial \mathbf{j}_{totalz}}{\partial t} = \frac{1}{\mu\mu_0}\left(\frac{\partial^2 E_z}{\partial x^2} + \frac{\partial^2 E_z}{\partial y^2} + \frac{\partial^2 E_z}{\partial z^2}\right) - \frac{1}{\mu\mu_0}\frac{\partial}{\partial z}\left(\frac{\partial E_x}{\partial x} + \frac{\partial E_y}{\partial y} + \frac{\partial E_z}{\partial z}\right) \tag{15}$$

At the interface, the following relation (Eremin,Y. & Wriedt,T., 2002) is also true:

$$div\mathbf{i}_\tau + I_{qx_1} - I_{qx_2} = -\frac{\partial\sigma}{\partial t} \tag{16}$$

Let us write conditions (see Equations 8–11) in the Cartesian coordinate system:

$$D_{x_1} - D_{x_2} = \sigma \tag{17}$$

$$E_{y_1} - E_{y_2} = 0 \tag{18}$$

$$E_{z_1} - E_{z_2} = 0 \tag{19}$$

$$B_{x_1} - B_{x_2} = 0 \tag{20}$$

$$H_{y_1} - H_{y_2} = i_z \tag{21}$$

$$H_{z_1} - H_{z_2} = i_y \tag{22}$$

where $\mathbf{i}_\tau = i_y\mathbf{j} + i_z\mathbf{k}$ is the surface-current density, and the coordinate x is directed along the normal to the interface. The densities i_y and i_z of the surface currents represent the electric charge carried in unit time by a segment of unit length positioned on the surface drawing the current perpendicularly to its direction.

The order of the system of differential equations (see Equations 13–15) is equal to 18. Therefore, at the interface S, it is necessary to set, by and large, nine boundary conditions. Moreover, the three additional conditions (see Equation 17, 21, and 22) containing (prior to the solution) unknown quantities should be fulfilled at this interface. Consequently, the total number of conjugation conditions at the boundary S should be equal to 12 for a correct solution of the problem.

Differentiating expression (see Equation 17) with respect to time and using relation (see Equation 16), we obtain the following condition for the normal components of the total current at the medium-medium interface:

$$div\mathbf{i}_{\tau} + \mathbf{j}_{totalx_1} = \mathbf{j}_{totalx_2} \tag{23}$$

that allows one to disregard the surface charge σ. Let us introduce the arbitrary function *f*: $[f]\big|_{x=\xi} = f_1\big|_{x=\xi+0} - f_2\big|_{x=\xi-0}$. In this case, expression (see Equation 23) will take the form

$$\left[div\mathbf{i}_{\tau} + \mathbf{j}_{totalx_1}\right]\Big|_{x=\xi} = 0 \tag{24}$$

It is assumed that, at the medium-medium interface, E_x is a continuous function of y and z. Then, differentiating Eq. (see Equation 23) with respect to y and z, we obtain

$$\left[\frac{\partial}{\partial y}\mathbf{j}_{totalx_1}\right]\Bigg|_{x=\xi} = -\frac{\partial\left(div\mathbf{i}_{\tau}\right)}{\partial y} \tag{25}$$

$$\left[\frac{\partial}{\partial z}\mathbf{j}_{totalx_1}\right]\Bigg|_{x=\xi} = -\frac{\partial\left(div\mathbf{i}_{\tau}\right)}{\partial z} \tag{26}$$

Let us differentiate conditions (see Equations 20–22) for the magnetic induction and the magnetic-field strength with respect to time. On condition that **B**=$\mu\mu_0$**H**

$$\left[\frac{\partial B_x}{\partial t}\right]\Bigg|_{x=\xi} = 0, \left[\frac{1}{\mu\mu_0}\frac{\partial B_y}{\partial t}\right]\Bigg|_{x=\xi} = \frac{\partial i_z}{\partial t}, \left[\frac{1}{\mu\mu_0}\frac{\partial B_z}{\partial t}\right]\Bigg|_{x=\xi} = \frac{\partial i_y}{\partial t} \tag{27}$$

Using Eq. (see Equation 7) and expressing (see Equation 27) in terms of projections of the electric-field rotor, we obtain

$$\left[rot_x\mathbf{E}\right]\big|_{x=\xi} = 0 \text{ and } \left[\frac{\partial E_z}{\partial y} - \frac{\partial E_y}{\partial z}\right]\Bigg|_{x=\xi} = 0 \tag{28}$$

$$\left[\frac{1}{\mu\mu_0}rot_y\mathbf{E}\right]\Bigg|_{x=\xi} = \frac{\partial i_z}{\partial t} \text{ or } \left[\frac{1}{\mu\mu_0}\left(\frac{\partial E_x}{\partial z} - \frac{\partial E_z}{\partial x}\right)\right]\Bigg|_{x=\xi} = \frac{\partial i_z}{\partial t} \tag{29}$$

$$\left[\frac{1}{\mu\mu_0}rot_z\mathbf{E}\right]\Bigg|_{x=\xi} = \frac{\partial i_z}{\partial t} \text{ or } \left[\frac{1}{\mu\mu_0}\left(\frac{\partial E_y}{\partial x} - \frac{\partial E_x}{\partial y}\right)\right]\Bigg|_{x=\xi} = \frac{\partial i_z}{\partial t} \tag{30}$$

Here, Eq. (see Equation 28) is the normal projection of the electric-field rotor, Eq. (see Equation 29) is the tangential projection of the rotor on y, and Eq. (see Equation 30) is the rotor projection on z.

Assuming that E_y and E_z are continuous differentiable functions of the coordinates y and z, from conditions (see Equations 18 and 19) we find

$$\left[\frac{\partial E_y}{\partial y}\right]\Bigg|_{x=\xi}=0,\left[\frac{\partial E_y}{\partial z}\right]\Bigg|_{x=\xi}=0$$
$$\left[\frac{\partial E_z}{\partial y}\right]\Bigg|_{x=\xi}=0,\left[\frac{\partial E_z}{\partial z}\right]\Bigg|_{x=\xi}=0 \tag{31}$$

In accordance with the condition that the tangential projections of the electric field on z and y are equal and in accordance with conditions (see Equations 18 and 19), the expressions for the densities of the surface currents i_z and i_y take the form

$$i_z=\bar{\lambda}E_z\big|_{x=\xi},\ i_y=\bar{\lambda}E_y\big|_{x=\xi} \tag{32}$$

where

$$\bar{\lambda}=\frac{1}{2}(\lambda_1+\lambda_2)\Big|_{x=\xi} \tag{33}$$

is the average value of the electrical conductivity at the interface between the adjacent media in accordance with the Dirichlet theorem for a piecewise-smooth, piecewise-differentiable function.

Consequently, formulas (see Equations 31–33) yield

$$\left[div\vec{i_\tau}\right]\Big|_{x=\xi}=0 \tag{34a}$$

Relation (see Equation 34) and hence the equality of the normal components of the total current were obtained (in a different manner) by G.A. Grinberg and V.A. Fok (Grinberg, G.A. & Fok, V.A., 1948). In this work, it has been shown that condition (34a) leads to the equality of the derivatives of the electric field strength along the normal to the surface

$$\left[\frac{\partial E_x}{\partial x}\right]\Bigg|_{x=\xi}=0 \tag{34b}$$

With allowance for the foregoing we have twelve conditions at the interface between the adjacent media that are necessary for solving the complete system of equations (see Equations 13–15):

a. the functions E_y and E_z are determined from Eqs. (see Equations 18 and 19);
b. E_x is determined from condition (see Equation 24);
c. the values of $\partial E_x/\partial y$, $\partial E_x/\partial z$, and $\partial E_x/\partial x$ are determined from relations (see Equations 25 and 26) with the use of the condition of continuity of the total-current normal component at the interface (see Equation 24) and the continuity of the derivative of the total current with respect to the coordinate *x*;
d. the values of $\partial E_y/\partial y$, $\partial E_y/\partial z$, and $\partial E_z/\partial z$ are determined from conditions (see Equations 31 and 32) in consequence of the continuity of the tangential components of the electric field along *y* and *z*;
e. the derivatives $\partial E_y/\partial x$ and $\partial E_z/\partial x$ are determined from conditions (see Equations 29 and 30) as a consequence of the equality of the tangential components of the electric-field rotor along *y* and *z*.

Note that condition (see Equation 23) was used by us in (Grinchik, N. N. & Dostanko, A. P., 2005) in the numerical simulation of the pulsed electrochemical processes in the one-dimensional case. Condition (see Equation 28) for the normal component of the electric-field rotor represents a linear combination of conditions (see Equations 31 and 32); therefore, $rot_x E = 0$ and there is no need to use it in the subsequent discussion. The specificity of the expression for the general law of electric-charge conservation at the interface is that the components $\partial E_y/\partial y$ and $\partial E_z/\partial z$ are determined from conditions (see Equations 31 and 32) that follow from the equality and continuity of the tangential components E_y and E_z at the boundary S.
Thus, at the interface between the adjacent media the following conditions are fulfilled: the equality of the total-current normal components; the equality of the tangential projections of the electric-field rotor; the electric-charge conservation law; the equality of the electric-field tangential components and their derivatives in the tangential direction; the equality of the derivatives of the total-current normal components in the direction tangential to the interface between the adjacent media, determined with account for the surface currents and without explicit introduction of a surface charge. They are true at each cross section of the sample being investigated.

2.1.2 Features of calculation of the propagation of electromagnetic waves in layered media

The electromagnetic effects arising at the interface between different media under the action of plane electromagnetic waves have a profound impact on the equipment because all real devices are bounded by the surfaces and are inhomogeneous in space. At the same time, the study of the propagation of waves in layered conducting media and, according to (Born, 1970), in thin films is reduced to the calculation of the reflection and transmission coefficients; the function $E(x)$ is not determined in the thickness of a film, i.e., the geometrical-optics approximation is used.
The physicomathematical model proposed allows one to investigate the propagation of an electromagnetic wave in a layered medium without recourse to the assumptions used in (Wei Hu & Hong Guo, 2002; Danae, D. et al., 2002; Larruquert, J. I., 2001; Ehlers, R. A. & Metaxas, A. C., 2003).
Since conditions (see Equations 23-32) are true at each cross section of a layered medium, we will use schemes of through counting without an explicit definition of the interface between the media. In this case, it is proposed to calculate E_x at the interface in the following way.
In accordance with Eq. (see Equation 17), $E_{x1} \neq E_{x2}$, i.e., $E_x(x)$ experiences a discontinuity of the first kind. Let us determine the strength of the electric field at the discontinuity point $x = \xi$ on condition that $E_x(x)$ is a piecewise-smooth, piecewise-differentiable function having finite one-sided derivatives $E'_{x^+}(x)$ and $E'_{x^-}(x)$. At the discontinuity points x_i,

$$E'_{x^+}(x_i) = \lim_{\Delta x_i \to +0} \frac{E(x_i + \Delta x_i) - E(x_i + 0)}{\Delta x_i} \tag{35}$$

$$E'_{x^-}(x_i) = \lim_{\Delta x_i \to -0} \frac{E(x_i + \Delta x_i) - E(x_i - 0)}{\Delta x_i} \tag{36}$$

In this case, in accordance with the Dirichlet theorem (Kudryavtsev, 1970), the Fourier series of the function $E(x)$ at each point x, including the discontinuity point ξ, converges and its sum is equal to

$$E_{x=\xi} = \frac{1}{2}\left[E(\xi - 0) + E(\xi + 0)\right] \tag{37}$$

The Dirichlet condition (see Equation 37) also has a physical meaning. In the case of contact of two solid conductors, e.g., dielectrics or electrolytes in different combinations (metal-electrolyte, dielectric-electrolyte, metal-vacuum, and so on), at the interface between the adjacent media there always arises an electric double layer (EDL) with an unknown (as a rule) structure that, however, substantially influences the electrokinetic effects, the rate of the electrochemical processes, and so on. It is significant that, in reality, the electrophysical characteristics λ, ε, and $E(x)$ change uninterruptedly in the electric double layer; therefore, (see Equation 37) is true for the case where the thickness of the electric double layer, i.e., the thickness of the interphase boundary, is much smaller than the characteristic size of a homogeneous medium. In a composite, e.g., in a metal with embedments of dielectric balls, where the concentration of both components is fairly large and their characteristic sizes are small, the interphase boundaries can overlap and condition (see Equation 37) can break down.

If the thickness of the electric double layer is much smaller than the characteristic size L of an object, (see Equation 37) also follows from the condition that $E(x)$ changes linearly in the EDL region. In reality, the thickness of the electric double layer depends on the kind of contacting materials and can comprise several tens of angstroms (Frumkin, 1987). In accordance with the modern views, the outer coat of the electric double layer consists of two parts, the first of which is formed by the ions immediately attracted to the surface of the metal (a "dense" or a "Helmholtz" layer of thickness h), and the second is formed by the ions separated by distances larger than the ion radius from the surface of the layer, and the number of these ions decreases as the distance between them and the interface (the "diffusion layer") increases. The distribution of the potential in the dense and diffusion parts of the electric double layer is exponential in actual practice (Frumkin, 1987), i.e., the condition that $E(x)$ changes linearly breaks down; in this case, the sum of the charges of the dense and diffusion parts of the outer coat of the electric double layer is equal to the charge of its inner coat (the metal surface). However, if the thickness of the electric double layer h is much smaller than the characteristic size of an object, the expansion of $E(x)$ into a power series is valid and one can restrict oneself to the consideration of a linear approximation. In accordance with the more general Dirichlet theorem (1829), a knowledge of this function in the EDL region is not necessary to substantiate Eq. (see Equation 37). Nonetheless, the above-indicated physical features of the electric double layer lend support to the validity of condition (see Equation 37).

The condition at interfaces, analogous to Eq. (see Equation 37), has been obtain earlier (Tikhonov, A. N. & Samarskii, A. A., 1977) for the potential field (where rot $\mathbf{E} = 0$) on the basis of introduction of the surface potential, the use of the Green formula, and the consideration of the discontinuity of the potential of the double layer. In (Tikhonov, A. N. & Samarskii, A. A., 1977), it is also noted that the consideration of the thickness of the double layer and the change in its potential at $h/L \ll 1$ makes no sense in general; therefore, it is advantageous to consider, instead of the volume potential, the surface potential of any density. Condition (see Equation 37) can be obtained, as was shown in (Kudryavtsev, 1970), from the more general Dirichlet theorem for a nonpotential vorticity field (Tikhonov, A. N. & Samarskii, A. A., 1977).

Thus, the foregoing and the validity of conditions (see Equations 17-19 and 25-.32) at each cross section of a layered medium show that, for numerical solution of the problem being considered it is advantageous to use schemes of through counting and make the discretization of the medium in such a way that the boundaries of the layers have common points.

The medium was divided into finite elements so that the nodes of a finite-element grid, lying on the separation surface between the media with different electrophysical properties, were shared by these media at a time. In this case, the total currents or the current flows at the interface should be equal if the Dirichlet condition (see Equation 37) is fulfilled.

2.1.3 Results of numerical simulation of the propagation of electromagnetic waves in layered media

Let us analyze the propagation of an electromagnetic wave through a layered medium that consists of several layers with different electrophysical properties in the case where an electromagnetic-radiation source is positioned on the upper plane of the medium. It is assumed that the normal component of the electric-field vector $E_x = 0$ and its tangential component $E_y = a \sin(\omega t)$, where a is the electromagnetic-wave amplitude (Fig. 2).

In this example, for the purpose of correct specification of the conditions at the lower boundary of the medium, an additional layer is introduced downstream of layer 6; this layer has a larger conductivity and, therefore, the electromagnetic wave is damped out rapidly in it. In this case, the condition $E_y = E_z = 0$ can be set at the lower boundary of the medium. The above manipulations were made to limit the size of the medium being considered because, in the general case, the electromagnetic wave is attenuated completely at an infinite distance from the electromagnetic-radiation source.

Numerical calculations of the propagation of an electromagnetic wave in the layered medium with electrophysical parameters $\varepsilon_1 = \varepsilon_2 = 1$, $\lambda_1 = 100$, $\lambda_2 = 1000$, and $\mu_1 = \mu_2 = 1$ were carried out. Two values of the cyclic frequency $\omega = 2\pi/T$ were used: in the first case, the electromagnetic-wave frequency was assumed to be equal to $\omega = 10^{14}$ Hz (infrared radiation), and, in the second case, the cyclic frequency was taken to be $\omega = 10^9$ Hz (radiofrequency radiation).

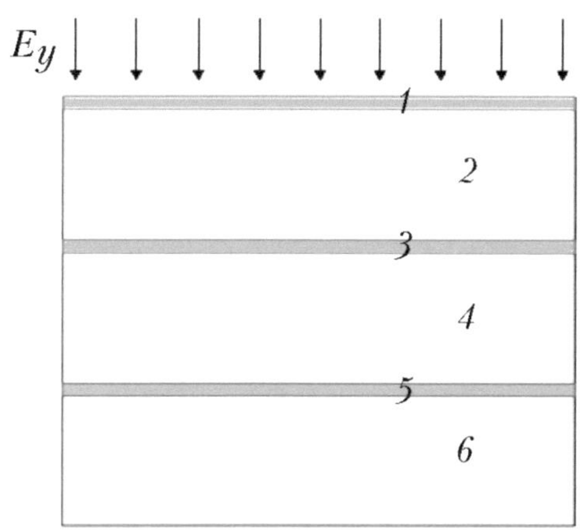

Fig. 2. Scheme of a layered medium: layers 1, 3, and 5 are characterized by the electrophysical parameters ε_1, λ_1, and μ_1, and layers 2, 4, and 6 — by ε_2, λ_2, μ_2.

As a result of the numerical solution of the system of equations (see Equations 13–15) with the use of conditions S (see Equations 24-34) at the interfaces, we obtained the time dependences of the electric-field strength at different distances from the surface of the layered medium (Fig. 3).

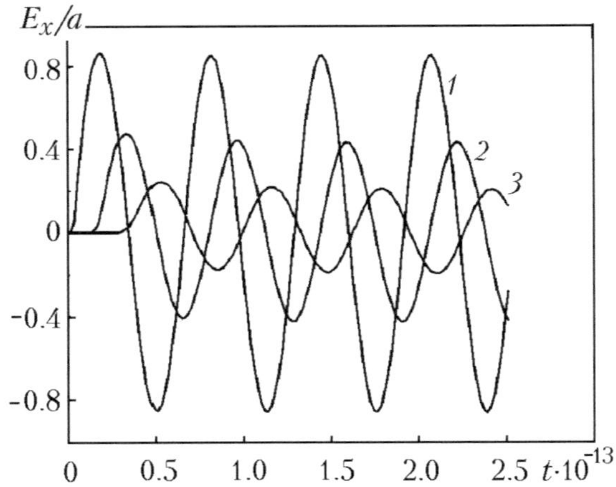

Fig. 3. Time change in the tangential component of the electric-field strength at a distance of 1 μm (1), 5 μm (2), and 10 μm (3) from the surface of the medium at $\lambda_1 = 100$, $\lambda_2 = 1000$, $\varepsilon_1 = \varepsilon_2 = 1$, $\mu_1 = \mu_2 = 1$, and $\omega = 10^{14}$ Hz. t, sec.

The results of our simulation (Fig. 4) have shown that a high-frequency electromagnetic wave propagating in a layered medium is damped out rapidly, whereas a low-frequency electromagnetic wave penetrates into such a medium to a greater depth. The model developed was also used for calculating the propagation of a modulated signal of frequency 20 kHz in a layered medium. As a result of our simulation (Fig. 5), we obtained changes in the electric-field strength at different depths of the layered medium, which points to the fact that the model proposed can be used to advantage for calculating the propagation of polyharmonic waves in layered media; such a calculation cannot be performed on the basis of the Helmholtz equation.

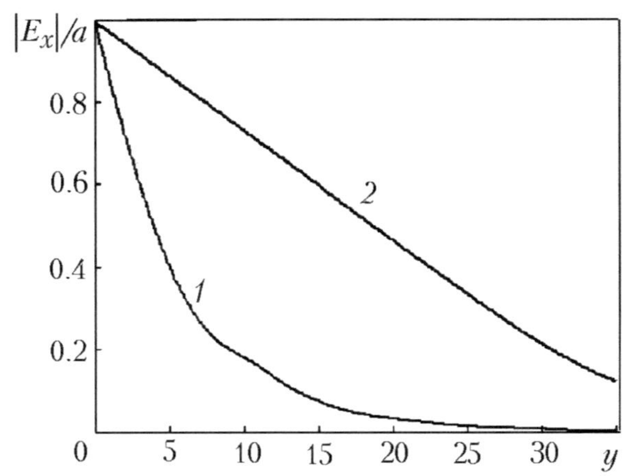

Fig. 4. Distribution of the amplitude of the electric-field-strength at the cross section of the layered medium: $\omega = 10^{14}$ (1) and 10^9 Hz (2). y, μm.

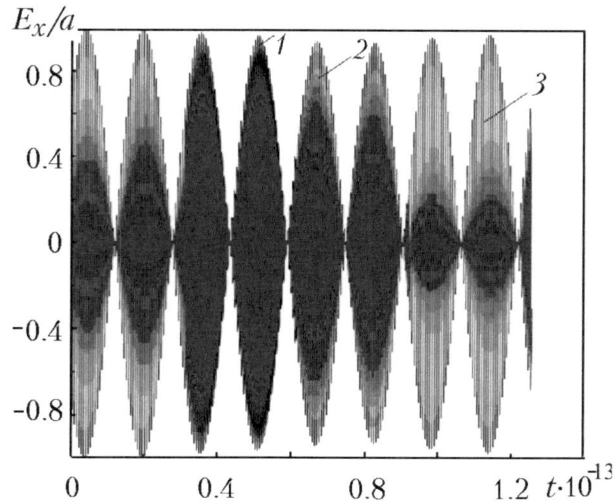

Fig. 5. Time change in the electric-field strength at a distance of 1 (1), 5 (2), and 10 µm (3) from the surface of the medium. t, sec.

The physicomathematical model developed can also be used to advantage for simulation of the propagation of electromagnetic waves in media with complex geometric parameters and large discontinuities of the electromagnetic field (Fig. 6).

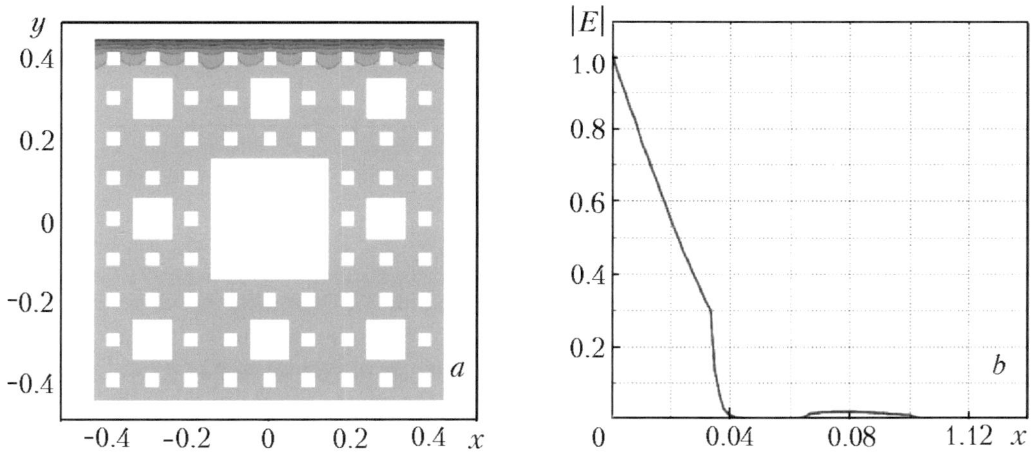

(a) Distribution of the amplitude of the electric-field strength in the two-dimensional medium

(b) Distribution of the amplitude of the electric-field strength in depth

Fig. 6. Distribution of the amplitude of the electric-field strength in the two-dimensional medium and in depth at $\varepsilon_1 = 15$, $\varepsilon_2 = 20$, $\lambda_1 = 10^{-6}$, $\lambda_2 = 10$, $\mu_1 = \mu_2 = 1$, and $\omega = 10^9$ Hz (the dark background denotes medium 1, and the light background – medium 2). x, y, mm; E, V/m.

Figure 6a shows the cross-sectional view of a cellular structure representing a set of parallelepipeds with different cross sections in the form of squares. The parameters of the materials in the large parallelepiped are denoted by index 1, and the parameters of the materials in the small parallelepipeds (the squares in the figure) are denoted by index 2.

An electromagnetic wave propagates in the parallelepipeds (channels) in the transverse direction. It is seen from Fig. 6b that, in the cellular structure there are "silence regions," where the amplitude of the electromagnetic-wave strength is close to zero, as well as inner regions where the signal has a marked value downstream of the "silence" zone formed as a result of the interference.

2.1.4 Results of numerical simulation of the scattering of electromagnetic waves in angular structures

It is radiolocation and radio-communication problems that are among the main challenges in the set of problems solved using radio-engineering devices.

Knowledge of the space-time characteristics of diffraction fields of electromagnetic waves scattered by an object of location into the environment is necessary for solving successfully any radiolocation problem. Irradiated object have a very intricate architecture and geometric shape of the surface consisting of smooth portions and numerous wedge-shaped for formations of different type-angular joints of smooth portions, surface fractures, sharp edges, etc. - with rounded radii much smaller than the probing-signal wavelength. Therefore, solution of radiolocation problems requires that the methods of calculation of the diffraction fields of electromagnetic waves excited and scattered by different surface portions of the objects, in particular, by wedge-shaped formations, be known, since the latter are among the main sources of scattered waves.

For another topical problem, i.e., radio communication effected between objects, the most difficult are the issues of designing of antennas arranged on an object, since their operating efficiency is closely related to the geometric and radiophysical properties of its surface.

The issues of diffraction of an electromagnetic wave in wedge-shaped regions are the focus of numerous of the problems for a perfectly conducting and impedance wedge for monochromatic waves is representation of the diffraction field in an angular region in the form of a Sommerfeld integral (Kryachko, A.F. et al., 2009).

Substitution of Sommerfeld integrals into the system of boundary conditions gives a system of recurrence functional equations for unknown analytical integrands. The system's coefficients are Fresnel coefficients defining the reflection of plane media or their refraction into the opposite medium. From the system of functional equations, one determines, in a recurrence manner, sequences of integrand poles and residues in these poles.

The edge diffraction field in both media is determined using a pair of Fredholm-type singular integral equations of the second kind which are obtained from the above-indicated systems of functional equations with subsequent computation of Sommerfeld integrals by the saddle-point approximation. The branching points of the integrands condition the presence of creeping waves excited by the edge of the dielectric wedge.

The proposed method is only true of monochromatic waves and of the approximate Leontovich boundary conditions, when the field of the electromagnetic wave slowly varies from point to point on a wavelength scale (Leontovich, 1948).

We note that the existing approximate Leontovich conditions have a number of other constants and should be used with caution (Leontovich, 1948).

In actual fact, the proposed calculation method does not work in the presence of, e.g., two wedges, when the sharp angles are pointed at each other, i.e., an optical knife, or in diffraction of the electromagnetic wave on a system of parallel lobes, when the gap between the lobes is in the region of microns, and the electromagnetic field is strongly "cut" throughout the space with a step much than the wavelength.

A) Optical Knife

Figure 7 shows the field of an electromagnetic wave in its diffraction on the optical knife. The parameters of the wave at entry and at exit are $E_x=10^4\sin(10^{10}t)$, $E_y=10^4\cos(10^{10}t)$.

The electrophysical characteristics are as follows: the wedge is manufactured from aluminum: $\varepsilon=1$; $\mu=1$; $\sigma=3.774\cdot10^7$ S/m; the ambient medium is air.

The dimensions of the computational domain are 0.1×0.05 m. The calculation time 10^{-9} sec, and the time step is 10^{-11} sec.

Numerical solution of the system of equations (13)-(15) yields the dependences of the distribution Ex (x,y) и Ey (x,y) on the optical knife. The calculations results are in good agreement with the existing experimental data and experiments specially conducted at the Department of the Physics and Chemistry of Nonequilibrium Media of the A.V. Luikov Heat and Mass Transfer Institute of the National Academy of Sciences by A.I. Bereznyak.

The experiments were carried out with an optical-range laser and were tentative in character but the obtained experimental photographs of diffraction fields and the calculated results turned out to be in good qualitative agreement. The authors express their thanks to A.I. Bereznyak for the conducting of the experiments.

B) Diffraction Grating

The parameters of the wave and the interfacial conditions are the same, as those for the case "optical knife". The electrophysical characteristics are as follows: 2D lobes, $\varepsilon=12$; and $\sigma=100$ S/m; the ambient medium is air; the characteristics of the prism and the square are identical to those of the lobes.

Figure 8 corresponds to a calculation time of 10^{-10} sec; the time step is 10^{-12} sec. Figure 9 corresponds to a calculation time of 10^{-9} sec; the time step is 10^{-11} sec.

It is seen from the modeling results that the proposed "comb" cab be used as a filter of a high-frequency signal. Furthermore, we carried out numerical calculations of a modulated signal at a frequency of 20 kHz. The results of the modulated-signal calculations are not given. To analyze the difference scheme for stability was analyzed by the initial data. When the time and space steps are large there appear oscillations of the grid solution and of its "derivatives" ("ripple") which strongly decrease the accuracy of the scheme. Undoubtedly, this issue calls for separate consideration. The proposed algorithm of solution of Maxwell equations allows circuitry-engineering modeling of high-frequency radio-engineering devices and investigation of the propagation of electromagnetic waves in media of intricate geometry in the presence of strong discontinuities of electromagnetic field.

The result of Para 2.1 were published in part (Grinchik, N.N et al., 2009).

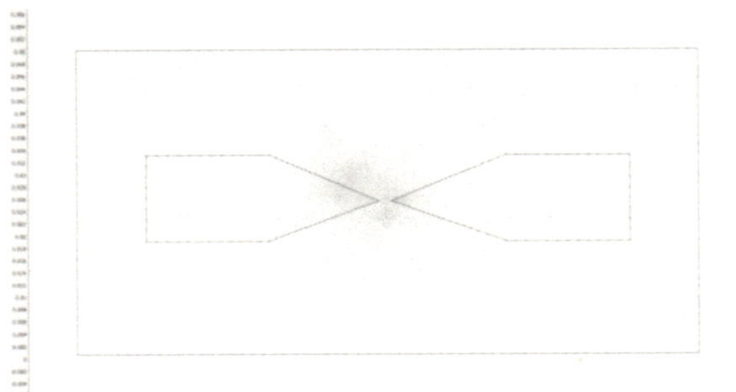

(a) Mesh

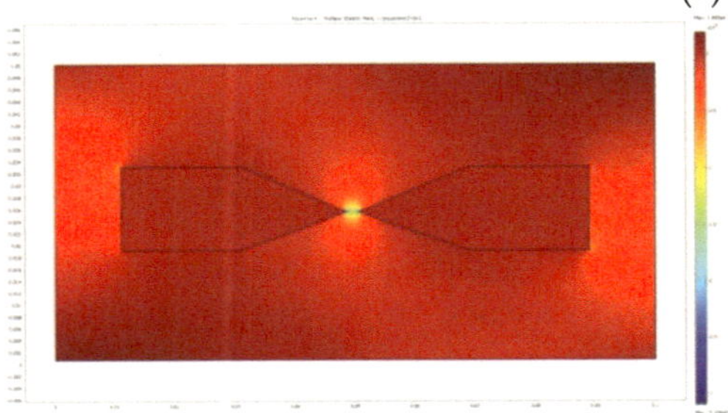

(b) Amplitude E_x

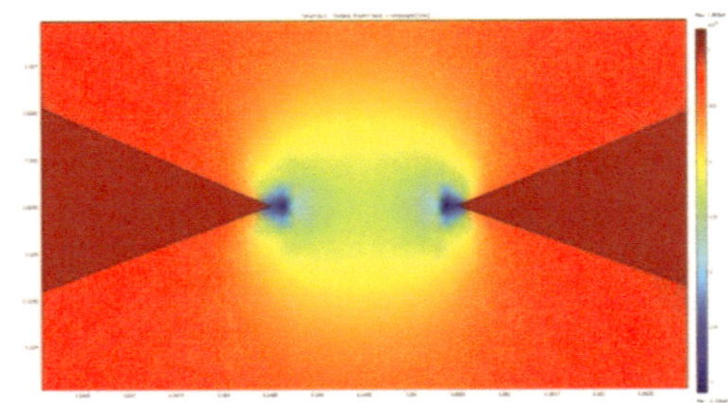

(c) Amplitude E_x

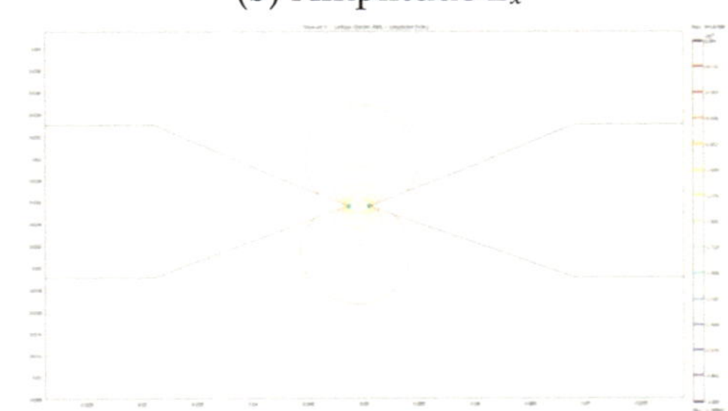

(d) Isolines E_x

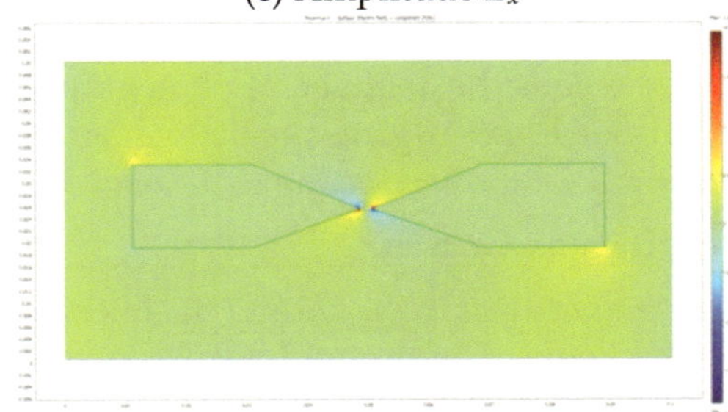

(e) Amplitude E_y

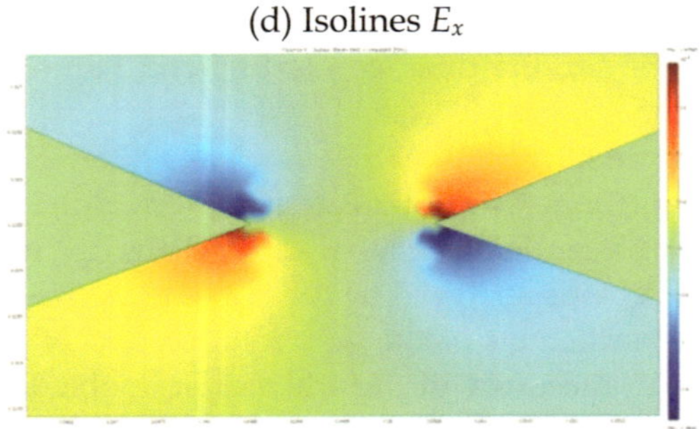

(f) Amplitude E_y

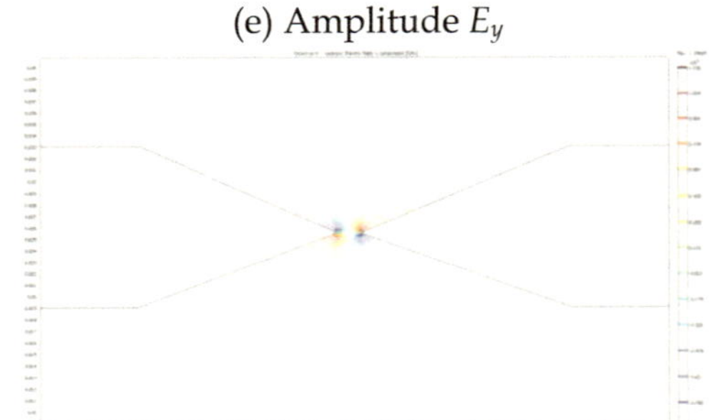

(g) Isolines E_y

Fig. 7. Mesh, amplitude E_x and isolines, amplitude E_y and isolines of the electromagnetic field strength

(a) Amplitude E_x

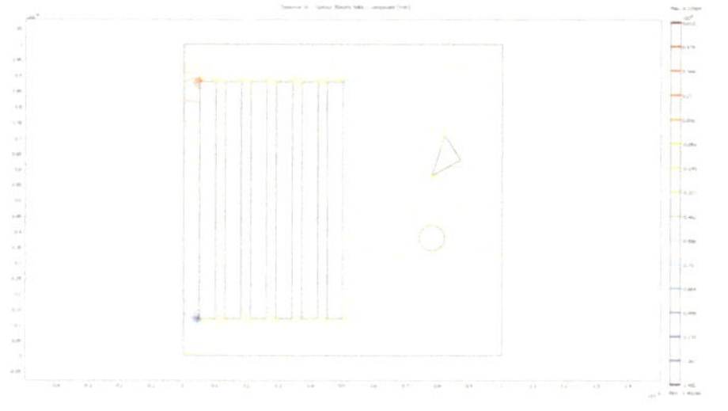

(b) Isolines E_x

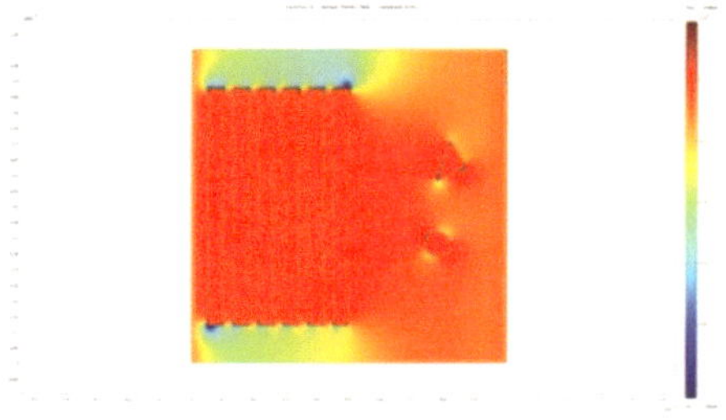

(c) Amplitude E_y

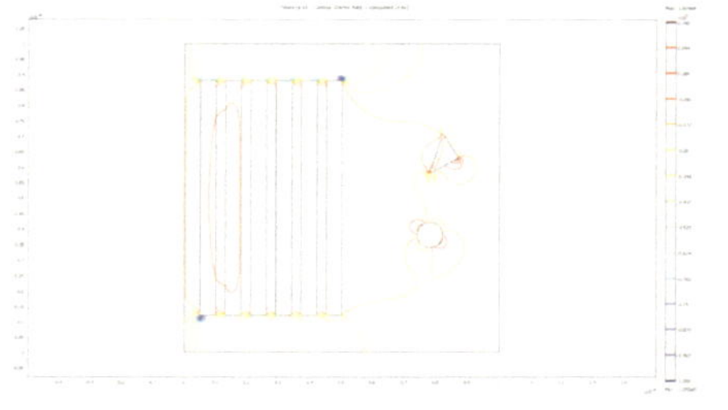

(d) Isolines E_y

Fig. 8. Amplitude E_x and isolines, amplitude E_y and isolines of the electromagnetic field strength

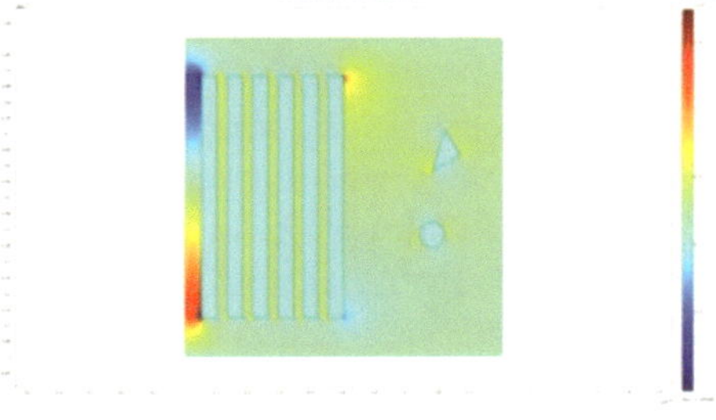

(a) Amplitude E_x

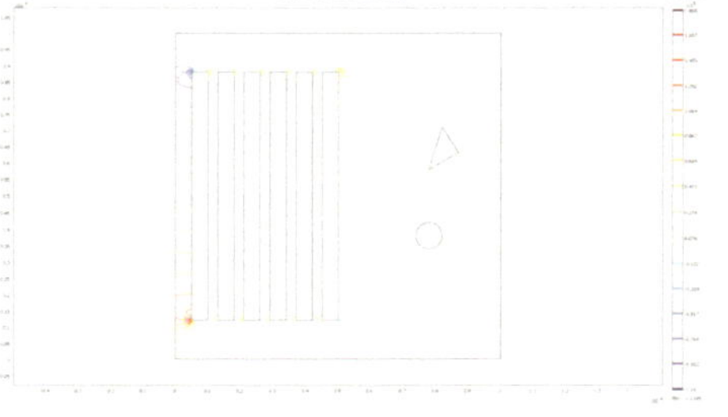

(b) Isolines E_x

(c) Amplitude E_y

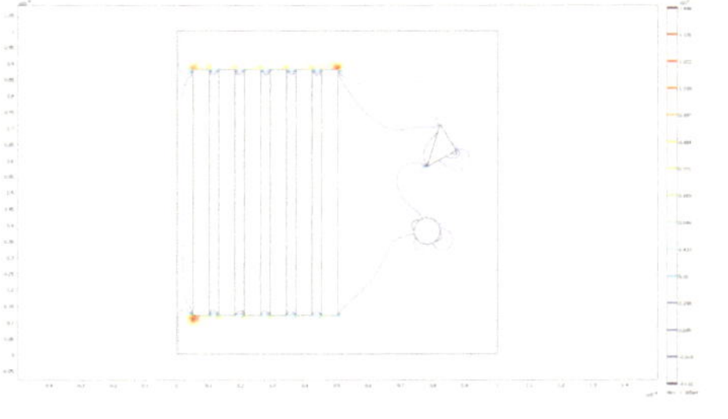

(d) Isolines E_y

Fig. 9. Amplitude E_x and isolines, amplitude E_y and isolines of the electromagnetic field strength

2.1.5 Conclusions

We were the first to construct a consistent physicomathematical model of propagation of electromagnetic waves in layered media without recourse to the matrices of the induced-surface-charge impedances. This model is based on the Maxwell equations, the electric-charge conservation law, the total-current continuity, and the Dirichlet theorem. Our numerical investigations have shown that the physical and mathematical model proposed can be used to advantage for simulation of the propagation of a high-frequency electromagnetic wave in a medium consisting of layers having different electrophysical properties.

2.2 Wave equation for $\vec{H}$ and conditions on the boundaries in the presence of strong discontinuities of the electromagnetic field. numerical modeling of electrodynamic processes in the surface layer

2.2.1 Introduction

During the interaction of an external magnetic field and magnetic abrasive particles, the particles are magnetized, and magnetic dipoles with the moment oriented predominantly along the field are formed. "Chains" along the force lines of the field (Shul'man, Z. P. & Kordonskii ,V. I., 1982; Khomich, 2006) appear that periodically act on the processable surface with a frequency $\omega=l/v$. A fixed elemental area of the material periodically experiences the effect of the magnetic field of one direction. Actually, the frequency and duration of the pulse will be still higher because of the rotation of the magnetic abrasive particle due to the presence of the moment of forces on contact and of the friction of the particle against the processable part. In what follows, we will not take into account the effect of rotation.

We assume that the particle velocity on the polisher is v. If the particle radius is r, then the angular frequency is $\omega=2\pi v/r$, and precisely this frequency determines the frequency of the effect of the variable magnetic field component due to the fact that for a ferromagnetic $\mu>1$. The magnetic permeability μ of ferromagnetics, which are usually used in magnetic abrasive polishing, is measured by thousands of units in weak fields. However, in polishing, the constant external magnetic field is strong and amounts to 10^5-10^6 A/m, and in this case the value of μ for compounds of iron and nickel and for Heusler alloy decreases substantially.

Because of the presence of a strong external magnetic field H_0 the "small" absolute value of μ of an abrasive particle leads to a periodic "increase" and "decrease" in the normal component of the magnetic induction near the processable surface. In the present work we used neodymium magnets (neodymium-iron-boron) with H_0>485,000 A/m. The magnetic permeability of a magnetic abrasive particle based on carbonyl iron was assumed in this case to be equal to $\mu_1=100$.

Due to the continuity of the normal magnetic induction component $B_{n1}=B_{n2}$, where $B_{n1}=\mu_1\mu_0H_1$; $B_{n2}=\mu_2\mu_0H_2$. For example, in glasses ($\mu_2=1$; therefore at the boundary of contact of the glass with the magnetic abrasive particle an additional variable magnetic field of strength $H_1 > H_0$ appears.

In (Levin, M. N. et al., 2003; Orlov, A. M. et al., 2001; Makara, V. A. et al., 2001; Rakomsin, 2000), magnetic field-induced effects in silicon are considered: a nonmonotonic change in the crystal lattice parameters in the surface layer of silicon, the gettering of defects on the surface, the change in the sorption properties of the silicon surface, and the change in the mobility of the edge dislocations and in the microhardness of silicon.

In (Golovin, Yu. I. et al.; Makara ,V. A. et al., 2008; Orlov, A. M. et al., 2003), the influence of an electromagnetic field on the domain boundaries, plasticity, strengthening, and on the reduction of metals and alloys was established.

In view of the foregoing, it is of interest to find the relationship between the discrete-impulse actions of a magnetic field of one direction on the surface layer of the processable material that contains domains. According to (Shul'man, Z. P. & Kordonskii ,V. I., 1982), the size of domains is as follows: 0.05 μm in iron, 1.5 μm in barium ferrite; 8 μm in the MnBi compound, and 0.5-1 μm in the acicular gamma ferric oxide. According to (Akulov, 1961), the size of a domain may reach 10^{-6} cm^3 (obtained by the method of magnetic metallography).

As a rule, an abrasive exhibits a distinct shape anisotropy, whereas the frequency of the effect is determined by the concentration of abrasive particles in a hydrophobic solution and by the velocity of its motion. We assume that on the surface of a processable crystal the magnetic field strength $H(t)=H_1 \sin^4(\omega t) + H_0$.

It is required to find the value of the magnetic field strength in the surface layer that has the characteristics λ_1, ε_1, and μ_1 and contains domains with electrophysical properties λ_2, ε_2, and μ_2. The domains may have the form of a triangular prism, a bar, a cylinder, etc.

2.2.2 Physicomathematical model. wave equation for $\vec{H}$

We will formulate a physicomathematical model of propagation of electromagnetic waves in a heterogeneous medium. The media in contact are considered homogeneous. We operate with the operator rot on the left- and right-hand sides of the first equation for the total current (see Equation 6) and multiply by $\mu_0\mu$; then we differentiate the second equation in Eq. (see Equation 7) with respect to time. Taking into consideration the solenoidality of the magnetic field (see Equation 7) and the rule of repeated application of the operator ∇ to the vector **H**, we obtain

$$\mu_0\varepsilon\varepsilon_0\frac{\partial^2\mathbf{H}}{\partial t^2}+\lambda\mu_0\frac{\partial\mathbf{H}}{\partial t}=\frac{1}{\mu}\nabla^2\mathbf{H} \tag{38}$$

In the Cartesian coordinates Eq. (see Equation 38) will have the form

$$\begin{aligned}
\varepsilon\varepsilon_0\frac{\partial^2 H_x}{\partial t^2}+\lambda\mu_0\frac{\partial H_x}{\partial t}&=\frac{1}{\mu}\left(\frac{\partial^2 H_x}{\partial x^2}+\frac{\partial^2 H_x}{\partial y^2}+\frac{\partial^2 H_x}{\partial z^2}\right)\\
\varepsilon\varepsilon_0\frac{\partial^2 H_y}{\partial t^2}+\lambda\mu_0\frac{\partial H_y}{\partial t}&=\frac{1}{\mu}\left(\frac{\partial^2 H_y}{\partial x^2}+\frac{\partial^2 H_y}{\partial y^2}+\frac{\partial^2 H_y}{\partial z^2}\right)\\
\varepsilon\varepsilon_0\frac{\partial^2 H_z}{\partial t^2}+\lambda\mu_0\frac{\partial H_z}{\partial t}&=\frac{1}{\mu}\left(\frac{\partial^2 H_z}{\partial x^2}+\frac{\partial^2 H_z}{\partial y^2}+\frac{\partial^2 H_z}{\partial z^2}\right)
\end{aligned} \tag{39}$$

One fundamental electromagnetic field equation is the equation $div\vec{B}=0$. The use of the Dirichlet theorem for approximation of the value of the magnetic field strength on the boundaries between adjacent media analogously to that of the electric field strength does not necessarily guarantees the observance of the condition of solenoidality of the magnetic field; furthermore, the magnetic properties of heterogeneous media were assumed constant in deriving generalized wave equations. The experience of numerical calculations has shown that when it is necessary to model nonstationary magnetic phenomena it is better in many cases to use a generalized wave equation for $\vec{E}$, accordingly expressing $\vec{H}(t,\vec{r})$ by $\vec{E}(t,\vec{r})$ and, if need be, to perform backward recalculation to $\vec{H}(t,\vec{r})$. This approach is difficult to apply to modeling of heterogeneous media with different magnetic properties, when the magnetic permeability μ is dependent on coordinates.

In media with a weak heterogeneity where μ (x, y, z) is a piecewise continuous quantity, the application of the proposed method of through counting is quite justified. Indeed, the system of equations (see Equations 13-15, 39) yields that the function's discontinuity on the boundaries between adjacent media is determined by the complexes which will be called the generalized permeability $\varepsilon^* \equiv \varepsilon\mu\varepsilon_0\mu_0$ and the generalized conductivity $\lambda^* \equiv \lambda\mu\mu_0$. Using the Direchlet theorem for ε^* and λ^*, we obtain their values on the boundaries between adjacent media and the values for the electric field strength at the discontinuity point (see Equation 37); here, we note that the value of the electric field strength is obtained without solving Maxwell equations. In fact, at the discontinuity point, we use linear interpolation of the function to obtain the values of ε^*, λ^*, and $E_{x=\xi} = \frac{1}{2}\left[E(\xi-0)+E(\xi+0)\right]$. Consequently, for piecewise continuous quantity μ (x, y, z), the application of the proposed method of through counting is justified. We note that the equality of the derivatives of the electric field strength along the normal to the surface at the discontinuity point according to Eq. (see Equation 34b) holds. When the wave equation for $\vec{H}$ is used for media with different magnetic permeabilities the condition of equality of the derivatives fails, i.e.,

$$\left.\frac{\partial H_x}{\partial x}\right|_{x=\xi-0} \neq \left.\frac{\partial H_x}{\partial x}\right|_{x=\xi+0}$$

which is a consequence of Eq. (see Equation 10); therefore, the use of through-counting schemes for the wave equation for $\vec{H}$ is difficult.

The generalized wave equation for $\vec{E}$ contains the term $grad\, div\vec{E}$ which directly allows for the influence of induced surface charges on the propagation of waves. We note that the proposed method of calculation can be used on condition that there are no built-in space charges and extraneous electromotive forces (Grinberg, G.A. & Fok, V.A., 1948).

By virtue of what has been stated above, for modeling of the propagation of electromagnetic waves in glasses having roughness and defects, we used system (see Equations 13-15) with boundary conditions (see Equations 24-34)

2.2.3 Results of numerical simulation

The physicomathematical model developed can also efficiently be used in modeling the propagation of electromagnetic waves in media with complex geometries and strong electromagnetic field discontinuities.

The transverse cut of a cellular structure represents a set of parallelepipeds and triangular prisms of various cross sections, as depicted in Fig. 10a and 11a. An electromagnetic wave propagates across the direction of parallelepipeds and triangular prisms (channels) along the coordinate x.

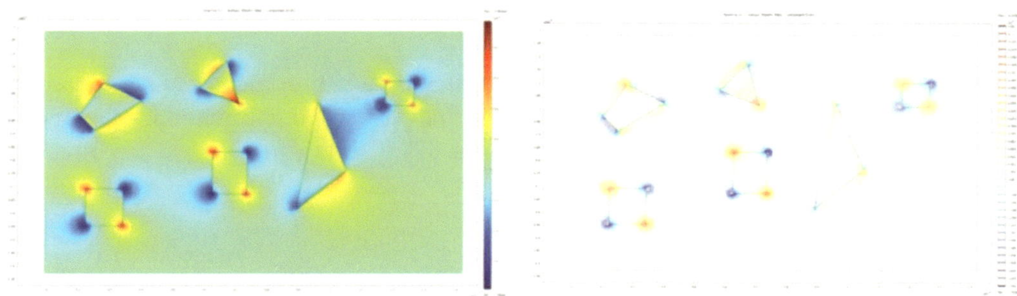

(a) Amplitude H_x (b) Isolines

Fig. 10. Amplitude H_x and isolines of the magnetic field strength

The size of the investigated two-dimensional object is $14\times20\cdot10^{-6}$ m, and the sizes of the domains are 2–4 μm. The frequency of the influence of the magnetic field is $\omega = 2\pi\cdot10^6$, and the strength of the field is

$$H_x = 21\cdot10^5 \sin^4\left(2\pi\cdot10^6 t\right) A/m \tag{40}$$

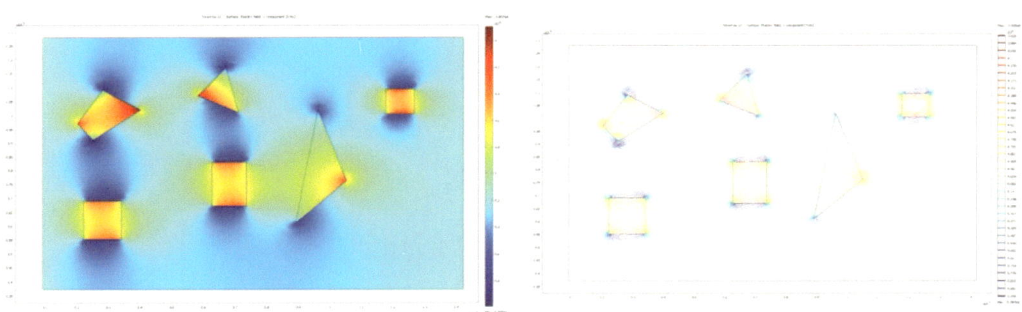

(a) Amplitude H_y (b) Isolines

Fig. 11. Amplitude H_y and isolines of the magnetic field strength

The electrophysical properties are as follows: of the large parallelepiped, $\mu = 1$, $\varepsilon = 8$, $\sigma = 10^{-9}$ Ω·m; of domains, $\mu = 1$, $\varepsilon = 6$, $\sigma = 10^{-8}$ Ω·m. They correspond to the electrophysical properties of glasses.

It was assumed that in a layer of thickness 15-20 μm an electromagnetic wave propagates without attenuation; therefore, on all the faces of the large parallelepiped the fulfillment of condition (see Equation 40) was considered valid. On the faces of the parallelepiped that are parallel to the OX axis condition (see Equation 40) corresponded to the "transverse" tangential component of the wave; on the faces parallel to OY condition (see Equation 40) corresponded to the normal component of the field.

The calculations were carried out with a time step of 10^{-13} sec up to a time instant of 10^{-10} sec. Figures 10a and 11a present the amplitude values of the magnetic field strength along H_x and H_y with a comparison scale, whereas Figs. 11b and 11b present the corresponding isolines. An analysis of these figures shows that at the places of discontinuity, on the wedges, force lines of the electromagnetic field concentrate. According to (Akulov, N. S.,

1939), precisely wedges are often the sources and sinks of the vacancies that determine, for example, the hardness and plasticity of a solid body.

Also, we modeled the propagation of waves in media, when domains possess magnetic properties. We assumed, in the calculations, that $\mu=100$; the remaining parameters correspond to the previous example of solution (Fig. 12)

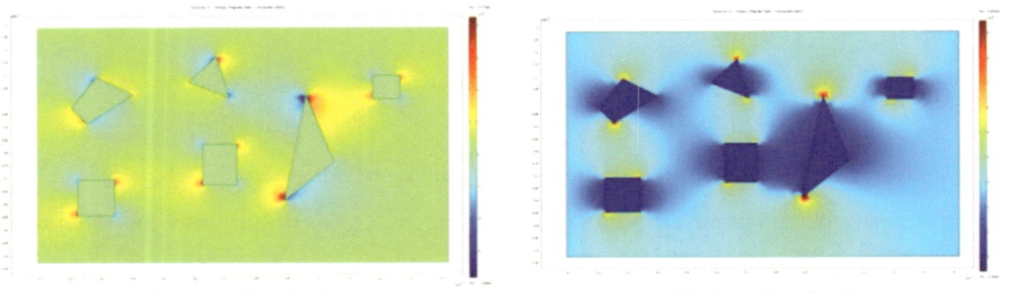

(a) Amplitude H_x (b) Amplitude H_y

Fig. 12. Amplitude H_x and H_y of the magnetic field strength

Of interest is the interaction of the electromagnetic wave with the rough surface shown in Fig. 13 and 14. As in the previous examples, we observe the concentration of electromagnetic energy on angular structures.

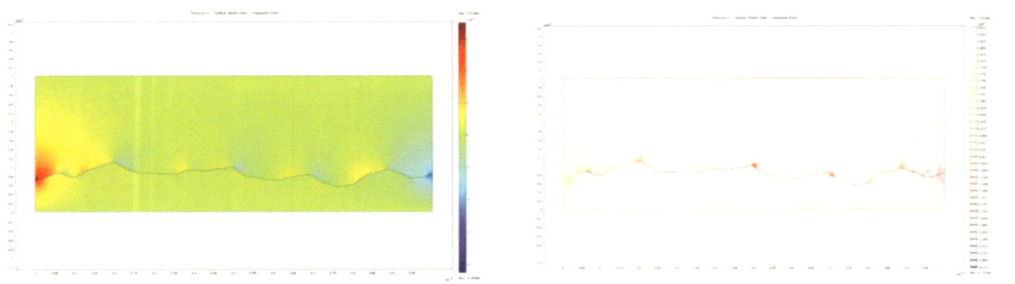

(a) Amplitude H_x (b) Isolines

Fig. 13. Amplitude H_x and isolines of the magnetic field strength

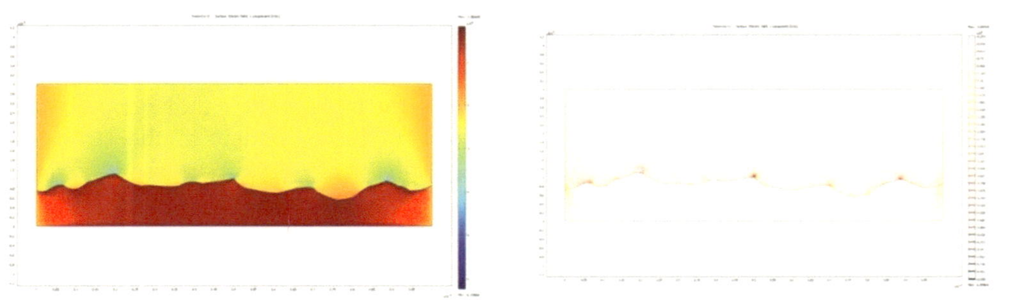

(a) Amplitude H_x (b) Isolines

Fig. 14. Amplitude H_y and isolines of the magnetic field strength

From Fig. 13 and 14, it is seen that electromagnetic heating of tapered structures may occur in addition to mechanical heating in magnetic abrasive machining.
As we have mentioned above, for investigation of the propagation of electromagnetic waves in nonmagnetic materials, it is more expedient to use the generalized equation for $\vec{E}$. For the purpose of illustration we give an example of numerical calculation of an optical knife with the wave equation for $\vec{H}$ (Fig. 15).
From Fig. 15, it is seen that the actual problem of diffraction on the optical knife remains to be solved, i.e., there is no "glow" on the optical-knife section, which is inconsistent with experimental data.

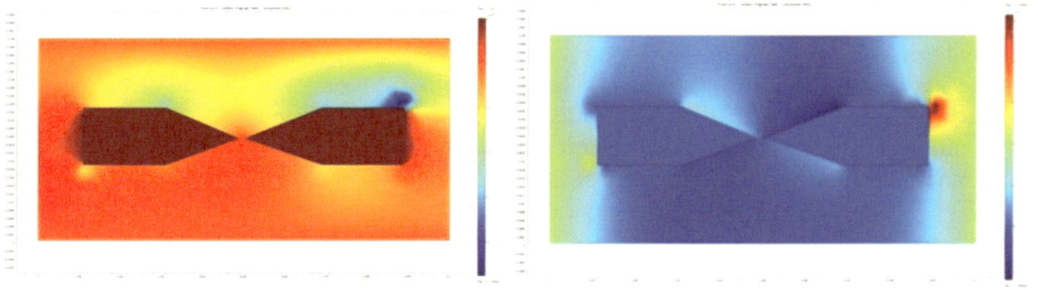

(a) Amplitude H_x (b) Amplitude H_y

Fig. 15. Amplitude H_x and H_y of the magnetic field strength

As is known (Bazarov, 1991), in thermodynamically equilibrium systems the temperature T and the electrical φ and chemical μ_c potentials are constant along the entire system:

$$grad\,T = 0,\; grad\,\varphi = 0,\; grad\,\mu_c = 0 \tag{41}$$

If these conditions are not fulfilled (grad $T \neq 0$, grad $\varphi \neq 0$, grad $\mu_c \neq 0$), irreversible processes of the transfer of mass, energy, electrical charge, etc. appear in the system.
The chemical potential of the j-th component is determined, for example, as a change of the free energy with a change in the number of moles:

$$\mu_{cj} = \left(\partial F / \partial n_j\right)_{T,V} \tag{42}$$

Where

$$dF = -SdT - PdV + HdB \tag{43}$$

The last term in Eq. (see Equation 43) takes into account the change in the free energy of a dielectric due to the change in the magnetic induction. The free energy of a unit volume of the dielectric in the magnetic field in this case has the form

$$F(T,D) = F_0 + \mu\mu_0 \frac{H^2}{2} \tag{44}$$

We assume that changes in the temperature and volume of the dielectric are small. Then the mass flux is determined by a quantity proportional to the gradient of the chemical potential or, according to Eq. (see Equation 43), we obtain

$$q_i = -D_{\mu_c} grad(HdB) = -D_{\mu_c} grad\, W \tag{45}$$

where $W = \mu\mu_0 \frac{H^2}{2}$ is the density of the magnetic field in the unit volume of the dielectric.
In magnetic abrasive polishing on the sharp protrusions of domains the gradients of magnetic energy are great, which can lead to the origination of vacancy flows.
An analysis of the results shows that the nonstationary component of the full electromagnetic energy is also concentrated in the region of fractures and wedges, i.e., at the sharp angles of domains, which may lead to the improvement of the structure of the sublayer of the treated surface due to the "micromagnetoplastic" effect. Maximum values of the nonstationary part of the total electromagnetic energy W_{max} in the sublayer correspond to a maximum value of the function sin $(2\pi{\cdot}10^6\, t)$ and occur for the time instants $t = (n/4)10^{-6}$ sec, where n is the integer, with the value of W_{max} for a neodymium magnet and a magnetoabrasive particle on the basis of carboxyl iron amounting to a value of the order of (see Equations 5-6)$\cdot 10^6$ J/m^3 . Having multiplied W_{max} by the volume of a domain, vacancy, or atom, we may approximately obtain the corresponding energy. The density of the electromagnetic energy in all of the cases is much smaller than the bonding energy of atoms, 10^{-18}-10^{-19} J. However, a periodic change in the magnetic field in one direction leads to a ponderomotive force that may influence the motion of various defects and dislocations to create a stable and equilibrium structure of atoms and molecules in magnetic abrasive polishing and, in the long run, in obtaining a surface with improved characteristics due to the "micromagnetoplastic" effect. The result of Para 2.2 were published in part (Grinchik, N.N. et al., 2010).

3. Interaction of nonstationary electric and thermal fields with allowance for relaxation processes

We investigate electric and thermal fields created by macroscopic charges and currents in continuous media. Of practical interest is modeling of local heat releases in media on exposure to a high-frequency electromagnetic field. We should take into account the influence of the energy absorption on the propagation of an electromagnetic wave, since the transfer processes are interrelated.
In an oscillatory circuit with continuously distributed parameters, the energy dissipation is linked (Kolesnikov, 2001) to the dielectric loss due to the dependence of the relative permittivity $\varepsilon(\omega)$ on frequency. In the general case ε is also complex, and the relationship between the electric displacement and electric field vectors has the form $D=\varepsilon(\omega)E$, where $\varepsilon(\omega)= \varepsilon'(\omega) - i\varepsilon''(\omega)$; here, ε' and ε'' are determined experimentally. As of now, the problems of dielectric heating of a continuous medium are reduced in many cases to consideration of an equivalent circuit based on lumped parameters, such as capacitance, inductance, loss angle, and relative-dielectric-loss factor (Skanavi, 1949; Perre P.; Turner I. W., 1996), that are established experimentally.
With this approach, there arise substantial difficulties in determining the temperature field of equivalent circuits. Also, we have polarization and the occurrence of an electric double layer of a prescribed electric moment in contact of media with different properties. Equivalent circuits in lamellar media additionally involve empirical lumped parameters: surface capacitance and surface resistance (Jaeger, 1977). The total current can always be separated into a dissipative, or conduction current which is in phase with the applied voltage and a displacement current shifted in time relative to the voltage. The exact physical

meaning of these components of the current is largely dependent on selection of an equivalent electric circuit. A unique equivalent circuit - series or parallel connection of the capacitor, the resistor, and the inductor - does not exist; it is determined by a more or less adequate agreement with experimental data.

In the case of electrolytic capacitors, the role of one plate is played by the electric double layer with a specific resistance much higher than the resistance of metallic plates. Therefore, decrease in the capacitance with frequency is observed, for such capacitors, even in the acoustic-frequency range (Jaeger, 1977). Circuits equivalent to an electrolytic capacitor are very bulky: up to 12 R, L, and C elements can be counted in them; therefore, it is difficult to obtain a true value of, e.g., the electrolyte capacitance. In (Jaeger, 1977) experimental methods of measurement of the dielectric properties of electrolyte solutions at different frequencies are given and ε' and ε" are determined. The frequency dependence of dispersion and absorption are essentially different consequences of one phenomenon: "dielectric-polarization inertia" (Jaeger, 1977). In actual fact, the dependence ε(ω) is attributable to the presence of the resistance of the electric double layer and to the electrochemical cell in the electrolytic capacitor being a system with continuously distributed parameters, in which the signal velocity is a finite quantity.

Actually, ε' and ε" are certain integral characteristics of a material at a prescribed constant temperature, which are determined by the geometry of the sample and the properties of the electric double layer. It is common knowledge that in the case of a field arbitrarily dependent on time any reliable calculation of the absorbed energy in terms of ε(ω) turns out to be impossible (Landau, L.D. & Lifshits, E.M., 1982). This can only be done for a specific dependence of the field E on time. For a quasimonochromatic field, we have (Landau, L.D. & Lifshits, E.M., 1982)

$$\mathbf{E}(t)=\frac{1}{2}\left[\mathbf{E}_0(t)e^{-i\omega t}+\mathbf{E}_0^*(t)e^{i\omega t}\right] \tag{46}$$

$$\mathbf{H}(t)=\frac{1}{2}\left[\mathbf{H}_0(t)e^{-i\omega t}+\mathbf{H}_0^*(t)e^{i\omega t}\right] \tag{47}$$

The values of $\mathbf{E}_0$(t) and $\mathbf{H}_0$(t), according to (Barash, Yu. & Ginzburg, V.L., 1976), must very slowly vary over the period T = 2π/ω. Then, for absorbed energy, on averaging over the frequency ω, we obtain the expression (Barash, Yu. & Ginzburg, V.L., 1976)

$$\begin{aligned}\frac{\partial \mathbf{D}(t)}{\partial t}E(t)=\frac{1}{4}\frac{d(\omega\varepsilon'(\omega))}{d(\omega)}\frac{\partial}{\partial t}\left[\mathbf{E}_0(t)\mathbf{E}_0^*(t)\right]+\frac{\omega\varepsilon''}{2}\mathbf{E}_0(t)\mathbf{E}_0^*(t)+\\ i\frac{d(\omega\varepsilon''(\omega))}{4d\omega}\left(\frac{\partial \mathbf{E}_0(t)}{\partial t}\mathbf{E}_0^*(t)-\frac{\mathbf{E}_0^*(t)}{\partial t}\mathbf{E}_0(t)\right)\end{aligned} \tag{48}$$

where the derivatives with respect to frequency are taken at the carrier frequency ω. We note that for an arbitrary function $\mathbf{E}$(t), it is difficult to represent it in the form

$$E(t)=a(t)\cos\varphi(t) \tag{49}$$

since we cannot unambiguously indicate the amplitude $a(t)$ and the phase φ (t). The manner in which $E(t)$ is decomposed into factors a and cosφ is not clear. Even greater difficulties appear in the case of going to the complex representation $W(t)=U(t)+iV(t)$ when the real oscillation $E(t)$ is supplemented with the imaginary part $V(t)$. The arising problems have been considered in (Vakman, D.E. & Vanshtein, L.A., 1977) in detail. In the indicated work, it has been emphasized that certain methods using a complex representation and claiming higher-than-average accuracy become trivial without an unambiguous determination of the amplitude, phase, and frequency.

Summing up the aforesaid, we can state that calculation of the dielectric loss is mainly empirical in character. Construction of the equivalent circuit and allowance for the influence of the electric double layer and for the dependence of electrophysical properties on the field's frequency are only true of the conditions under which they have been modeled; therefore, these are fundamental difficulties in modeling the propagation and absorption of electromagnetic energy.

As we believe, the release of heat in media on exposure to nonstationary electric fields can be calculated on the basis of allowance for the interaction of electromagnetic and thermal fields as a system with continuously distributed parameters from the field equation and the energy equation which take account of the distinctive features of the boundary between adjacent media. When the electric field interacting with a material medium is considered we use Maxwell equations (see Equations 6–7). We assume that space charges are absent from the continuous medium at the initial instant of time and they do not appear throughout the process. The energy equation will be represented in the form

$$\rho C_p \frac{dT}{dt} = div\left[k(T)\, grad(T)\right] + Q \tag{50}$$

where Q is the dissipation of electromagnetic energy.

According to (Choo, 1962), the electromagnetic energy converted to heat is determined by the expression

$$Q = \rho\left[\mathbf{E}\frac{d}{dt}\left(\frac{\mathbf{D}}{\rho}\right) + H\frac{d}{dt}\left(\frac{\mathbf{B}}{\rho}\right)\right] + \mathbf{J}_q\mathrm{E} \tag{51}$$

In deriving this formula, we used the nonrelativistic approximation of Minkowski's theory. If ε, μ, and ρ = const, there is no heat release; therefore, the intrinsic dielectric loss is linked to the introduction of $\varepsilon'(\omega)$ and $\varepsilon''(\omega)$. The quantity Q is affected by the change in the density of the substance $\rho(T)$.

A characteristic feature of high frequencies is the lag of the polarization field behind the charge in the electric field in time. Therefore, the electric-polarization vector is expediently determined by solution of the equation $\mathbf{P}(t+\tau_e)=(\varepsilon-1)\varepsilon_0\mathbf{E}(t)$ with allowance for the time of electric relaxation of dipoles τ_e. Restricting ourselves to the first term of the expansion $\mathbf{P}(t+\tau_e)$ in a Taylor series, from this equation, we obtain

$$\mathbf{P}(t) + \tau_e \frac{d\mathbf{P}(t)}{dt} = (\varepsilon - 1)\varepsilon_0 \mathbf{E}(t) \tag{52}$$

The solution (see Equation 52), on condition that **P**=0 at the initial instant of time, will take the form

$$\mathbf{P} = \frac{(\varepsilon - 1)\varepsilon_0}{\tau_e} \int_{t_0}^{t} \mathbf{E}(\tau) e^{-(t-\tau)/\tau_e} d\tau \tag{53}$$

It is noteworthy that Eq. (see Equation 52) is based on the classical Debay model. According to this model, particles of a substance possess a constant electric dipole moment. The indicated polarization mechanism involves partial arrangement of dipoles along the electric field, which is opposed by the process of disorientation of dipoles because of thermal collisions. The restoring "force", in accordance with Eq. (see Equation 52), does not lead to oscillations of electric polarization. It acts as if constant electric dipoles possessed strong damping.

Molecules of many liquids and solids possess the Debay relaxation polarizability. Initially polarization aggregates of Debay oscillators turn back to the equilibrium state $\mathbf{P}(t)=\mathbf{P}(0)\exp(-t/\tau_e)$.

A dielectric is characterized, as a rule, by a large set of relaxation times with a characteristic distribution function, since the potential barrier limiting the motion of weakly coupled ions may have different values (Skanavi, 1949); therefore, the mean relaxation time of the ensemble of interacting dipoles should be meant by τ_e in Eq. (see Equation 52).

To eliminate the influence of initial conditions and transient processes we set $t_0 = -\infty$, $\mathbf{E}(\infty)=0$, $\mathbf{H}(\infty)=0$, as it is usually done. If the boundary regime acts for a fairly long time, the influence of initial data becomes weaker with time owing to the friction inherent in every real physical system. Thus, we naturally arrive at the problem without the initial conditions:

$$\mathbf{P} = \frac{(\varepsilon - 1)\varepsilon_0}{\tau_e} \int_{-\infty}^{t} \mathbf{E}(\tau) e^{-(t-\tau)/\tau_e} d\tau \tag{54}$$

Let us consider the case of the harmonic field $\mathbf{E} = \mathbf{E}_0 \sin\omega t$; then, using Eq. (see Equation 54) we have, for the electric induction vector

$$\begin{aligned} \mathbf{D} = \varepsilon_0 \mathbf{E} + \mathbf{P} = \frac{(\varepsilon - 1)\varepsilon_0}{\tau_e} \int_{-\infty}^{t} \mathbf{E}(\tau) e^{-(t-\tau)/\tau_e} d\tau + \varepsilon_0 \mathbf{E}_0 \sin\omega t = \\ \frac{\mathbf{E}_0 (\varepsilon - 1)\varepsilon_0}{1+\omega^2\tau_e^2}(\sin\omega t - \omega\tau_e \cos\omega t) + \varepsilon_0 \mathbf{E}_0 \sin\omega t \end{aligned} \tag{55}$$

The electric induction vector is essentially the sum of two absolutely different physical quantities: the field strength and the polarization of a unit volume of the medium.

If the change in the density of the substance is small, we obtain, from formula (see Equation 51), for the local instantaneous heat release

$$\mathbf{Q} = \mathbf{E}\frac{d\mathbf{D}}{dt} = \frac{E_0^2 (\varepsilon - 1)\varepsilon_0}{1+\omega^2\tau_e^2}\left(\omega \sin\omega t \cos\omega t + \omega^2 \tau_e \sin^2 \omega t\right) \tag{56}$$

when we write the mean value of Q over the total period T:

$$\mathbf{Q} = \frac{1}{2}\frac{\mathbf{E}_0^2 (\varepsilon - 1)\varepsilon_0}{1+\omega^2\tau_e^2}\omega^2\tau_e + \lambda \mathbf{E}^2/2 \tag{57}$$

For high frequencies ($\omega\to\infty$), heat release ceases to be dependent on frequency, which is consistent with formula (see Equation 57) and experiment (Skanavi, 1949).
When the relaxation equation for the electric field is used we must also take account of the delay of the magnetic field, when the magnetic polarization lags behind the change in the strength of the external magnetic field:

$$\mathbf{I}(t)+\tau_i\frac{d\mathbf{I}(t)}{dt}=\mu\mu_0\mathbf{H}(t) \tag{58}$$

Formula (see Equation 57) is well known in the literature; it has been obtained by us without introducing complex parameters. In the case of "strong" heating of a material where the electrophysical properties of the material are dependent on temperature expression (see Equation 52) will have a more complicated form and the expression for Q can only be computed by numerical methods. Furthermore, in the presence of strong field discontinuities, we cannot in principle obtain the expression for Q because of the absence of closing relations for the induced surface charge and the surface current on the boundaries of adjacent media; therefore, the issue of energy relations in macroscopic electrodynamics is difficult, particularly, with allowance for absorption.
Energy relations in a dispersive medium have repeatedly been considered; nonetheless, in the presence of absorption, the issue seems not clearly understood (or at least not sufficiently known), particularly in the determination of the expression of released heat on the boundaries of adjacent media.
Indeed, it is known from the thermodynamics of dielectrics that the differential of the free energy F has the form

$$dF=-SdT-pdV+\mathbf{E}d\mathbf{D} \tag{59}$$

If the relative permittivity and the temperature and volume of the dielectric are constant quantities, from Eq. (see Equation 59) we have

$$F(T,\mathbf{D})=F_0+D^2/2 \tag{60}$$

where F_0 is the free energy of the dielectric in the absence of the field.
The change of the internal energy of the dielectric during its polarization at constant temperature and volume can be found from the Gibbs-Helmholtz equation, in which the external parameter $\mathbf{D}$ is the electric displacement. Disregarding F_0 which is independent of the field strength, we can obtain

$$U(T,\mathbf{D})=F(T,\mathbf{D})-T(dF/dT)_D \tag{61}$$

If the relative dielectric constant is dependent on temperature ($\varepsilon(T)$), we obtain

$$U(T,\mathbf{D})=\varepsilon_0 E^2/2(\varepsilon+T(d\varepsilon/dT)_D) \tag{62}$$

Expression (see Equation 62) determines the change in the internal energy of the dielectric in its isothermal polarization but with allowance for the energy transfer to a thermostat, if the polarization causes the dielectric temperature to change. A more detailed substantiation of Eq. (see Equation 62) will be given in the book. In the works on microwave heating, that we know, expression (see Equation 62) is not used.

A characteristic feature of high frequencies is that the polarization field lags behind the change in the external field in time; therefore, the polarization vector is expediently determined by solution of the equation

$$\mathbf{P}(t+\tau_e)=\left(\varepsilon-1+T\left(d\varepsilon/dT\right)_D\right)\varepsilon_0\mathbf{E}(t) \tag{63}$$

With allowance for the relaxation time, i.e., restricting ourselves to the first term of the expansion $\mathbf{P}(t+\tau_e)$ in a Taylor series, we obtain

$$\mathbf{P}(t)+\tau_e d\mathbf{P}(t)/dT=\left(\varepsilon-1+T\left(d\varepsilon/dT\right)_D\right)\varepsilon_0\mathbf{E}(t) \tag{64}$$

In the existing works on microwave heating with the use of complex parameters, they disregard the dependence ε"(T). In (Antonets, I.V.; Kotov, L.N.; Shavrov, V.G. & Shcheglov, V.I., 2009), consideration has been given to the incidence of a one-dimensional wave from a medium with arbitrary complex parameters on one or two boundaries of media whose parameters are also arbitrary. The amplitudes of waves reflected from and transmitted by each boundary have been found. The refection, transmission, and absorption coefficients have been obtained from the wave amplitudes. The well-known proposition that a traditional selection of determinations of the reflection, transmission, and absorption coefficients from energies (reflectivity, transmissivity, and absorptivity) in the case of complex parameters of media comes into conflict with the law of conservation of energy has been confirmed and exemplified. The necessity of allowing for ε"(T) still further complicates the problem of computation of the dissipation of electromagnetic energy in propagation of waves through the boundaries of media with complex parameters.

The proposed method of computation of local heat release is free of the indicated drawbacks and makes it possible, for the first time, to construct a consistent model of propagation of nonmonochromatic waves in a heterogeneous medium with allowance for frequency dispersion without introducing complex parameters.

In closing, we note that a monochromatic wave is infinite in space and time, has infinitesimal energy absorption in a material medium, and transfers infinitesimal energy, which is the idealization of real processes. However with these stringent constraints, too, the problem of propagation of waves through the boundary is open and far from being resolved even when the complex parameters of the medium are introduced and used. In reality, the boundary between adjacent media is not infinitely thin and has finite dimensions of the electric double layers; therefore, approaches based on through-counting schemes for a hyperbolic equation without explicit separation of the boundary between adjacent media are promising.

6. Conclusion

The consistent physicomathematical model of propagation of an electromagnetic wave in a heterogeneous medium has been constructed using the generalized wave equation and the Dirichlet theorem. Twelve conditions at the interfaces of adjacent media were obtained and justified without using a surface charge and surface current in explicit form. The conditions are fulfilled automatically in each section of the heterogeneous medium and are conjugate, which made it possible to use through-counting schemes for calculations. For the first time

the effect of concentration of "medium-frequency" waves with a length of the order of hundreds of meters at the fractures and wedges of domains of size 1-3 μm has been established. Numerical calculations of the total electromagnetic energy on the wedges of domains were obtained. It is shown that the energy density in the region of wedges is maximum and in some cases may exert an influence on the motion, sinks, and the source of dislocations and vacancies and, in the final run, improve the near-surface layer of glass due to the "micromagnetoplastic" effect.
The results of these calculations are of special importance for medicine, in particular, when microwaves are used in the therapy of various diseases. For a small, on the average, permissible level of electromagnetic irradiation, the concentration of electromagnetic energy in internal angular structures of a human body (cells, membranes, neurons, interlacements of vessels, etc) is possible.

7. Acknowledgment

The authors express their gratitude to Corresponding Member of the National Academy of Sciences of Belarus N.V. Pavlyukevich, Corresponding Member of the National Academy of Sciences of Belarus Prof. V.I. Korzyuk and Dr. R. Wojnar for a useful discussion of the work. This work war carried out with financial support from the Belarusian Republic Foundation for Basic Research (grant T10P-122) and from the Science Support Foundation of Poland "Kassa im. Myanowski" (2005).

8. References

Akulov, N. S. (1961). *Dislocations and Plasticity [in Russian].* Minsk: Izd. AN BSSR.
Akulov, N. S. (1939). *Ferromagnetism [in Russian].* Moscow–Leningrad: ONTI.
Antonets, I.V.; Kotov, L.N.; Shavrov, V.G. & Shcheglov, V.I. (2009). Energy characteristics of propagation of a wave through the boundaries of media with complex parameters. *Radiotekhnika i Elektronika , 54* (10), 1171-1183.
Barash, Yu. & Ginzburg, V.L. (1976). *Usp.Fiz.Nauk , 118* (3), 523.
Barta, O.; Pistora, I.; Vesec, I. et al. (2001). Magneto-optics in bi-gyrotropic garnet waveguide. *Opto-Electronics Review , 9* (3), 320–325.
Bazarov, I. P. (1991). *Thermodynamics: Textbook for Higher Educational Establishments [in Russian].* Moscow : Vysshaya Shkola.
Born, M. &. (1970). *Principles of Optics [Russian translation].* Moscow: Mir.
Broe, I. & Keller, O. (2002). Quantum-well enhancement of the Goos–Hanchen shift for p-polarized beams in a two-prism configuration. *J. Opt. Soc. Am. B , 19* (6), 1212–1221.
Choo, B.-T. (1962). *Plasma in a Magnetic Field and Direct Thermal-to-Electric Energy Conversion [Russian translation].* Moscow.
Danae, D. et al. (2002). Rigorous electromagnetic analysis of dipole emission in periodically corrugated layers: the grating-assisted resonant-cavity light-emitting diode. *J. Opt. Soc. Am. B , 19* (5), 871–881.
Ehlers, R. A. & Metaxas, A. C. (2003). 3-DFE Discontinuous sheet for microwave heating. *IEEE Trans. Microwave Theory Tech. , 51* (3), 718–726.
Eremin,Y. & Wriedt,T. (2002). Large dielectric non-spherical particle in an evanescent wave field near a plane surface. *Optics Communications* (214), 34–45.
Frumkin, A. (1987). *Electrode Processes [in Russian].* Moscow: Nauka.

Golovin, Yu. I. et al. Influence of weak magnetic fields on the dynamics of changes in the microhardness of silicon initiated by low-intensity beta-irradiation. *Fiz. Tverd. Tela*, *49* (5).

Grinberg, G.A. & Fok, V.A. (1948). *On the theory of Coastal Refraction of Electromagnetic Waves [in Russian].* (In Collected Papers "Investigations on Propagation of Radio Waves" (B.A. Vvedenskii (ed.) Ausg., Bd. 2). M-L., AN SSSR.

Grinchik, N. N. & Dostanko, A. P. (2005). *Influence of Thermal and Diffusional Processes on the Propagation of Electromagnetic Waves in Layered Materials [in Russian].* Minsk: ITMO im. A. V. Luikova, NAN Belarusi.

Grinchik, N.N et al. (2009). Electrodynamics of layered media with boundary conditions corresponding to the total-current continuum. *Journal of Engeneering Physics and Thermodynamics*, *82* (4), 810-819.

Grinchik, N.N. et al. (2010). Electrodynamic processes in a surface layer in magnetoabrasive polishing. *Journal of Engeneering Physics and Thermodynamics*, *83* (3), 638-649.

Jaeger, J. (1977). *Methods of Measurement in Electrochemistry [Russian translation]* (Bd. 2). Moscow.

Keller, O. (1997). Local fields in linear and nonlinear optics of mesoscopic system. *Prog. Opt.* (37), 257–343.

Keller, O. (1995). Optical response of a quantum-well sheet: internal electrodynamics. *J. Opt. Soc. Am. B*, *12* (6), 997–1005.

Keller, O. (1995). Sheet-model description of the linear optical response of quantum wells. *J. Opt. Soc. Am. B*, *12* (6), 987–997.

Khomich, M. (2006). *Magnetic-abrasive machining of the manufactured articles [in Russian].* Minsk: BNTU.

Kolesnikov, P. (2001). *Theory and Calculation of Waveguides, lightguides, and integral-optoelectronics elements. Electrodynamics and Theory of Waveguides [in Russian]* (Bd. 1). Minsk: ITMO NAN Belarusi.

Koludzija, B. M. (1999). Electromagnetic modeling of composite metallic and dielectric structures, IEEE Trans. Microwave Theory Tech. *47* (7), 1021-1029.

Kryachko, A.F. et al. (2009). *Theory of scattering of electromagnetic waves in the angular structure.* Nauka.

Kudryavtsev, L. (1970). *Mathematical Analysis [in Russian]* (Bd. 2). Moscow: Mir.

Landau, L.D. & Lifshits, E.M. (1982). *Theoretical Physics. Vol. 8. Electrodynamics of Continuous Media [in Russian].* Moscow.

Larruquert, J. I. (2001). Reflectance enhancement with sub-quarterwave multilayers of highly absorbing materials. *J. Opt. Soc. Am. B*, *18* (6), 1406–1415.

Leontovich, M. (1948). *On the approximate boundary conditions for the electromagnetic field on the surface of well conducting bodies.* Moscow: Academy of Science of USSR.

Levin, M. N. et al. (2003). Activation of the surface of semiconductors by the effect of a pulsed magnetic field. *Zh. Tekh. Fiz*, *73* (10), 85–87.

Makara ,V. A. et al. (2008). Magnetic field-induced changes in the impurity composition and microhardness of the near-surface layers of silicon crystals. *Fiz. Tekh. Poluprovadn.*, *42* (9), 1061-1064.

Makara, V. A. et al. (2001). On the influence of a constant magnetic field on the electroplastic effect in silicon crystals. *Fiz. Tverd. Tela* (3), 462–465.

Monzon, I.; Yonte,T.; Sanchez-Soto, L. (2003). Characterizing the reflectance of periodic lasered media. *Optics Communications* (218), 43–47.

Orlov, A. M. et al. (2003). Dynamics of the surface dislocation ensembles in silicon in the presence of mechanical and magnetic perturbation. *Fiz. Tverd. Tela* , *45* (4), 613–617.

Orlov, A. M. et al. (2001). Magnetic- stimulated alteration of the mobility of dislocations in the plastically deformed silicon of n-type. *Fiz. Tverd. Tela* , *43* (7), 1207–1210.

Perre P.; Turner I. W. (1996). 10 Int. Druing Sympos. IDS 96., (p. 183). Krakow, Poland.

Rakomsin, A. P. (2000). *Strengthening and Restoration of Items in an Electromagnetic Field [in Russian]*. Minsk : Paradoks.

Shul'man, Z. P. & Kordonskii ,V. I. (1982). *Magnetorheological Effect [in Russian]*. Minsk: Nauka i Tekhnika.

Skanavi, T. (1949). *Dielectric Physics (Region of Weak Fields) [in Russian]*. Moscow: Gostekhizdat.

Tikhonov, A. N. & Samarskii, A. A. (1977). *Equations of Mathematical Physics [in Russian]*. Moscow: Nauka.

Vakman, D.E. & Vanshtein, L.A. (1977). *Usp.Fiz.Nauk* , *123* (4), 657.

Wei Hu & Hong Guo. (2002). Ultrashort pulsed Bessel beams and spatially induced group-velocity dispersio. *J. Opt. Soc. Am. B* , *19* (1), 49–52.

3

Nonlinear Propagation of Electromagnetic Waves in Antiferromagnet

Xuan-Zhang Wang and Hua Li
School of Physics and Electronic Engineering, Harbin Normal University
China

1. Introduction

The nonlinearities of common optical materials result from the nonlinear response of their electric polarization to the electric field of electromagnetic waves (EMWs), or $\vec{P}^{NL} = \ddot{\chi}^{(1)} \cdot \vec{E} + \ddot{\chi}^{(2)} : \vec{E}\vec{E} + \ddot{\chi}^{(3)} : \vec{E}\vec{E}\vec{E} + \cdots$. From the Maxwell equations and related electromagnetic boundary conditions including this nonlinear polarization, one can present the origin of most nonlinear optical phenomena.

However, the magnetically optical nonlinearities of magnetic materials come from the nonlinear response of their dynamical magnetization to the magnetic field of EWMs, or the magnetization $\vec{m}^{NL} = \ddot{\chi}^{(1)} \cdot \vec{H} + \ddot{\chi}^{(2)} : \vec{H}\vec{H} + \ddot{\chi}^{(3)} : \vec{H}\vec{H}\vec{H} + \ldots$. From these one can predict or explain various magnetic optical nonlinear features of magnetic materials. The magnetic mediums are optical dispersive, which originates from the magnetic permeability as a function of frequency. Since various nonlinear phenomena from ferromagnets and ferrimagnets almost exist in the microwave region, these phenomena are important for the microwave technology.

In the concept of ferromagnetism(Morrish, 2001), there is such a kind of magnetic ordering media, named antiferromagnets (AFs), such as NiO, MnF_2, FeF_2, and CoF_2 *et. al*. This kind of materials may possess two or more magnetic sublattices and all lattice points on any sublattice have the same magnetic moment, but the moments on adjacent sublattices are opposite in direction and counteract to each other. We here present an example in Fig.1, a bi-sublattice AF structure. In contrast to the ferromagnets or ferrimagnets, it is very difficult to magnetize AFs by a magnetic field of ordinary intensity since very intense AF exchange interaction exists in them, so they are almost not useful in the fields of electronic and electric engineering. But the dynamical properties of AFs should be paid a greater attention to. The resonant frequencies of the AFs usually fall in millimeter or far infrared (IR) frequency regime. Therefore the experimental methods to study AFs optical properties are optical or quasi-optical ones. In addition, these frequency regions also are the working frequency regions of the THz technology, so the AFs may be available to make new elements in the field of THz technology.

The propagation of electromagnetic waves in AFs can be divided into two cases. In the first case, the frequency of an EMW is far to the AF resonant frequency and then the AF can be optically considered as an ordinary dielectric. The second case means that the wave frequency is situated in the vicinity of the AF resonant frequency and the dynamical magnetization of the AF then couples with the magnetic field of the EMW. Consequently, modes of EMW propagation in this frequency region are some AF polaritons. In the linear case, the AF polaritons in AF films, multilayers and superlattices had been extendedly discussed before the year 2000 (Stamps & Camley, 1996; Camley & Mills, 1982; Zhu & Cao, 1987; Oliveros, et. al., 1992; Camley, 1992; Raj & Tilley, 1987; Wang & Tilley, 1987; Almeida & Tilley, 1990).

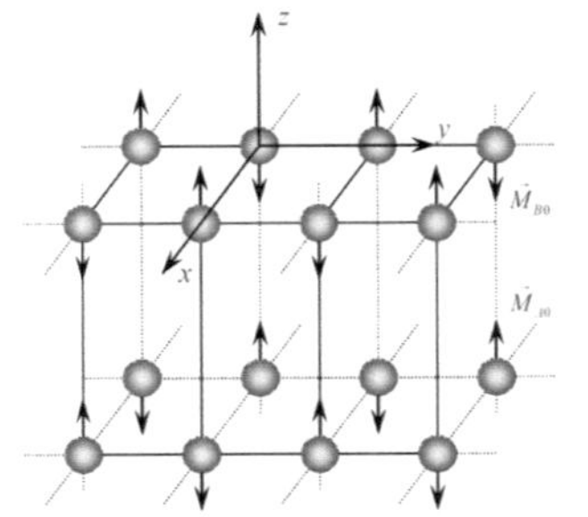

Fig. 1. The sketch of a bi-sublattice AF structure.

The magnetically nonlinear investigation of AF systems was not given great attention until the 1990s. In the recent years, many progresses have been made in understanding the magnetic dynamics of AF systems (Costa, et. al. ,1993; Balakrishnan, et. al.,1990, 1992; Daniel & Bishop,1992; Daniel & Amuda,1994; Balakrishnan & Blumenfeld,1997). Many investigations have been carried out on nonlinear guided and surface waves (Wang & Awai,1998; Almeida & Mills, 1987; Kahn, et. al., 1988; Wright & Stegeman, 1992; Boardman & Egan,1986), second-harmonic generation (Lim, 2002, 2006; Fiebig et. al, 1994, 2001, 2005), bistability (Vukovic, 1992) and dispersion properties (Wang,Q, 2000). Almeida and Mills first discussed the nonlinear infrared responses of the AFs and explore the field-dependent of transmission through thin AF films and superlattices, where the third-order approximation of dynamical magnetization was used, but no analytical expressions of nonlinear magnetic susceptibilities in the AF films or layers were obtained (Almeida & Mills, 1987; Kahn, et. al., 1988). Lim first obtained the expressions of the susceptibilities in the third-order approximation, in a special situation where a circularly polarized magnetic field and the cylindrical coordinate system were applied in the derivation process (Lim, et. al., 2000). It is obvious that those expressions cannot be conveniently used in various geometries and boundaries of different shape. In analogue to what done in the ordinary nonlinear optics, the nonlinear magnetic susceptibilities were presented in the Cartesian coordinate system by Wang et. al. (Wang & Fu, 2004; Zhou, et. al., 2009), and were used to discuss the nonlinear polaritons of AF superlattices and the second-harmonic generation (SHG) of AF films (Wang & Li, 2005; Zhou & Wang, 2008), as well as transmission and reflection bi-stability (Bai, et. al., 2007; Zhou, 2010).

2. Nonlinear susceptibilities of antiferromagnets

AF susceptibility is considered as one important physical quantity to describe the response of magnetization in AFs to the driving magnetic filed. It is also a basis of investigating dynamic properties and magneto-optical properties. In this section, the main steps and results of deriving nonlinear magnetic susceptibilities of AFs will be presented in the right-angled coordinate system, or the Cartesian system. The detail mathmetical procedure can be found from our previous works (Wang & Fu, 2004; Zhou, et. al., 2009). The used bi-sublattice AF structure and coordinate system are shown in Fig.1, where we take the AF anisotropy axis and the external magnetic field H_0 along the z axis. The sublattic magnetzation M_0 is the absolute projection value of each total sublattice magnatization to the anisotropy axis. The driving magnetic field $\vec{H}$ changes with time, according to $\exp(-i\omega t)$.

2.1 Basical assumptions, definitions and the first-order susceptibilities

We begin with the assumption that this AF crystal is at a low temperature, or the temperature is much lower than its Neel temperature and the magnetic ordering is properly preserved. Then the magnetization on each sublattice is regarded as saturated without a driving field. In the alternating driving field $\vec{H}$, each sublattice magnetization deviates the AF anisotropy axis and makes a precession with respect to the effective field acting on it. This precession is described by the Bloch's equation with damping,

$$\frac{\partial}{\partial t}\vec{M}_{A(B)} = \gamma\vec{M}_{A(B)} \times \vec{H}^{eff}_{A(B)} - \frac{\tau\vec{M}_{A(B)}}{M_0} \times \frac{\partial}{\partial t}\vec{M}_{A(B)} \tag{2-1a}$$

where γ is the gyromagnetic ratio and τ the damping coefficient, $\vec{M}_A$ and $\vec{M}_B$ are the total sublattice magnetizations and contain two parts, the static part $\pm M_0$ and changing part with time $\vec{m}_{A(B)}$, produced by the driving field,

$$\vec{M}_A = M_0\vec{e}_z + \vec{m}_A\,,\ \vec{M}_B = -M_0\vec{e}_z + \vec{m}_B \tag{2-1b}$$

$\vec{H}^{eff}_A$ and $\vec{H}^{eff}_B$ are the effective fields acting on sublattices A and B, respectively, and are given by

$$\vec{H}^{eff}_A = (H_0 + \frac{H_a M^z_A}{M_0})\vec{e}_z - \frac{H_e\vec{M}_B}{M_0} + \vec{H}\,,\ \vec{H}^{eff}_B = (H_0 + \frac{H_a M^z_B}{M_0})\vec{e}_z - \frac{H_e\vec{M}_A}{M_0} + \vec{H} \tag{2-1c}$$

where H_a is the AF anisotropy field, H_e is the AF exchange field and $\vec{H}$ indicates the driving field. Substituting (2-1b) and (2-1c) into (2-1a), we have

$$\frac{\partial\vec{M}_A}{\partial t} = \gamma\{m_{Ay}[H_0 + \frac{H_a M_{Az}}{M_0} + H_z - \frac{H_e(m_{Bz}-M_0)}{M_0}] - (M_0 + m_{Az})[H_y - \frac{H_e}{M_0}m_{By}]\}\vec{e}_x$$

$$+\gamma\{(M_0 + m_{Az})[H_x - \frac{H_e}{M_0}m_{Bx}] - m_{Ax}[H_0$$

$$+\frac{H_a M_{Az}}{M_0} + H_z - \frac{H_e(m_{Bz}-M_0)}{M_0}]\}\vec{e}_y + \gamma\{m_{Ax}[H_y(\omega) - \frac{H_e}{M_0}m_{By}]$$

$$-m_{Ay}[H_x(\omega) - \frac{H_e}{M_0}m_{Bx}]\}\vec{e}_z + \frac{\tau}{M_0}\{[m_{Ay}\frac{\partial}{\partial t}(M_0 + m_{Az})$$

$$-(M_0 + m_{Az})\frac{\partial}{\partial t}m_{Ay}]\vec{e}_x + [(M_0 + m_{Az})\frac{\partial}{\partial t}m_{Ax} - m_{Ax}\frac{\partial}{\partial t}(M_0 + m_{Az})]\vec{e}_y$$
$$+[m_{Ax}\frac{\partial}{\partial t}m_{Ay} - m_{Ay}\frac{\partial}{\partial t}m_{Ax}]\vec{e}_z\} \tag{2-2a}$$

$$\frac{\partial}{\partial t}\vec{M}_B = \gamma\{m_{By}[H_0 + \frac{H_a M_{Bz}}{M_0} + H_z - \frac{H_e(M_0 + m_{Az})}{M_0}] - (m_{Bz} - M_0)[H_y - \frac{H_e}{M_0}m_{Ay}]\}\vec{e}_x$$
$$+\gamma\{(m_{Bz} - M_0)[H_x - \frac{H_e}{M_0}m_{Ax}]$$
$$-m_{Bx}[H_0 + \frac{H_a M_{Bz}}{M_0} + H_z - \frac{H_e(M_0 + m_{Az})}{M_0}]\}\vec{e}_y + \gamma\{m_{Bx}[H_y - \frac{H_e}{M_0}m_{Ay}]$$
$$-m_{By}[H_x - \frac{H_e}{M_0}m_{Ax}]\}\vec{e}_z + \frac{\tau}{M_0}\{[m_{By}\frac{\partial}{\partial t}(-M_0 + m_{Bz})$$
$$-(-M_0 + m_{Bz})\frac{\partial}{\partial t}m_{By}]\vec{e}_x + [(-M_0 + m_{Bz})\frac{\partial}{\partial t}m_{Bx} - m_{Bx}\frac{\partial}{\partial t}(-M_0 + m_{Bz})]\vec{e}_y$$
$$+(m_{Bx}\frac{\partial}{\partial t}m_{By} - m_{By}\frac{\partial}{\partial t}m_{Bx})\vec{e}_z\}. \tag{2-2b}$$

We shall use the perturbation expansion method to derive nonlinear magnetizations and susceptibilies of various orders. We take M_0 and H_0 as the 0-order magnetization and the 0-order field. $\vec{H}$ is considered as the first-order field and we note that the complex conjugation of this field should be included in higher-order mathmetical procesures higher than the first-order one. In the third-order aproximation, the induced magnetizations $\vec{m}_{A(B)}$ are divided into the first-, second- and third-order parts, or

$$\vec{m}_{A(B)} = \vec{m}^{(1)}_{A(B)} + \vec{m}^{(2)}_{A(B)} + \vec{m}^{(3)}_{A(B)} + c.c. \tag{2-3}$$

where *c.c.* indicates the complex conjugation. In practice, one needs the AF magnetization rather than the lattice magnetizations, so we define $\vec{m} = \vec{m}_A + \vec{m}_B$ as the AF magnetization and $\vec{n} = \vec{m}_A - \vec{m}_B$ as its supplemental quantity. In the linear case, $M_{Az} = M_0$, $M_{Bz} = -M_0$ and considering that the linear magnetizations should change with time according to $\exp(-i\omega t)$, Eqs.(2-2) can be simplified as

$$-i\omega\vec{m}^{(1)}_A = [(\omega_0 + \omega_a + \omega_e + i\tau\omega)m^{(1)}_{Ay} + \omega_e m^{(1)}_{By} - \omega_m H_y]\vec{e}_x$$
$$-[(\omega_0 + \omega_a + \omega_e + i\tau\omega)m^{(1)}_{Ax} + \omega_e m^{(1)}_{Bx} - \omega_m H_x]\vec{e}_y \tag{2-4a}$$

$$-i\omega\vec{m}^{(1)}_B = [(\omega_0 - \omega_a - \omega_e - i\tau\omega)m^{(1)}_{By} - \omega_e m^{(1)}_{Ay} + \omega_m H_y]\vec{e}_x$$
$$-[(\omega_0 - \omega_a - \omega_e - i\tau\omega)m^{(1)}_{Bx} - \omega_e m^{(1)}_{Ax} + \omega_m H_x]\vec{e}_y \tag{2-4b}$$

where the special frequencies are defined with $\omega_0 = \gamma H_0$, $\omega_a = \gamma H_a$, $\omega_e = \gamma H_e$ and $\omega_m = 4\pi\gamma M_0$. The first-order z-components of the sublattice magnetizations are vanishing. From the definitions $m_i^{(1)} = \sum_j \chi_{ij}^{(1)} H_j$ and $n_i^{(1)} = \sum_j N_{ij}^{(1)} H_j$, we have the nonzero elements of the first-order magnetic susceptibilty and supplementary susceptibility

$$\chi_{xx}^{(1)} = \chi_{yy}^{(1)} = \chi_1 = 2A\omega_m \omega_a' Z_{--}(\omega)\,,\ \chi_{xy}^{(1)} = -\chi_{yx}^{(1)} = i\chi_2 = -4iA\omega_m \omega_a' \omega_0 \omega \tag{2-5a}$$

$$N_{xx}^{(1)} = N_{yy}^{(1)} = -2A\omega_0 \omega_m Z_{-+}(\omega)\,,\ N_{xy}^{(1)} = -N_{yx}^{(1)} = 2iA\omega\omega_m Z_{+-}(\omega) \tag{2-5b}$$

where $\omega_r'^2 = \omega_a'(2\omega_e + \omega_a')$ and $Z_{\pm\pm}(\omega) = \omega_r'^2 \pm \omega_0^2 \pm \omega^2$ with $\omega_a' = \omega_a + i\tau\omega$, and $A = \{[\omega_r'^2 - (\omega - \omega_0)^2][\omega_r'^2 - (\omega + \omega_0)^2]\}^{-1}$. The linear magnetic permeability often used in the past is $\vec{\mu} = \mu_0[1 + \vec{\chi}^{(1)}]$, or $\mu_{xx} = \mu_{yy} = \mu_0(1 + \chi_1) = \mu_0\mu_1$ and $\mu_{xy} = -\mu_{yx} = i\mu_0\chi_2 = i\mu_0\mu_2$.

2.2 The second-order approximation

Similar to the second-order electric polarization in the nonlinear optics, the second-order magneizations also are divided into the dc part unchanging with time and the second-harmonic part varying with time according to $\exp(-2i\omega t)$. Here we first derive the dc susceptibility, which will appear in the third-order ones. Neglecting the linear, third-order terms and the second-harmonic terms in (2-2), reserving only the second-order 0-frequency terms, we obtain the following equations

$$0 = \omega_0 m_y^{(2)}(0) + \omega_a n_y^{(2)}(0) + \gamma H_z m_y^{(1)*} + \gamma m_y^{(1)} H_z^* \tag{2-6a}$$

$$0 = \omega_0 n_y^{(2)}(0) + (\omega_a + 2\omega_e) m_y^{(2)}(0) + \gamma H_z n_y^{(1)*} + \gamma n_y^{(1)} H_z^* \tag{2-6b}$$

$$0 = -\omega_0 m_x^{(2)}(0) - \omega_a n_x^{(2)}(0) - \gamma H_z^* m_x^{(1)} - \gamma H_z m_x^{(1)*} \tag{2-6c}$$

$$0 = -\omega_0 n_x^{(2)}(0) - (\omega_a + 2\omega_e) m_x^{(2)}(0) - \gamma H_z n_x^{(1)*} - \gamma H_z^* n_x^{(1)} \tag{2-6d}$$

In addition, the z component of the dc magnetization can be obtained from the conservation of each sulattice magnetic moment, and we see

$$m_{Az}^{(2)}(0) = -\frac{1}{M_0}[m_{Ax}^{(1)}(\omega) m_{Ax}^{(1)*}(\omega) + m_{Ay}^{(1)}(\omega) m_{Ay}^{(1)*}(\omega)] \tag{2-7a}$$

$$m_{Bz}^{(2)}(0) = \frac{1}{M_0}[m_{Bx}^{(1)}(\omega) m_{Bx}^{(1)*}(\omega) + m_{By}^{(1)}(\omega) m_{By}^{(1)*}(\omega)] \tag{2-7b}$$

These lead directly to the z component to be

$$m_z^{(2)}(0) = -\frac{1}{2M_0}(m_x^{(1)} n_x^{(1)*} + n_x^{(1)} m_x^{(1)*} + m_y^{(1)} n_y^{(1)*} + n_y^{(1)} m_y^{(1)*}) \tag{2-8}$$

Here we have used $m_i^{(2)}(0)$ and $n_i^{(2)}(0)$ directly to represent $m_i^{(2)}(0) + m_i^{(2)*}(0)$ and $n_i^{(2)}(0) + n_i^{(2)*}(0)$ for simplicity. Substituting the linear results into (2-6) and (2-8), and using the definitions of $n_i^{(2)}(0)=\sum_{jk} N_{ijk}^{(2)}(0)H_jH_k^*$ and $m_i^{(2)}(0)=\sum_{jk}\chi_{ijk}^{(2)}(0)H_j(\omega)H_k^*(\omega)$, we find the corresponding nonzero elements

$$\chi_{xxz}^{(2)}(0)=\chi_{yyz}^{(2)}(0)=\chi_{yzy}^{(2)}(0)^*=\chi_{xzx}^{(2)}(0)^*=\omega_m(\omega_0\chi_{xx}^{(1)}-\omega_a N_{xx}^{(1)})/[M_0(\omega_r^2-\omega_0^2)] \tag{2-9a}$$

$$\chi_{xyz}^{(2)}(0)=\chi_{xzy}^{(2)}(0)^*=-\chi_{yxz}^{(2)}(0)=-\chi_{yzx}^{(2)}(0)^*=\omega_m(\omega_0\chi_{xy}^{(1)}-\omega_a N_{xy}^{(1)})/[M_0(\omega_r^2-\omega_0^2)] \tag{2-9b}$$

$$\chi_{zxx}^{(2)}(0)=\chi_{zyy}^{(2)}(0)=-(\chi_{xx}^{(1)}N_{xx}^{(1)*}+\chi_{xy}^{(1)}N_{xy}^{(1)*}+c.c)/4M_0 \tag{2-9c}$$

$$\chi_{zyx}^{(2)}(0)=-\chi_{zxy}^{(2)}(0)=(N_{xx}^{(1)}\chi_{xy}^{(1)*}+\chi_{xx}^{(1)}N_{xy}^{(1)*}-c.c.)/4M_0 \tag{2-9d}$$

$$\begin{aligned}&N_{xxz}^{(2)}(0)=N_{yyz}^{(2)}(0)=N_{xzx}^{(2)}(0)^*=N_{yzy}^{(2)}(0)^*\\&=\omega_m[\omega_0 N_{xx}^{(1)}-(\omega_a+2\omega_e)\chi_{xx}^{(1)}]/[M_0(\omega_r^2-\omega_0^2)]\end{aligned} \tag{2-9e}$$

$$\begin{aligned}&N_{xyz}^{(2)}(0)=N_{xzy}^{(2)}(0)^*=-N_{yxz}^{(2)}(0)=-N_{yzx}^{(2)}(0)^*\\&=\omega_m[\omega_0 N_{xy}^{(1)}-(\omega_a+2\omega_e)\chi_{xy}^{(1)}]/[M_0(\omega_r^2-\omega_0^2)]\end{aligned} \tag{2-9f}$$

$$N_{zxx}^{(2)}(0)=N_{zyy}^{(2)}(0)=-(N_{xy}^{(1)}N_{xy}^{(1)*}+\chi_{xx}^{(1)}\chi_{xx}^{(1)*}+\chi_{xy}^{(1)}\chi_{xy}^{(1)*}+N_{xx}^{(1)}N_{xx}^{(1)*})/4M_0 \tag{2-9g}$$

$$N_{zyx}^{(2)}(0)=-N_{zxy}^{(2)}(0)=-(\chi_{xy}^{(1)}\chi_{xx}^{(1)*}+N_{xy}^{(1)}N_{xx}^{(1)*}-c.c.)/4M_0 \tag{2-9h}$$

Next, we are going to derive the second-harmonic (SH) magnetization and susceptibility. They will not be used only in the third-order susceptibility, but also be applied to describe the SH generation in various AF systems. In equations (2-2), reserving only the SH terms, we obtain the following equations

$$-2i\omega m_x^{(2)}(2\omega)=\omega_0 m_y^{(2)}(2\omega)+\omega_a'' n_y^{(2)}(2\omega)+\omega_m h_z m_y^{(1)}/M_0 \tag{2-10a}$$

$$-2i\omega n_x^{(2)}(2\omega)=\omega_0 n_y^{(2)}(2\omega)+(\omega_a''+2\omega_e)m_y^{(2)}(2\omega)+\omega_m h_z n_y^{(1)}/M_0 \tag{2-10b}$$

$$-2i\omega m_y^{(2)}(2\omega)=-\omega_0 m_x^{(2)}(2\omega)-\omega_a'' n_x^{(2)}(2\omega)-\omega_m h_z m_x^{(1)}/M_0 \tag{2-10c}$$

$$-2i\omega n_y^{(2)}(2\omega)=-\omega_0 n_x^{(2)}(2\omega)-(\omega_a''+2\omega_e)m_x^{(2)}(2\omega)-\omega_m h_z n_x^{(1)}/M_0 \tag{2-10d}$$

with $\omega_a''=\omega_a+2i\omega\tau$. Meanwhile the conservation of each sulattice magnetic moment results in

$$m_z^{(2)}(2\omega)=-\frac{1}{2M_0}[m_x^{(1)}n_x^{(1)}+m_y^{(1)}n_y^{(1)}] \tag{2-10e}$$

Applying the expressions of the first-order components and the expressions

$$m_i^{(2)}(2\omega) = \sum_{jk} \chi_{ijk}^{(2)}(2\omega) H_j H_k \tag{2-11a}$$

$$n_i^{(2)}(2\omega) = \sum_{jk} N_{ijk}^{(2)}(2\omega) H_j H_k \tag{2-11b}$$

one finds

$$\chi_{xxz}^{(2)}(2\omega) = \chi_{xzx}^{(2)}(2\omega) = \chi_{yyz}^{(2)}(2\omega) = \chi_{yzy}^{(2)}(2\omega) = AB\omega_m^2\omega_0\{\omega_a' Z_{--}(\omega) \\ Z_{-+}(2\omega) + \omega_a'' Z_{-+}(\omega) Z_{--}(2\omega) + 4\omega^2[\omega_a' Z_{+-}(2\omega) + \omega_a'' Z_{+-}(\omega)]\} / M_0 \tag{2-12a}$$

$$\chi_{xyz}^{(2)}(2\omega) = \chi_{xzy}^{(2)}(2\omega) = -\chi_{yxz}^{(2)}(2\omega) = -\chi_{yzx}^{(2)}(2\omega) = -iAB\omega\omega_m^2\{2\omega_0^2[\omega_a' \\ Z_{-+}(2\omega) + 2\omega_a'' Z_{-+}(\omega)] + \omega_a'' Z_{+-}(\omega) Z_{--}(2\omega) + 2\omega_a' Z_{--}(\omega) Z_{+-}(2\omega)\} / M_0 \tag{2-12b}$$

$$\chi_{zxx}^{(2)}(2\omega) = \chi_{zyy}^{(2)}(2\omega) = 2A\omega_m^2\omega_a'\omega_0 / M_0 \tag{2-12c}$$

$$N_{xxz}^{(2)}(2\omega) = N_{xzx}^{(2)}(2\omega) = N_{yyz}^{(2)}(2\omega) = N_{yzy}^{(2)}(2\omega) = -AB\omega_m^2\{\omega_a'(\omega_a'' + 2\omega_e) \\ [Z_{--}(2\omega) Z_{--}(\omega) + 8\omega_0^2\omega^2] + \omega_0^2 Z_{-+}(2\omega) Z_{-+}(\omega) + 2\omega^2 Z_{+-}(2\omega) Z_{+-}(\omega)\} / M_0 \tag{2-12d}$$

$$N_{xyz}^{(2)}(2\omega) = N_{xzy}^{(2)}(2\omega) = -N_{yxz}^{(2)}(2\omega) = -N_{yzx}^{(2)}(2\omega) = iAB\omega\omega_0\omega_m^2\{2\omega_a'(\omega_a'' + 2\omega_e) \\ [Z_{--}(2\omega) + 2Z_{--}(\omega)] + 2Z_{+-}(2\omega) Z_{-+}(\omega) + Z_{-+}(2\omega) Z_{+-}(\omega)\} / M_0 \tag{2-12e}$$

$$N_{zxx}^{(2)}(2\omega) = N_{zyy}^{(2)}(2\omega) = \frac{A^2\omega_m^2}{M_0}[\omega^2 Z_{+-}(\omega)^2 - \omega_a'^2 Z_{--}(\omega)^2 - \omega_0^2 Z_{-+}(\omega)^2 + 4\omega_a'^2\omega^2\omega_0^2] \tag{2-12f}$$

where $B = 1 / [\omega_r''^2 - (2\omega + \omega_0)^2][\omega_r''^2 - (2\omega - \omega_0)^2]$.

2.3 The third-order approximation

The third-order magnetization also contains two part, or one varies with time according to $\exp(-i\omega t)$ and the orther is the third-harmonic part with $\exp(-3i\omega t)$. Because we do not consern with the third-harmonic (TH) generation, so the first-order, and second-order and TH terms in equations (2-2) all are ignored. Thus we have

$$-i\omega m_x^{(3)}(\omega) = \omega_0 m_y^{(3)}(\omega) + \omega_a' n_y^{(3)}(\omega) + \eta_x \tag{2-13a}$$

$$-i\omega n_x^{(3)}(\omega) = \omega_0 n_y^{(3)}(\omega) + (\omega_a' + 2\omega_e) m_y^{(3)}(\omega) + \eta_x' \tag{2-13b}$$

$$-i\omega m_y^{(3)}(\omega) = -\omega_0 m_x^{(3)}(\omega) - \omega_a' n_x^{(3)}(\omega) + \eta_y \tag{2-13c}$$

$$-i\omega n_y^{(3)}(\omega) = -\omega_0 n_x^{(3)}(\omega) - (\omega_a' + 2\omega_e) m_x^{(3)}(\omega) + \eta_y' \tag{2-13d}$$

$$
\begin{aligned}
&-i\omega m_z^{(3)}(\omega)=\frac{\omega_m}{M_0}[m_x^{(2)}(0)H_y+m_x^{(2)}(2\omega)H_y^*-m_y^{(2)}(0)H_x \\
&-m_y^{(2)}(2\omega)H_x^*]+\frac{\tau}{M_0}[-2i\omega m_{Ax}^{(1)*}(\omega) \\
&m_{Ay}^{(2)}(2\omega)-i\omega m_{Ax}^{(2)}(0)m_{Ay}^{(1)}(\omega)+i\omega m_{Ax}^{(2)}(2\omega)m_{Ay}^{(1)*}(\omega) \\
&+2i\omega m_{Ay}^{(1)*}(\omega)m_{Ax}^{(2)}(2\omega)+i\omega m_{Ay}^{(2)}(0)m_{Ax}^{(1)}(\omega) \\
&-i\omega m_{Ay}^{(2)}(2\omega)m_{Ax}^{(1)*}(\omega)]-2i\omega m_{Bx}^{(1)*}(\omega)m_{By}^{(2)}(2\omega) \\
&-i\omega m_{Bx}^{(2)}(0)m_{By}^{(1)}(\omega)+i\omega m_{Bx}^{(2)}(2\omega)m_{By}^{(1)*}(\omega) \\
&+2i\omega m_{By}^{(1)*}(\omega)m_{Bx}^{(2)}(2\omega)+i\omega m_{By}^{(2)}(0)m_{Bx}^{(1)}(\omega) \\
&-i\omega m_{By}^{(2)}(2\omega)m_{Bx}^{(1)*}(\omega)]
\end{aligned}
\tag{2-13e}
$$

where

$$
\begin{aligned}
&\eta_x=\eta_{Ax}+\eta_{Bx}=\frac{\omega_m}{M_0}[m_y^{(2)}(0)H_z+m_y^{(2)}(2\omega)H_z^*-m_z^{(2)}(0)H_y \\
&-m_z^{(2)}(2\omega)H_y^*]+\frac{\omega_a}{M_0}[m_{Ay}^{(1)}(\omega)m_{Az}^{(2)}(0) \\
&+m_{Ay}^{(1)*}(\omega)m_{Az}^{(2)}(2\omega)+m_{By}^{(1)}(\omega)m_{Bz}^{(2)}(0)+m_{By}^{(1)*}(\omega)m_{Bz}^{(2)}(2\omega)] \\
&+\frac{\tau}{M_0}[-2i\omega m_{Ay}^{(1)*}(\omega)m_{Az}^{(2)}(2\omega)+i\omega m_{Az}^{(2)}(0) \\
&m_{Ay}^{(1)}(\omega)-i\omega m_{Az}^{(2)}(2\omega)m_{Ay}^{(1)*}(\omega)-2i\omega m_{By}^{(1)*}(\omega)m_{Bz}^{(2)}(2\omega) \\
&+i\omega m_{Bz}^{(2)}(0)m_{By}^{(1)}(\omega)-i\omega m_{Bz}^{(2)}(2\omega)m_{By}^{(1)*}(\omega)]
\end{aligned}
\tag{2-14a}
$$

$$
\begin{aligned}
&\eta_y=\eta_{Ay}+\eta_{By}=\frac{\omega_m}{M_0}[m_z^{(2)}(0)H_x+m_z^{(2)}(2\omega)H_x^*-m_x^{(2)}(0)H_z \\
&-m_x^{(2)}(2\omega)H_z^*]-\frac{\omega_a}{M_0}[m_{Az}^{(2)}(0)m_{Ax}^{(1)}(\omega) \\
&+m_{Az}^{(2)}(2\omega)m_{Ax}^{(1)*}(\omega)+m_{Bz}^{(2)}(0)m_{Bx}^{(1)}(\omega)+m_{Bz}^{(2)}(2\omega)m_{Bx}^{(1)*}(\omega)] \\
&+\frac{\tau}{M_0}[-i\omega m_{Az}^{(2)}(0)m_{Ax}^{(1)}(\omega)+i\omega m_{Az}^{(2)}(2\omega) \\
&m_{Ax}^{(1)*}(\omega)+2i\omega m_{Ax}^{(1)*}(\omega)m_{Az}^{(2)}(\omega)-i\omega m_{Bz}^{(2)}(0)m_{Bx}^{(1)}(\omega) \\
&+i\omega m_{Bz}^{(2)}(2\omega)m_{Bx}^{(1)*}(\omega)+2i\omega m_{Bx}^{(1)}(\omega)m_{Bz}^{(2)}(2\omega)]
\end{aligned}
\tag{2-14b}
$$

$$
\begin{aligned}
&\eta_x'=\eta_{Ax}-\eta_{Bx}=\frac{\omega_m}{M_0}[n_y^{(2)}(0)H_z+n_y^{(2)}(2\omega)H_z^*-n_z^{(2)}(0)H_y \\
&-n_z^{(2)}(2\omega)H_y^*]+\frac{\omega_a}{M_0}[m_{Ay}^{(1)}(\omega)m_{Az}^{(2)}(0)
\end{aligned}
$$

$$
\begin{aligned}
&+m_{Ay}^{(1)^*}(\omega)m_{Az}^{(2)}(2\omega)-m_{By}^{(1)}(\omega)m_{Bz}^{(2)}(0)-m_{By}^{(1)^*}(\omega)m_{Bz}^{(2)}(2\omega)]\\
&+\frac{2\omega_e}{M_0}[m_{By}^{(1)}(\omega)m_{Az}^{(2)}(0)+m_{By}^{(1)^*}(\omega)m_{Az}^{(2)}(2\omega)\\
&-m_{Ay}^{(1)}(\omega)m_{Bz}^{(2)}(0)-m_{Ay}^{(1)^*}(\omega)m_{Bz}^{(2)}(2\omega)]\\
&+\frac{\tau}{M_0}[-2i\omega m_{Ay}^{(1)^*}(\omega)m_{Az}^{(2)}(2\omega)+i\omega m_{Az}^{(2)}(0)m_{Ay}^{(1)}(\omega)\\
&-i\omega m_{Az}^{(2)}(2\omega)m_{Ay}^{(1)^*}(\omega)+2i\omega m_{By}^{(1)^*}(\omega)m_{Bz}^{(2)}(2\omega)\\
&-i\omega m_{Bz}^{(2)}(0)m_{By}^{(1)}(\omega)+i\omega m_{Bz}^{(2)}(2\omega)m_{By}^{(1)^*}(\omega)]
\end{aligned}
\tag{2-14c}
$$

$$
\begin{aligned}
&\eta'_y=\eta_{Ay}-\eta_{By}=\frac{\omega_m}{M_0}[n_z^{(2)}(0)H_x+n_z^{(2)}(2\omega)H_x^*-n_x^{(2)}(0)H_z\\
&-n_x^{(2)}(2\omega)H_z^*]-\frac{\omega_a}{M_0}[m_{Az}^{(2)}(0)m_{Ax}^{(1)}(\omega)+\\
&m_{Az}^{(2)}(2\omega)m_{Ax}^{(1)^*}(\omega)-m_{Bz}^{(2)}(0)m_{Bx}^{(1)}(\omega)-m_{Bz}^{(2)}(2\omega)m_{Bx}^{(1)^*}(\omega)]\\
&+\frac{2\omega_e}{M_0}[m_{Bz}^{(2)}(0)m_{Ax}^{(1)}(\omega)+m_{Bz}^{(2)}(2\omega)m_{Ax}^{(1)^*}(\omega)\\
&-m_{Az}^{(2)}(0)m_{Bx}^{(1)}(\omega)-m_{Az}^{(2)}(2\omega)m_{Bx}^{(1)^*}(\omega)]+\frac{\tau}{M_0}\\
&[-i\omega m_{Az}^{(2)}(0)m_{Ax}^{(1)}(\omega)+i\omega m_{Az}^{(2)}(2\omega)m_{Ax}^{(1)^*}(\omega)\\
&+2i\omega m_{Ax}^{(1)^*}(\omega)m_{Az}^{(2)}(\omega)+i\omega m_{Bz}^{(2)}(0)m_{Bx}^{(1)}(\omega)\\
&-i\omega m_{Bz}^{(2)}(2\omega)m_{Bx}^{(1)^*}(\omega)-2i\omega m_{Bx}^{(1)}(\omega)m_{Bz}^{(2)}(2\omega)]
\end{aligned}
\tag{2-14d}
$$

Substituting the definitions of $m_i^{(1)}$, $m_i^{(2)}$, $n_i^{(1)}$ and $n_i^{(2)}$ into equations (2-13,2-14), and after some complicated algebra, we finally obtain

$$\chi_{xxxx}^{(3)}(\omega)=A[\omega_0 Z_{-+}(\omega)f-\omega'_a Z_{--}(\omega)f'-i\omega Z_{+-}(\omega)a+2i\omega\omega_0\omega'_a a'] \tag{2-15a}$$

$$\chi_{xyyx}^{(3)}(\omega)=A[\omega_0 Z_{-+}(\omega)e-\omega'_a Z_{--}(\omega)e'-i\omega Z_{+-}(\omega)b+2i\omega\omega_0\omega'_a b'] \tag{2-15b}$$

$$\chi_{xzzx}^{(3)}(\omega)=A[\omega_0 Z_{-+}(\omega)g-\omega'_a Z_{--}(\omega)g'-i\omega Z_{+-}(\omega)c+2i\omega\omega_0\omega'_a c'] \tag{2-15c}$$

$$\chi_{xxyx}^{(3)}(\omega)=\frac{A}{2}[-\omega_0 Z_{-+}(\omega)h+\omega'_a Z_{--}(\omega)h'-i\omega Z_{+-}(\omega)d+2i\omega\omega_0\omega'_a d'] \tag{2-15d}$$

$$\chi_{xxxy}^{(3)}(\omega)=A[-\omega_0 Z_{-+}(\omega)b+\omega'_a Z_{--}(\omega)b'-i\omega Z_{+-}(\omega)e+2i\omega\omega_0\omega'_a e'] \tag{2-15e}$$

$$\chi_{xyyy}^{(3)}(\omega)=A[-\omega_0 Z_{-+}(\omega)a+\omega'_a Z_{--}(\omega)a'-i\omega Z_{+-}(\omega)f+2i\omega\omega_0\omega'_a f'] \tag{2-15f}$$

$$\chi_{xzzy}^{(3)}(\omega)=A[-\omega_0 Z_{-+}(\omega)c+\omega'_a Z_{--}(\omega)c'-i\omega Z_{+-}(\omega)g+2i\omega\omega_0\omega'_a g'] \tag{2-15g}$$

$$\chi^{(3)}_{xxyy}(\omega)=\frac{A}{2}[\omega_0 Z_{-+}(\omega)d-\omega'_a Z_{--}(\omega)d'-i\omega Z_{+-}(\omega)h+2i\omega\omega_0\omega'_a h'] \tag{2-15h}$$

$$\chi^{(3)}_{xxzz}(\omega)=\frac{A}{2}[\omega_0 Z_{-+}(\omega)p-\omega'_a Z_{--}(\omega)p'-i\omega Z_{+-}(\omega)l+2i\omega\omega_0\omega'_a l'] \tag{2-15i}$$

$$\chi^{(3)}_{xyzz}(\omega)=\frac{A}{2}[-\omega_0 Z_{-+}(\omega)l+\omega'_a Z_{--}(\omega)l'-i\omega Z_{+-}(\omega)p+2i\omega\omega_0\omega'_a p'] \tag{2-15j}$$

$$\begin{aligned}\chi^{(3)}_{zxzx}(\omega)&=\frac{\tau}{4M_0}[6\chi^{(1)*}_{xy}\chi^{(2)}_{xxz}(2\omega)-6\chi^{(1)*}_{xx}\chi^{(2)}_{xyz}(2\omega)\\&+3N^{(1)*}_{xy}N^{(2)}_{xxz}(2\omega)-3N^{(1)*}_{xx}N^{(2)}_{xyz}(2\omega)+\\&\chi^{(1)}_{xx}\chi^{(2)}_{xzy}(0)-\chi^{(1)}_{xy}\chi^{(2)}_{xzx}(0)+N^{(1)}_{xx}N^{(2)}_{xzy}(0)-N^{(1)}_{xy}N^{(2)}_{xzx}(0)]\\&+\frac{i\omega_m}{2\omega M_0}[\chi^{(2)}_{xzy}(0)+2\chi^{(2)}_{xyz}(2\omega)]\end{aligned} \tag{2-15k}$$

$$\begin{aligned}\chi^{(3)}_{zyzx}(\omega)&=\frac{\tau}{4M_0}[6\chi^{(1)*}_{xy}\chi^{(2)}_{xyz}(2\omega)+6\chi^{(1)*}_{xx}\chi^{(2)}_{xxz}(2\omega)\\&+3N^{(1)*}_{xy}N^{(2)}_{xyz}(2\omega)+3N^{(1)*}_{xx}N^{(2)}_{xxz}(2\omega)\\&+\chi^{(1)}_{xx}\chi^{(2)}_{xzx}(0)+\chi^{(1)}_{xy}\chi^{(2)}_{xzy}(0)+N^{(1)}_{xx}N^{(2)}_{xzx}(0)\\&+N^{(1)}_{xy}N^{(2)}_{xzy}(0)]+\frac{i\omega_m}{2\omega M_0}[\chi^{(2)}_{xzx}(0)-2\chi^{(2)}_{xxz}(2\omega)]\end{aligned} \tag{2-15l}$$

$$\begin{aligned}\chi^{(3)}_{zxxz}(\omega)&=\frac{\tau}{2M_0}[\chi^{(1)}_{xx}\chi^{(2)}_{xyz}(0)+N^{(1)}_{xx}N^{(2)}_{xyz}(0)-N^{(1)}_{xy}N^{(2)}_{xxz}(0)\\&-\chi^{(1)}_{xy}\chi^{(2)}_{xxz}(0)]+\frac{i\omega_m}{\omega M_0}\chi^{(2)}_{xyz}(0)\end{aligned} \tag{2-15m}$$

$$\begin{aligned}\chi^{(3)}_{zyyz}(\omega)&=\frac{\tau}{2M_0}[\chi^{(1)}_{xx}\chi^{(2)}_{xyz}(0)+N^{(1)}_{xx}N^{(2)}_{xyz}(0)-N^{(1)}_{xy}N^{(2)}_{xxz}(0)\\&-\chi^{(1)}_{xy}\chi^{(2)}_{xxz}(0)]+\frac{i\omega_m}{\omega M_0}\chi^{(2)}_{xyz}(0)\end{aligned} \tag{2-15n}$$

$$\begin{aligned}\chi^{(3)}_{zxzy}(\omega)&=-\frac{\tau}{4M_0}[6\chi^{(1)*}_{xx}\chi^{(2)}_{xxz}(2\omega)+6\chi^{(1)*}_{xy}\chi^{(2)}_{xyz}(2\omega)\\&+3N^{(1)*}_{xx}N^{(2)}_{xxz}(2\omega)+3N^{(1)*}_{xy}N^{(2)}_{xyz}(2\omega)+\\&\chi^{(1)}_{xx}\chi^{(2)}_{xzx}(0)+\chi^{(1)}_{xy}\chi^{(2)}_{xzy}(0)+N^{(1)}_{xx}N^{(2)}_{xzx}(0)+N^{(1)}_{xy}N^{(2)}_{xzy}(0)]\\&+\frac{i\omega_m}{2\omega M_0}[2\chi^{(2)}_{xxz}(2\omega)-\chi^{(2)}_{xzx}(0)]\end{aligned} \tag{2-15o}$$

$$\chi_{zyzy}^{(3)}(\omega)=\frac{\tau}{4M_0}[6\chi_{xy}^{(1)*}\chi_{xxz}^{(2)}(2\omega)-6\chi_{xx}^{(1)*}\chi_{xyz}^{(2)}(2\omega)$$
$$+3N_{xy}^{(1)*}N_{xxz}^{(2)}(2\omega)-3N_{xx}^{(1)*}N_{xyz}^{(2)}(2\omega)+\chi_{xx}^{(1)}\chi_{xzy}^{(2)}(0)$$
$$-\chi_{xy}^{(1)}\chi_{xzx}^{(2)}(0)+N_{xx}^{(1)}N_{xzy}^{(2)}(0)-N_{xy}^{(1)}N_{xzx}^{(2)}(0)]$$
$$+\frac{i\omega_m}{2\omega M_0}[\chi_{xzy}^{(2)}(0)+2\chi_{xyz}^{(2)}(2\omega)] \tag{2-15p}$$

with the coefficients

$$a=\frac{i\omega\tau}{2M_0}[3\chi_{zxx}^{(2)}(2\omega)\chi_{xy}^{(1)*}+3N_{zxx}^{(2)}(2\omega)N_{xy}^{(1)*}$$
$$-\chi_{xy}^{(1)}\chi_{zxx}^{(2)}(0)-N_{xy}^{(1)}N_{zxx}^{(2)}(0)]$$
$$-\frac{\omega_a}{2M_0}[\chi_{xy}^{(1)}\chi_{zxx}^{(2)}(0)+N_{xy}^{(1)}N_{zxx}^{(2)}(0)$$
$$+N_{zxx}^{(2)}(2\omega)N_{xy}^{(1)*}+\chi_{zxx}^{(2)}(2\omega)\chi_{xy}^{(1)*}] \tag{2-16a}$$

$$b=\frac{i\omega\tau}{2M_0}[3\chi_{zxx}^{(2)}(2\omega)\chi_{xy}^{(1)*}+3N_{zxx}^{(2)}(2\omega)N_{xy}^{(1)*}$$
$$+\chi_{xx}^{(1)}\chi_{zyx}^{(2)}(0)+N_{xx}^{(1)}N_{zyx}^{(2)}(0)]$$
$$+\frac{1}{2M_0}\{\omega_a[\chi_{xx}^{(1)}\chi_{zyx}^{(2)}(0)+N_{xx}^{(1)}N_{zyx}^{(2)}(0)-\chi_{zxx}^{(2)}(2\omega)\chi_{xy}^{(1)*}$$
$$-N_{zxx}^{(2)}(2\omega)N_{xy}^{(1)*}]-2\omega_m\chi_{zyx}^{(2)}(0)\} \tag{2-16b}$$

$$c=-\frac{\omega_m}{M_0}\chi_{xzy}^{(2)}(0) \tag{2-16c}$$

$$d=\frac{i\omega\tau}{2M_0}[\chi_{xx}^{(1)}\chi_{zxx}^{(2)}(0)+N_{xx}^{(1)}N_{zxx}^{(2)}(0)$$
$$-\chi_{xy}^{(1)}\chi_{zyx}^{(2)}(0)-N_{xy}^{(1)}N_{zyx}^{(2)}(0)]$$
$$+\frac{1}{2M_0}\{\omega_a[\chi_{xx}^{(1)}\chi_{zxx}^{(2)}(0)-\chi_{xy}^{(1)}\chi_{zyx}^{(2)}(0)+N_{xx}^{(1)}N_{zxx}^{(2)}(0)$$
$$-N_{xy}^{(1)}N_{zyx}^{(2)}(0)]-2\omega_m\chi_{zxx}^{(2)}(0)\} \tag{2-16d}$$

$$e=\frac{i\omega\tau}{2M_0}[-3\chi_{zxx}^{(2)}(2\omega)\chi_{xx}^{(1)*}-3N_{zxx}^{(2)}(2\omega)N_{xx}^{(1)*}$$
$$+\chi_{xy}^{(1)}\chi_{zyx}^{(2)}(0)+N_{xy}^{(1)}N_{zyx}^{(2)}(0)]$$

$$+\frac{1}{2M_0}\{\omega_a[\chi_{xy}^{(1)}\chi_{zyx}^{(2)}(0)+N_{xy}^{(1)}N_{zyx}^{(2)}(0)+\chi_{zxx}^{(2)}(2\omega)\chi_{xx}^{(1)*}$$
$$+N_{zxx}^{(2)}(2\omega)N_{xx}^{(1)*}]-2\omega_m\chi_{zxx}^{(2)}(2\omega)\} \tag{2-16e}$$

$$f = \frac{i\omega\tau}{2M_0}[-3N_{zxx}^{(2)}(2\omega)N_{xx}^{(1)*} - 3\chi_{zxx}^{(2)}(2\omega)\chi_{xx}^{(1)*}$$
$$+\chi_{xx}^{(1)}\chi_{zxx}^{(2)}(0) + N_{xx}^{(1)}N_{zxx}^{(2)}(0)]$$
$$+\frac{1}{2M_0}\{\omega_a[\chi_{xx}^{(1)}\chi_{zxx}^{(2)}(0) + N_{xx}^{(1)}N_{zxx}^{(2)}(0) + N_{zxx}^{(2)}(2\omega)N_{xx}^{(1)*} \tag{2-16f}$$
$$+\chi_{zxx}^{(2)}(2\omega)\chi_{xx}^{(1)*}] - 2\omega_m[\chi_{zxx}^{(2)}(2\omega) + \chi_{zxx}^{(2)}(0)]\}$$

$$g = \frac{\omega_m}{M_0}\chi_{xzx}^{(2)}(0) \tag{2-16g}$$

$$h = -\frac{i\omega\tau}{2M_0}[\chi_{xx}^{(1)}\chi_{zyx}^{(2)}(0) + N_{xx}^{(1)}N_{zyx}^{(2)}(0)$$
$$+\chi_{xy}^{(1)}\chi_{zxx}^{(2)}(0) + N_{xy}^{(1)}N_{zxx}^{(2)}(0)] \tag{2-16h}$$
$$-\frac{1}{2M_0}\{\omega_a[\chi_{xx}^{(1)}\chi_{zyx}^{(2)}(0) + N_{xx}^{(1)}N_{zyx}^{(2)}(0)$$
$$+\chi_{xy}^{(1)}\chi_{zxx}^{(2)}(0) + N_{xy}^{(1)}N_{zxx}^{(2)}(0)] - 2\omega_m\chi_{zyx}^{(2)}(0)\}$$

$$l = -\frac{\omega_m}{M_0}[2\chi_{xyz}^{(2)}(2\omega) + \chi_{xyz}^{(2)}(0)] \tag{2-16i}$$

$$p = \frac{\omega_m}{M_0}[2\chi_{xxz}^{(2)}(2\omega) + \chi_{xxz}^{(2)}(0)] \tag{2-16j}$$

$$a' = \frac{i\omega\tau}{2M_0}[3\chi_{xy}^{(1)*}N_{zxx}^{(2)}(2\omega) + 3N_{xy}^{(1)*}\chi_{zxx}^{(2)}(2\omega)$$
$$-\chi_{xy}^{(1)}N_{zxx}^{(2)}(0) - N_{xy}^{(1)}\chi_{zxx}^{(2)}(0)] - \frac{1}{2M_0}\{2\omega_e[\chi_{xy}^{(1)}N_{zxx}^{(2)}(0)$$
$$+\chi_{xy}^{(1)*}N_{zxx}^{(2)}(2\omega) - N_{xy}^{(1)}\chi_{zxx}^{(2)}(0) - N_{xy}^{(1)*}\chi_{zxx}^{(2)}(2\omega)] \tag{2-17a}$$
$$+\omega_a[\chi_{xy}^{(1)}N_{zxx}^{(2)}(0) + \chi_{xy}^{(1)*}N_{zxx}^{(2)}(2\omega)$$
$$+N_{xy}^{(1)*}\chi_{zxx}^{(2)}(2\omega) + N_{xy}^{(1)}\chi_{zxx}^{(2)}(0)]\}$$

$$b' = \frac{i\omega\tau}{2M_0}[3\chi_{xy}^{(1)*}N_{zxx}^{(2)}(2\omega) + 3N_{xy}^{(1)*}\chi_{zxx}^{(2)}(2\omega) + N_{xx}^{(1)}\chi_{zyx}^{(2)}(0)$$
$$+\chi_{xx}^{(1)}N_{zyx}^{(2)}(0)] + \frac{1}{2M_0}\{2\omega_e[\chi_{xx}^{(1)}N_{zyx}^{(2)}(0)$$
$$-\chi_{xy}^{(1)*}N_{zxx}^{(2)}(2\omega) - N_{xx}^{(1)}\chi_{zyx}^{(2)}(0) + N_{xy}^{(1)*}\chi_{zxx}^{(2)}(2\omega)] \tag{2-17b}$$
$$+\omega_a[\chi_{xx}^{(1)}N_{zyx}^{(2)}(0) - \chi_{xy}^{(1)*}N_{zxx}^{(2)}(2\omega)$$
$$+N_{xx}^{(1)}\chi_{zyx}^{(2)}(0) - N_{xy}^{(1)*}\chi_{zxx}^{(2)}(2\omega)] - 2\omega_m N_{zyx}^{(2)}(0)\}$$

$$c' = -\frac{\omega_m}{M_0} N_{xzy}^{(2)}(0) \tag{2-17c}$$

$$\begin{aligned} d' = &\frac{i\omega\tau}{2M_0}[\chi_{xx}^{(1)} N_{zxx}^{(2)}(0) - \chi_{xy}^{(1)} N_{zyx}^{(2)}(0) + N_{xx}^{(1)} \chi_{zxx}^{(2)}(0) \\ &-N_{xy}^{(1)} \chi_{zyx}^{(2)}(0)] + \frac{1}{2M_0}\{2\omega_e[\chi_{xx}^{(1)} N_{zxx}^{(2)}(0) \\ &-\chi_{xy}^{(1)} N_{zyx}^{(2)}(0) - N_{xx}^{(1)} \chi_{zxx}^{(2)}(0) + N_{xy}^{(1)} \chi_{zyx}^{(2)}(0)] + \omega_a[\chi_{xx}^{(1)} N_{zxx}^{(2)}(0) \\ &-\chi_{xy}^{(1)} N_{zyx}^{(2)}(0) - N_{xy}^{(1)} \chi_{zyx}^{(2)}(0) \\ &+N_{xx}^{(1)} \chi_{zxx}^{(2)}(0)] - 2\omega_m N_{zxx}^{(2)}(0)\} \end{aligned} \tag{2-17d}$$

$$\begin{aligned} e' = &\frac{i\omega\tau}{2M_0}[-3\chi_{xx}^{(1)*} N_{zxx}^{(2)}(2\omega) - 3N_{xx}^{(1)*} \chi_{zxx}^{(2)}(2\omega) + \chi_{xy}^{(1)} N_{zyx}^{(2)}(0) \\ &+N_{xy}^{(1)} \chi_{zyx}^{(2)}(0)] + \frac{1}{2M_0}\{2\omega_e[\chi_{xx}^{(1)*} N_{zxx}^{(2)}(2\omega) \\ &+\chi_{xy}^{(1)} N_{zyx}^{(2)}(0) - N_{xx}^{(1)*} \chi_{zxx}^{(2)}(2\omega) - N_{xy}^{(1)} \chi_{zyx}^{(2)}(0)] + \omega_a[\chi_{xx}^{(1)*} N_{zxx}^{(2)}(2\omega) \\ &+\chi_{xy}^{(1)} N_{zyx}^{(2)}(0) + N_{xx}^{(1)*} \chi_{zxx}^{(2)}(2\omega) \\ &+N_{xy}^{(1)} \chi_{zyx}^{(2)}(0)] - 2\omega_m N_{zxx}^{(2)}(2\omega)\} \end{aligned} \tag{2-17e}$$

$$\begin{aligned} f' = &-\frac{i\omega\tau}{2M_0}[3\chi_{xx}^{(1)*} N_{zxx}^{(2)}(2\omega) + 3N_{xx}^{(1)*} \chi_{zxx}^{(2)}(2\omega) - \chi_{xx}^{(1)} N_{zxx}^{(2)}(0) \\ &-N_{xx}^{(1)} \chi_{zxx}^{(2)}(0)] + \frac{1}{2M_0}\{2\omega_e[\chi_{xx}^{(1)} N_{zxx}^{(2)}(0) \\ &+\chi_{xx}^{(1)*} N_{zxx}^{(2)}(2\omega) - N_{xx}^{(1)} \chi_{zxx}^{(2)}(0) - N_{xx}^{(1)*} \chi_{zxx}^{(2)}(2\omega)] + \omega_a[\chi_{xx}^{(1)} N_{zxx}^{(2)}(0) \\ &+\chi_{xx}^{(1)*} N_{zxx}^{(2)}(2\omega) + N_{xx}^{(1)} \chi_{zxx}^{(2)}(0) \\ &+N_{xx}^{(1)*} \chi_{zxx}^{(2)}(2\omega)] - 2\omega_m (N_{zxx}^{(2)}(0) + N_{zxx}^{(2)}(2\omega)]\} \end{aligned} \tag{2-17f}$$

$$g' = \frac{\omega_m}{M_0} N_{xzx}^{(2)}(0) \tag{2-17g}$$

$$\begin{aligned} h' = &-\frac{i\omega\tau}{2M_0}[\chi_{xx}^{(1)} N_{zyx}^{(2)}(0) + \chi_{xy}^{(1)} N_{zxx}^{(2)}(0) + N_{xx}^{(1)} \chi_{zyx}^{(2)}(0) \\ &+N_{xy}^{(1)} \chi_{zxx}^{(2)}(0)] - \frac{1}{2M_0}\{2\omega_e[\chi_{xx}^{(1)} N_{zyx}^{(2)}(0) \\ &+\chi_{xy}^{(1)} N_{zxx}^{(2)}(0) - N_{xx}^{(1)} \chi_{zyx}^{(2)}(0) - N_{xy}^{(1)} \chi_{zxx}^{(2)}(0)] + \omega_a[\chi_{xx}^{(1)} N_{zyx}^{(2)}(0) \\ &+\chi_{xy}^{(1)} N_{zxx}^{(2)}(0) + N_{xx}^{(1)} \chi_{zyx}^{(2)}(0) \\ &+N_{xy}^{(1)} \chi_{zxx}^{(2)}(0)] - 2\omega_m N_{zyx}^{(2)}(0)\} \end{aligned} \tag{2-17h}$$

$$l' = -\frac{\omega_m}{M_0}[2N^{(2)}_{xyz}(2\omega) + N^{(2)}_{xyz}(0)] \quad \text{(2-17i)}$$

$$p' = \frac{\omega_m}{M_0}[2N^{(2)}_{xxz}(2\omega) + N^{(2)}_{xxz}(0)] \quad \text{(2-17j)}$$

The symmetry relations among the third-order elements are found to be

$\chi^{(3)}_{xxxx}(\omega) = \chi^{(3)}_{yyyy}(\omega)$, $\chi^{(3)}_{xyyx}(\omega) = \chi^{(3)}_{yxxy}(\omega)$

$\chi^{(3)}_{xzzx}(\omega) = \chi^{(3)}_{yzzy}(\omega)$, $\chi^{(3)}_{xxxy}(\omega) = -\chi^{(3)}_{yyyx}(\omega)$, $\chi^{(3)}_{xyyy}(\omega) = -\chi^{(3)}_{yxxx}(\omega)$

$\chi^{(3)}_{xxyx}(\omega) = \chi^{(3)}_{xyxx}(\omega) = -\chi^{(3)}_{yxyy}(\omega) = -\chi^{(3)}_{yyxy}(\omega)$, $\chi^{(3)}_{xzzy}(\omega) = -\chi^{(3)}_{yzzx}(\omega)$, $\chi^{(3)}_{zxzy}(\omega) = \chi^{(3)}_{zzxy}(\omega)$,

$\chi^{(3)}_{zxzx}(\omega) = \chi^{(3)}_{zzxx}(\omega)$, $\chi^{(3)}_{xxyy}(\omega) = \chi^{(3)}_{xyxy}(\omega) = \chi^{(3)}_{yxyx}(\omega) = \chi^{(3)}_{yyxx}(\omega)$, $\chi^{(3)}_{zyzx}(\omega) = \chi^{(3)}_{zzyx}(\omega)$,

$\chi^{(3)}_{xxzz}(\omega) = \chi^{(3)}_{xzxz}(\omega) = \chi^{(3)}_{yyzz}(\omega) = \chi^{(3)}_{yzyz}(\omega)$, $\chi^{(3)}_{zyzy}(\omega) = \chi^{(3)}_{zzyy}(\omega)$,

$\chi^{(3)}_{xyzz}(\omega) = \chi^{(3)}_{xzyz}(\omega) = -\chi^{(3)}_{yxzz}(\omega) = -\chi^{(3)}_{yzxz}(\omega)$.

Although there are 81 elements of the third-order susceptibility tensor and their expressions are very complicated, but many among them may not be applied due to the plane or line polarization of used electromagnetic waves. for example when the magnetic field $\vec{H}$ is in the x-y plane, the third-order elements with only subscripts x and y, such as $\chi^{(3)}_{xxxx}(\omega)$, $\chi^{(3)}_{xxyx}(\omega)$, $\chi^{(3)}_{xyyx}(\omega)$ and $\chi^{(3)}_{xyyy}(\omega)$ *et. al.*, are usefull. In addition, if the external magnetic field H_0 is removed, many the first- second- and third-order elements will disappear, or become 0. In the following sections, when one discusses AF polaritons the damping is neglected, but when investigating transmission and reflection the damping is considered.

3. Linear polaritons in antiferromagnetic systems

The linear AF polaritons of AF systems (AF bulk, AF films and superlattices) are eigen modes of electromagnetic waves propagating in the systems. The features of these modes can predicate many optical and electromagnetic properties of the systems. There are two kinds of the AF polaritons, the surface modes and bulk modes. The surface modes propagate along a surface of the systems and exponentially attenuate with the increase of distance to this surface. For these AF systems, an optical technology was applied to measure the AF polariton spectra (Jensen, 1995). The experimental results are completely consistent with the theoretical predications. In this section, we take the Voigt geometry usually used in the experiment and theoretical works, where the waves propagate in the plane normal to the AF anisotropy axis and the external magnetic field is pointed along this anisotropy axis.

3.1 Polaritons in AF bulk and film

Bulk AF polaritons can be directly described by the wave equation of EMWs in an AF crystal,

$$\nabla(\nabla \cdot \vec{H}) - \nabla^2\vec{H} - \varepsilon_a\omega^2\vec{\mu} \cdot \vec{H} = 0 \quad \text{(3-1)}$$

where ε_a is the AF dielectric constant and $\vec{\mu}$ is the magnetic permeability tensor. It is interesting that the magnetic field of AF polaritons vibrates in the x-y plane since the field does not couple with the AF magnetization for it along the z axis. We take the magnetic field as $\vec{H} = \vec{A}\exp(i\vec{k}\cdot\vec{r} - i\omega t)$ with the amplitude $\vec{A}$. Thus applying equation (3-1) we find directly the dispersion relation of bulk polaritons

$$k_x^2 + k_y^2 = \varepsilon_a \mu_v \omega^2 \tag{3-2}$$

with $\mu_v = [\mu_1^2 - \mu_2^2] / \mu_1$ the AF effective permeability. Equation (3-2) determines the continuums of AF polaritons in the $k - \omega$ figure (see Fig.2).

The best and simplest example available to describe the surface AF polariton is a semi-infinite AF. We assume the semi-infinite AF occupies the lower semi-space and the upper semi-space is of vacuum. The y axis is normal to the surface. The surface polariton moves along the x axis. The wave field in different spaces can be shown by

$$\vec{H} = \begin{cases} \vec{A}_0 \exp(-\alpha_0 y + ik_x x - i\omega t), & (\text{in the vaccum}) \\ \vec{A}\exp(\alpha y + ik_x x - i\omega t), & (\text{in the AF}) \end{cases} \tag{3-3}$$

where α_0 and α are positive attenuation factors . From the magnetic field (3-3) and the Maxwell equation $\nabla \times \vec{H} = \partial\vec{D} / \partial t$, we find the corresponding electric field

$$\vec{E} = \vec{e}_z \begin{cases} \dfrac{i}{\varepsilon_0 \omega}[ik_x A_{0y} + \alpha_0 A_{0x}]\exp(-\alpha_0 y + ik_x x - i\omega t) \\ \dfrac{i}{\varepsilon_a \omega}[ik_x A_y - \alpha A_x]\exp(\alpha y + ik_x x - i\omega t), \end{cases} \tag{3-4}$$

Here there are 4 amplitude components, but we know from equation $\nabla \cdot (\ddot{\mu} \cdot \vec{H}) = 0$ that only two are independent. This bounding equation leads to

$$A_{0y} = ik_x A_{0x} / \alpha_0 , \; A_y = i(k_x\mu_1 - \alpha\mu_2)A_x / (k_x\mu_2 - \alpha\mu_1) \tag{3-5}$$

The wave equation (3-1) shows that

$$\alpha_0^2 = k_x^2 - (\omega / c)^2 , \alpha^2 = k_x^2 - \mu_v(\omega / c)^2 \tag{3-6}$$

determining the two attenuation constants. The boundary conditions of H_x and E_z continuous at the interface (y=0) lead to the dispersion relation

$$\mu_1(\alpha_0\mu_v + \varepsilon_a\alpha) = \varepsilon_a\mu_2 k_x \tag{3-7}$$

where the permeability components and dielectric constants all are their relative values. Equation (3-7) describes the surface AF polariton under the condition that the attenuation factors both are positive. In practice, Eq.(3-6) also shows the dispersion relation of bulk modes as that attenuation factor is vanishing.

We illustrate the features of surface and bulk AF polaritons in Fig.2. There are three bulk continua where electromagnetic waves can propagate. Outside these regions, one sees the surface modes, or the surface polariton. The surface polariton is non-reciprocal, or the polariton exhibits completely different properties as it moves in two mutually opposite directions, respectively. This non-reciprocity is attributed to the applied external field that breaks the magnetic symmetry of the AF. If we take an AF film as example to discuss this subject, we are easy to see that the surface mode is changed only in quantity, but the bulk modes become so-called guided modes, which no longer form continua and are some separated modes (Cao & Caillé, 1982).

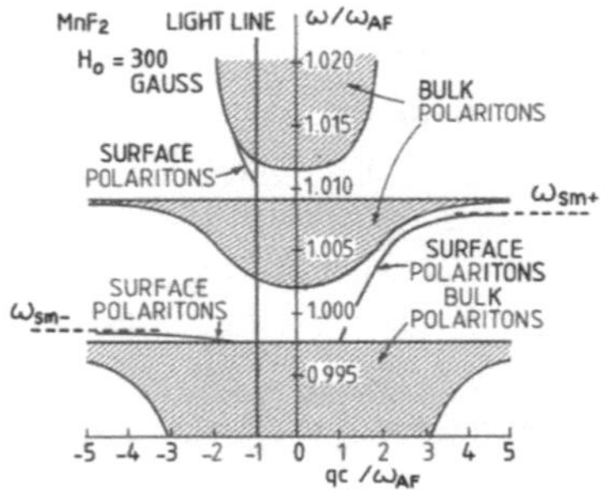

Fig. 2. Surface polariton dispersion curves and bulk continua on the MnF_2 in the geometry with an applied external field. After Camley & Mills,1982

3.2 Polaritons in antiferromagnetic multilayers and superlattices

There have been many works on the magnetic polaritons in AF multilayers or superlattices. This AF structure is the one-dimension stack, commonly composed of alternative AF layers and dielectric (DE) layers, as illustrated in Fig.3.

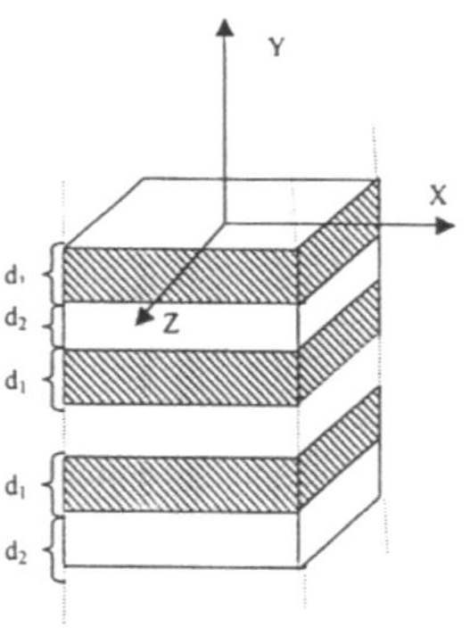

Fig. 3. The structure of AF superlattice and selected coordinate system.

In the limit case of small stack period, the effective-medium method was developed (Oliveros, et. al., 1992; Camley, 1992; Raj & Tilley, 1987; Almeida & Tilley, 1990; Cao & Caillé, 1982; Almeida & Mills,1988; Dumelow & Tilley,1993; Elmzughi, 1995a, 1995b). According to this method, one can consider these structures as some homogeneous films or bulk media with effective magnetic permeability and dielectric constant. This method and its results are very simple in mathematics. Of course, this is an approximate method. The other method is called as the transfer-matrix method (Born & Wolf, 1964; Raj & Tilley, 1989), where the electromagnetic boundary conditions at one interface set up a matrix relation between field amplitudes in the two adjacent layers, or adjacent media. Thus amplitudes in any layer can be related to those in another layer by the product of a series of matrixes. For an infinite AF superlattice, the Bloch's theorem is available and can give an additional relation between the corresponding amplitudes in two adjacent periods. Using these matrix relations, bulk AF polaritons in the superlattices can be determined. For one semi-finite structure with one surface, the surface mode can exist and also will be discussed with the method.

3.2.1 The limit case of short period, effective-medium method

Now we introduce the effective-medium method, with the condition of the wavelength λ much longer than the stack period $D = d_1 + d_2$ (d_1 and d_2 are the AF and DE thicknesses). The main idea of this method is as follows. We assume that there are an effective relation $\vec{B} = \ddot{\mu}_{eff} \cdot \vec{H}$ between effective magnetic induction and magnetic field, and an effective relation $\vec{D} = \ddot{\varepsilon}_{eff} \cdot \vec{E}$ between effective electric field and displacement, where these fields are considered as the wave fields in the structures. But $\vec{b} = \ddot{\mu} \cdot \vec{h}$ and $\vec{d} = \varepsilon\vec{e}$ in any layer, where $\ddot{\mu}$ is given in section 2 for AF layers and $\ddot{\mu} = 1$ for DE layers. These fields are local fields in the layers. For the components of magnetic induction and field continuous at the interface, one assumes

$$H_x = h_{1x} = h_{2x}, H_z = h_{1z} = h_{2z}, B_y = b_{1y} = b_{2y} \tag{3-8a}$$

and for those components discontinuous at the interface, one assumes

$$B_x = f_1 b_{1x} + f_2 b_{2x}, B_z = f_1 b_{1z} + f_2 b_{2z}, H_y = f_1 H_{1y} + f_2 H_{2y} \tag{3-8b}$$

where the AF ratio $f_1 = d_1 / (d_1 + d_2)$ and the DE ratio $f_2 = 1 - f_1$. Thus the effective magnetic permeability is obtained from equations (3-8) and its definition $\vec{B} = \ddot{\mu}_{eff} \cdot \vec{H}$,

$$\ddot{\mu}_{eff} = \begin{pmatrix} \mu_{xx}^e & i\mu_{xy}^e & 0 \\ -i\mu_{xy}^e & \mu_{yy}^e & 0 \\ 0 & 0 & 1 \end{pmatrix} \tag{3-9}$$

with the elements

$$\mu_{xx}^e = f_1\mu_1 + f_2 - \frac{f_1 f_2 \mu_2^2}{f_1 + f_2\mu_1}, \mu_{yy}^e = \frac{\mu_1}{f_1 + f_2\mu_1}, \mu_{xy}^e = \frac{f_1\mu_2}{f_1 + f_2\mu_1} \tag{3-10}$$

On the similar principle, we can find that the effective dielectric permittivity tensor is diagonal and its elements are

$$\varepsilon_{xx}^e = \varepsilon_{zz}^e = f_1\varepsilon_1 + f_2\varepsilon_2, \varepsilon_{yy}^e = \varepsilon_1\varepsilon_2 / (f_1\varepsilon_2 + f_2\varepsilon_1) \tag{3-11}$$

On the base of these effective permeability and permittivity, one can consider the AF multilayers or superlattices as homogeneous and anisotropical AF films or bulk media, so the same theory as that in section 3.1 can be used. Magnetic polaritons of AF multilayers (Oliveros, et.al., 1992; Raj & Tilley, 1987), AF superlattices with parallel or transverse surfaces (Camley, et. al., 1992; Barnas, 1988) and one-dimension AF photonic crystals (Song, et.al., 2009; Ta, et. al.,2010) have been discussed with this method.

3.2.2 Polaritons and transmission of AF multilayers: transfer-matrix method

If the wavelength is comparable to the stack period, the effective-medium method is no longer available so that a strict method is necessary. The transfer-matrix method is such a method. In this subsection, we shall present magnetic polaritons of AF multilayers or superlattices with this method. We introduce the wave magnetic field in two layers in the *l*th stack period as follows.

$$\vec{H} = \mathrm{e}^{ik_x x - i\omega t}\begin{cases}(\vec{A}_+^l \mathrm{e}^{ik_1 y} + \vec{A}_-^l \mathrm{e}^{-ik_1 y}) & (\text{in the AF layer}) \\ (\vec{B}_+^l \mathrm{e}^{ik_2 y} + \vec{B}_-^l \mathrm{e}^{-ik_2 y}) & (\text{in the DE layer})\end{cases} \tag{3-12}$$

where k_1 and k_2 are determined with $k_1^2 + k_x^2 = \varepsilon_1 \mu_v \omega^2$ and $k_2^2 + k_x^2 = \varepsilon_2 \mu_0 \omega^2$. Similar to Eq. (3-4) in subsection 3.1, the corresponding electric field in this period is written as

$$\vec{E} = \vec{e}_z e^{ik_x x - i\omega t}\begin{cases}\dfrac{i}{\varepsilon_1 \omega}[(ik_x A_{+y}^l - ik_1 A_{+x}^l)\mathrm{e}^{ik_1 y} + (ik_x A_{-y}^l + ik_1 A_{-x}^l)\mathrm{e}^{-ik_1 y}] \\ \dfrac{i}{\varepsilon_2 \omega}[(ik_x B_{+y}^l - ik_2 B_{+x}^l)\mathrm{e}^{ik_2 y} + (ik_x B_{-y}^l + ik_2 B_{-x}^l)\mathrm{e}^{-ik_2 y}]\end{cases} \tag{3-13}$$

Here there is a relation between per pair of amplitude components, or

$$A_{\pm y}^l = i(k_x \mu_1 \mp ik_1 \mu_2) A_{\pm x}^l / (k_x \mu_2 \mp ik_1 \mu_1) = \lambda_\pm A_{\pm x}^l, \; B_{\pm y}^l = \mp k_x B_{\pm x}^l / k_2 \tag{3-14}$$

As a result, we can take $A_{\pm x}^l$ and $B_{\pm x}^l$ as 4 independent amplitude components. Next, according to the continuity of electromagnetic fields at that interface in the period, we find

$$A_{+x}^l e^{ik_1 d_1} + A_{-x}^l e^{-ik_1 d_1} = B_{+x}^l + B_{-x}^l \tag{3-15a}$$

$$\frac{1}{\varepsilon_1}[(-k_x A_{+y}^l + k_1 A_{+x}^l)\mathrm{e}^{ik_1 d_1} - (k_x A_{-y}^l + k_1 A_{-x}^l)\mathrm{e}^{-ik_1 d_1}] = \frac{\omega \mu_0}{k_2}(B_{+x}^l - B_{-x}^l) \tag{3-15b}$$

At the interface between the *l*th and *l+1*th periods, one see

$$(A_{+x}^{l+1} + A_{-x}^{l+1}) = B_{+x}^l e^{ik_2 d_2} + B_{-x}^l e^{-ik_2 d_2} \tag{3-15c}$$

$$\frac{1}{\varepsilon_1}[(k_1 A_{+x}^{l+1} - k_x A_{+y}^{l+1}) - (k_1 A_{-x}^{l+1} + k_x A_{-y}^{l+1})] = \frac{\omega \mu_0}{k_2}(B_{+x}^l \mathrm{e}^{ik_2 d_2} - B_{-x}^l \mathrm{e}^{-ik_2 d_2}) \tag{3-15d}$$

Thus the matrix relation between the amplitude components in the same period is introduced as

$$\begin{pmatrix} B_{+x}^l \\ B_{-x}^l \end{pmatrix} = \begin{pmatrix} \Gamma_{11} & \Gamma_{12} \\ \Gamma_{21} & \Gamma_{22} \end{pmatrix}\begin{pmatrix} A_{+x}^l \\ A_{-x}^l \end{pmatrix} \tag{3-16}$$

where the matrix elements are given by

$$\Gamma_{11} = \frac{e^{ik_1 d_1}}{2}(1 + \Delta_+), \Gamma_{12} = \frac{e^{-ik_1 d_1}}{2}(1 - \Delta_-), \Gamma_{21} = \frac{e^{ik_1 d_1}}{2}(1 - \Delta_+), \; \Gamma_{22} = \frac{e^{-ik_1 d_1}}{2}(1 + \Delta_-) \tag{3-17}$$

with $\Delta_{\pm} = k_2(k_1 \mp \lambda_{\pm} k_x) / \omega\mu_0\varepsilon_1$. From (3-15), the other relation also is obtained, or

$$\begin{pmatrix} B_{+x}^{l} \\ B_{-x}^{l} \end{pmatrix} = \begin{pmatrix} \Lambda_{11} & \Lambda_{12} \\ \Lambda_{21} & \Lambda_{22} \end{pmatrix} \begin{pmatrix} A_{+x}^{l+1} \\ A_{-x}^{l+1} \end{pmatrix} \tag{3-18}$$

with

$$\Lambda_{11} = \frac{e^{-ik_2d_2}}{2}(1+\Delta_+)\,,\ \Lambda_{12} = \frac{e^{-ik_2d_2}}{2}(1-\Delta_-)\,,\ \Lambda_{21} = \frac{e^{ik_2d_2}}{2}(1-\Delta_+)\,,\ \Lambda_{22} = \frac{e^{ik_2d_2}}{2}(1+\Delta_-) \tag{3-19}$$

Commonly, the matrix relation between the amplitude components in the lth and $l+1$th periods is written as

$$\begin{pmatrix} A_{+x}^{l} \\ A_{-x}^{l} \end{pmatrix} = \Gamma^{-1}\Lambda \begin{pmatrix} A_{+x}^{l+1} \\ A_{-x}^{l+1} \end{pmatrix} = T \begin{pmatrix} A_{+x}^{l+1} \\ A_{-x}^{l+1} \end{pmatrix} \tag{3-20}$$

In order to discuss bulk AF polaritons, an infinite AF superlattice should be considered. Then the Bloch's theorem is available so that $A_{\pm x}^{l+1} = gA_{\pm x}^{l}$ with $g = \exp(-iQD)$, and then the dispersion relation of bulk magnetic polaritons just is

$$\cos(QD) = \frac{1}{2}(T_{11} + T_{22}) \tag{3-21}$$

It can be reduced into a more clearly formula, or

$$\cos(QD) = \cos(k_1d_1)\cos(k_2d_2) - \frac{k_1^2 + k_2^2\mu_v^2 - k_x^2\mu_2^2 / \mu_1^2}{2k_1k_2\mu_v}\sin(k_1d_1)\sin(k_2d_2) \tag{3-22}$$

When one wants to discuss the surface polariton, the semi-infinite system is the best and simplest example. In this situation, the Bloch's theorem is not available and the polariton wave attenuates with the distance to the surface, according to $\exp(-\alpha lD)$, where lD is the distance and α is the attenuation coefficient and positive. As a result,

$$\cosh(\alpha D) = \frac{1}{2}(T_{11} + T_{22}) \tag{3-23}$$

It should remind that equation (3-23) cannot independently determine the dispersion of the surface polariton since the attenuation coefficient is unknown, so an additional equation is necessary. We take the wave function outside this semi-infinite structure as $\vec{H} = \vec{A}_0 \exp(-\alpha_0 y + ik_x x - i\omega t)$ with α_0 the vacuum attenuation constant. The two components of the amplitude vector are related with $A_{0y} = ik_x A_{0x} / \alpha_0$ and $k_x^2 - \alpha_0^2 = (\omega / c)^2$. The corresponding electric field is $E_z = -(i\omega\mu_0 / \alpha_0)H_x$. The boundary conditions of field components H_x and E_z continuous at the surface lead to

$$A_{0x} = A_{+x} + A_{-x} \tag{3-24a}$$

$$-(i\omega\mu_0\varepsilon_1 / \alpha_0)A_{0x} = (k_1A_{+x} - k_xA_{+y}) - (k_1A_{-x} + k_xA_{-y}) \quad (3\text{-}24b)$$

$$A_{+x} = g'(T_{11}A_{+x} + T_{12}A_{-x}) \quad (3\text{-}24c)$$

with $g' = \exp(-\alpha D)$. These equations result in another relation,

$$g'T_{12}(k_x\lambda_+ - k_1 - \Delta_0) + (1 - g'T_{11})(k_x\lambda_- + k_1 - \Delta_0) = 0 \quad (3\text{-}25)$$

Eqs. (3-23) and (3-25) jointly determine the dispersion properties of the surface polariton under the conditions of $\alpha, \alpha_0 > 0$.

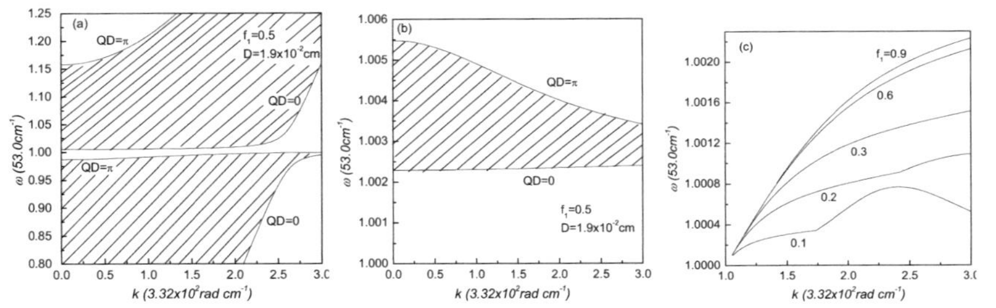

Fig. 4. Frequency spectrum of the polaritons of the FeF_2/ZnF_2 superalttice. (a) shows the top and bottom bands, and (b) presents the middle band. The surface mode is illustrated in (c). f_1 denotes the ratio of the FeF_2 in one period of the superlattice. After Wang & Li, 2005.

We present a figure example to show features of bulk and surface polaritons, as shown in Fig.4. Because of the symmetry of dispersion curves with respective to *k=0,* we present only the dispersion pattern in the range of *k>0.* The bulk polaritons form several separated continuums, and the surface mode exists in the bulk-polariton stop-bands. The bulk polaritons are symmetrical in the propagation direction, or possess the reciprocity, but is not the surface mode. These properties also can be found from the dispersion relations. For the bulk polaritons, the wave vector appears in dispersion equation (3-22) in its k_x^2 style, but for the surface mode, k_x and k_x^2 both are included dispersion equation (3-25).

3.2.3 Transmission of AF multilayers

In practice, infinite AF superlattices do not exist, so the conclusions from them are approximate results. For example, if the incident-wave frequency falls in a bulk-polariton stop-band of infinite AF superlattice, the transmission of the corresponding AF multilayer must be very weak, but not vanishing. Of course, it is more intensive in the case of frequency in a bulk-polariton continuum. Based on the above results, we derive the transmission ratio of an AF multilayer, where this structure has two surfaces, the upper surface and lower surface. We take a TE wave as the incident wave, with its electric component normal to the incident plane (the x-y plane) and along the z axis. The incident wave illuminates the upper surface and the transmission wave comes out from the lower surface. We set up the wave function above and below the multilayer as

$$\vec{H} = [\vec{I}_0 \exp(-ik_0y) + \vec{R}_0 \exp(ik_0y)]\exp(ik_xx), \text{(above the system)} \quad (3\text{-}26a)$$

$$\vec{H} = \vec{T}_0 \exp(-ik_0 y + ik_x x) \text{ (below the system)} \tag{3-26b}$$

The wave function in the multilayer has been given by (3-12) and (3-13). By the mathematical process similar to that in subsection 3.2.2, we can obtain the transmission and reflection of the multilayer with N periods from the following matrix relation,

$$\begin{pmatrix} I_0 \\ R_0 \end{pmatrix} = \Lambda_0 T^{N-1} \Gamma^{-1} \Lambda_1 \begin{pmatrix} T_0 \\ T_0 \end{pmatrix} \tag{3-27}$$

in which two new matrixes are shown with

$$\Lambda_0 = \begin{pmatrix} 1+\Delta'_+ & 1+\Delta'_- \\ 1-\Delta'_+ & 1-\Delta'_- \end{pmatrix}, \Lambda_1 = \begin{pmatrix} 1-k_2/k_0 & 0 \\ 0 & 1+k_2/k_0 \end{pmatrix} \tag{3-28}$$

with $k_0 = [(\omega/c)^2 - k_x^2]^{1/2}$ and $\Delta'_\pm = k_0(k_x\lambda_\pm \mp k_1)/\omega\mu_0\varepsilon_1$. Thus the reflection and transmission are determined with equation (3-27). In numerical calculations, the damping in the permeability cannot is ignored since it implies the existence of absorption. We have obtained the numerical results on the AF multilayer, and transmission spectra are consistent with the polariton spectra (Wang, J. J. et. al, 1999), as illustrated in Fig.5.

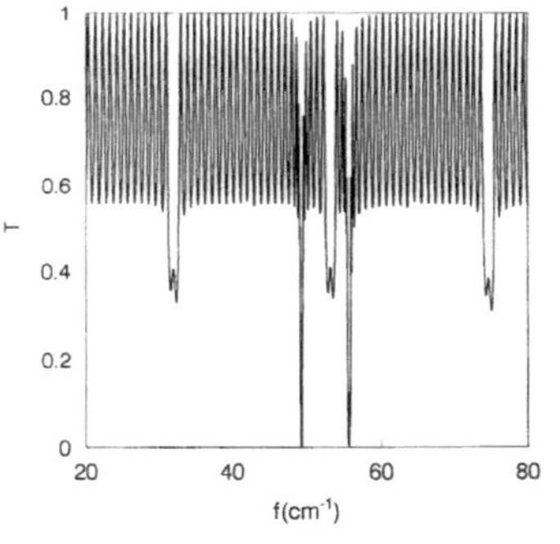

Fig. 5. Transmission curve for FeF_2 multilayer in Voigt geometry. After Wang, J. J. et. al, 1999.

4. Nonlinear surface and bulk polaritons in AF superlattices

In the previous section, we have discussed the linear propagation of electromagnetic waves in various AF systems, including the transmission and reflection of finite thickness multilayer. The results are available to the situation of lower intensity of electromagnetic waves. If the intensity is very high, the nonlinear response of magnetzation in AF media to the magnetic component of electromagnetic waves cannot be neglected. Under the present laser technology, this case is practical. Because we have found the second- and third-order magnetic susceptibilities of AF media, we can directly derive and solve nonlinear dispersion equations of electromagnetic waves in various AF systems. There also are two situations to be discussed. First, if the wavelenght λ is much longer than the superlattice period L ($\lambda \gg L$), the superlattice behaves like an anisotropic bulk medium(Almeida & Mills,1988; Raj & Tilley,1987), and the effective-medium approch is reasonable. We have introduced a nonlinear effective-medium theory(Wang & Fu, 2004), to solve effective susceptibilities of magnetic superlattices or multilayers. This method has a key point that the effective second- and third-order magnetizations come from the contribution of AF layers or $\vec{m}_e^{(2)} = f_1\vec{m}^{(2)}$ and $\vec{m}_e^{(3)} = f_1\vec{m}^{(3)}$.

4.1 Polaritons in AF superlattice

In this section we shall use a stricter method to deal with nonlinear propagation of AF polaritons in AF superlattices. In section 2, we have obtained various nonlinear susceptibilities of AF media, which means that one has obtained the expressions of $\vec{m}^{(2)}$ and $\vec{m}^{(3)}$. In AF layers, the polariton wave equation is

$$\nabla(\nabla\cdot\vec{H}^{NL}) - \nabla^2\vec{H}^{NL} - k_0^2\vec{\mu}\cdot\vec{H}^{NL} = k_0^2\vec{m}^{(3)}, \qquad k_0^2 = \varepsilon_1(\omega/c)^2, \tag{4-1}$$

where $\vec{\mu}$ is the linear permeability of antiferromagnetic layers given in section 2, and the nonzeroelements $\mu_{yy} = \mu_{xx} = \mu$, $\mu_{zz} = 1$. The third-order magnetization is indicated by $m_i^{(3)} = \sum_{jkl}\chi_{ijkl}^{(3)}H_jH_kH_l^*$ with the nonlinear susceptibility elements presened in section 2. As an approximation, we consider the field components H_i in $m_i^{(3)}$ as linear ones to find the nonlinear solution of $\vec{H}^{NL}$ included in wave equanion (4-1). For the linear surface wave propagating along the x-axis and the linear bulk waves moving in the x-y plane, $\partial/\partial z = 0$. Thus the wave equation is rewritten as

$$ik_x\frac{\partial}{\partial y}H_y^{NL} - \frac{\partial^2}{\partial y^2}H_x^{NL} - \varepsilon_1\omega^2\mu_1H_x^{NL} = \varepsilon_1\omega^2\Gamma(y)(\chi_{xxxy}^{(3)}H_x^{NL} - \chi_{xyyx}^{(3)}H_y^{NL}) \tag{4-2a}$$

$$ik_x\frac{\partial}{\partial y}H_x^{NL} + (k_x^2 - \varepsilon_1\omega^2\mu)H_y^{NL} = \varepsilon_1\omega^2\Gamma(y)(\chi_{xxxy}^{(3)}H_y^{NL} - \chi_{xyyx}^{(3)}H_x^{NL}) \tag{4-2b}$$

$$(k^2 - \frac{\partial^2}{\partial y^2} - \varepsilon_1\omega^2)H_z^{NL} = \varepsilon_1\omega^2 m_z^{(3)} \tag{4-2c}$$

with $\Gamma(y) = (H_xH_y^* - H_x^*H_y)$. Eq.(4-2c) implies that H_z is a third-order small quantity and equal to zero in the circumstance of linearity (TM waves). We begin from the linear wave solution that has been given section 3 to look for the nonlinear wave solution in AF layers. In the case of linearity, the relations among the wave amplitudes, $A_{\pm y} = \mp ik_xA_{\pm x}/\alpha_1$ with $\alpha_1 = [k_x^2 - \varepsilon_1\mu(\omega/c)^2]^{1/2}$. The nonlinear terms in equations (4-2) should contain a factor $F(m) = \exp(-mn\beta D)$ with $m = 3$ and β is defined as the attenuation constant for the surface modes, and m=1 and $\beta = iQ$ with Q the Bloch's wavwnumber for the bulk modes. $A_1 \sim D_1$ and $A_2 \sim D_2$ are nonlinear coefficients. After solving the derivation of equation (4-2b) with respect to y, substituting it into (4-2a) leads to the wave solutions

$$\begin{aligned}H_x = A_{+x}e^{i(k_xx-\omega t)}e^{-n\beta D}\{e^{\alpha_1(y+nD)} + \alpha' e^{-\alpha_1(y+nD)} + f_n[(y+nD)\alpha_1L_1e^{\alpha_1(y+nD)} \\ +\alpha(y+nD)\alpha_1L_2e^{-\alpha_1(y+nD)} + L_3e^{3\alpha_1(y+nD)} + L_4e^{-3\alpha_1(y+nD)}]\}\end{aligned} \tag{4-3a}$$

and

$$H_y = -\frac{ik_x}{\alpha_1} A_{+x} e^{-\beta nD} e^{i(k_x x - \omega t)} \{ e^{\alpha_1 (y+nD)} - \alpha' e^{-\alpha_1 (y+nD)} + f_n [((y+nD)\alpha_1 L_1 + S) \\ e^{\alpha_1 (y+nD)} + \alpha(-(y+nD)\alpha_1 L_2 + T) e^{-\alpha_1 (y+nD)} + L_3' e^{3\alpha_1 (y+nD)} + L_4' e^{-3\alpha_1 (y+nD)}]\} \quad (4\text{-}3b)$$

in which $f_n = 1$ for the bulk modes and $f_n = \exp(-2n\beta D)$ for the surface modes. The expressions of coefficients in Eqs.(4-3a) and (4-3b) are presented as follows:

1. When α_1 is a real number, the coefficients in Eq.(4-3a) are

$$A_1 = 2\alpha k_x A_m \chi_+ ,\ B_1 = -2|\alpha|^2 k_x A_m \chi_- ,\ C_1 = 2k_x A_m \chi_- ,\ D_1 = -2\alpha |\alpha|^2 k_x A_m \chi_+ \quad (4\text{-}4a)$$

$$A_2 = 2i\alpha k_x A_m \delta_+ ,\ B_2 = 2ik_x |\alpha|^2 A_m \delta_- ,\ C_2 = -2ik_x A_m \delta_- ,\ D_2 = -2ik_x \alpha |\alpha|^2 A_m \delta_+ \quad (4\text{-}4b)$$

where $\chi_\pm = i\alpha_1 \chi^{(3)}_{xxxy} \pm k_x \chi^{(3)}_{xyyx}$, $\delta_\pm = ik_x \chi^{(3)}_{xxxy} \pm \alpha_1 \chi^{(3)}_{xyyx}$, $A_m = \varepsilon_1 \omega^2 |A|^2 / [k_x^2 + |\alpha_1|^2]$. The field strength $|A|^2 = |A_x|^2 + |A_y|^2 = [|\alpha_1|^2 + k_x^2] |A_x|^2 / |\alpha_1|^2$. From the boundary conditions of the linear field, one also can easily prove that α included in the formulae is

$$\alpha = A_{-x} / A_{+x} = (\alpha_0 \mu + \alpha_1) / (\alpha_0 \mu - \alpha_1) \quad (4\text{-}5a)$$

for the surface modes and

$$\alpha = \frac{e^{-\alpha_1 d_1} [\alpha_1 \cosh(\alpha_2 d_2) - \mu \alpha_2 \sinh(\alpha_2 d_2)] - \alpha_1 e^{-iQD}}{\alpha_1 e^{-iQD} - e^{\alpha_1 d_1} [\alpha_1 \cosh(\alpha_2 d_2) + \mu \alpha_2 \sinh(\alpha_2 d_2)]} \quad (4\text{-}5b)$$

for the bulk modes. The coefficients in Eq.(4-3) can written as

$$L_1 = \frac{1}{2\varepsilon_1 \omega^2 \mu} (A_1 - \frac{ik_x}{\alpha_1} A_2) = \frac{A_m k_x \alpha}{\mu \varepsilon_1 \omega^2} (\chi_+ + \frac{k_x}{\alpha_1} \delta_+)$$
$$L_2 = -\frac{1}{2\varepsilon_1 \omega^2 \mu \alpha} (B_1 + \frac{ik_x}{\alpha_1} B_2) = \frac{A_m \alpha^* k_x}{\mu \varepsilon_1 \omega^2} (\chi_- + \frac{k_x}{\alpha_1} \delta_-) \quad (4\text{-}6a)$$

$$L_3 = \frac{1}{8\varepsilon_1 \omega^2 \mu} (C_1 - \frac{3ik_x}{\alpha_1} C_2) = \frac{A_m k_x}{4\mu \varepsilon_1 \omega^2} (\chi_- - \frac{3k_x}{\alpha_1} \delta_-)$$
$$L_4 = \frac{1}{8\varepsilon_1 \omega^2 \mu} (D_1 + \frac{3ik_x}{\alpha_1} D_2) = -\frac{A_m \alpha |\alpha|^2 k_x}{4\mu \varepsilon_1 \omega^2} (\chi_+ - \frac{3k_x}{\alpha_1} \delta_+) \quad (4\text{-}6b)$$

$$L_3' = 3L_3 + \frac{i}{\alpha_1 k_x} C_2 = \frac{A_m}{4\mu \varepsilon_1 \omega^2} [3k_x \chi_- - \frac{\delta_-}{\alpha_1} (k_x^2 + 8\alpha_1^2)]$$
$$L_4' = -3L_4 + \frac{i}{\alpha_1 k_x} D_2 = \frac{A_m \alpha |\alpha|^2}{4\mu \varepsilon_1 \omega^2} [3k_x \chi_+ - \frac{\delta_+}{\alpha_1} (k_x^2 + 8\alpha_1^2)] \quad (4\text{-}6c)$$

$$S = L_1 + \frac{i}{\alpha_1 k_x} A_2 = \frac{A_m \alpha}{\mu \varepsilon_1 \omega^2} [k_x \chi_+ + \frac{\delta_+}{\alpha_1}(2\alpha_1^2 - k_x^2)]$$
$$T = k_x L_2 + \frac{i}{\alpha_1 \alpha k_x} B_2 = \frac{k_x A_m \alpha^*}{\mu \varepsilon_1 \omega^2} [k_x \chi_- + \frac{\delta_-}{\alpha_1}(2\alpha_1^2 - k_x^2)] \quad (4\text{-}6d)$$

2. If α_1 is imaginary, i.e. $\alpha_1 = i\lambda$, these coefficients should be changed into

$$A_1 = -2|\alpha|^2 k_x A_m \chi_+ ,\ B_1 = 2\alpha k_x A_m \chi_- ,\ C_1 = -2k_x \alpha^* A_m \chi_- ,\ D_1 = 2k_x \alpha^2 A_m \chi_+ \quad (4\text{-}7a)$$

$$A_2 = -2ik_x |\alpha|^2 A_m \delta_+ ,\ B_2 = -2ik_x \alpha A_m \delta_- ,\ C_2 = 2ik_x \alpha^* A_m \delta_- ,\ D_2 = 2ik_x \alpha^2 A_m \delta_+ \quad (4\text{-}7b)$$

$$L_1 = -\frac{A_m |\alpha|^2 k_x}{\varepsilon_1 \omega^2 \mu}(\chi_+ + \frac{k_x}{\alpha_1}\delta_+) ,\ L_2 = -\frac{A_m k_x}{\varepsilon_1 \omega^2 \mu}(\chi_- + \frac{k_x}{\alpha_1}\delta_-) \quad (4\text{-}8a)$$

$$L_3 = -\frac{A_m \alpha^* k_x}{4\varepsilon_1 \omega^2 \mu}(\chi_- - \frac{3k_x}{\alpha_1}\delta_-) ,\ L_4 = \frac{A_m \alpha^2 k_x}{4\varepsilon_1 \omega^2 \mu}(\chi_+ - \frac{3k_x}{\alpha_1}\delta_+) \quad (4\text{-}8b)$$

$$L_3' = -\frac{A_m \alpha^*}{4\varepsilon_1 \omega^2 \mu}[3k_x \chi_- - \frac{\delta_-}{\alpha_1}(k_x^2 + 8\alpha_1^2)] ,\ L_4' = -\frac{A_m \alpha^2}{4\varepsilon_1 \omega^2 \mu}[3k_x \chi_+ - \frac{\delta_+}{\alpha_1}(k_x^2 + 8\alpha_1^2)] \quad (4\text{-}8c)$$

$$S = -\frac{A_m |\alpha|^2}{\varepsilon_1 \omega^2 \mu}[k_x \chi_+ + \frac{\delta_+}{\alpha_1}(2\alpha_1^2 - k_x^2)] ,\ T = -\frac{A_m}{\varepsilon_1 \omega^2 \mu}[k_x \chi_- + \frac{\delta_-}{\alpha_1}(2\alpha_1^2 - k_x^2)] \quad (4\text{-}8d)$$

Note that all these coefficients contain implicitly the factor $\Delta = |A|^2 / 4\pi M_0^2$, so we say that they are of the second-order. For simplicity in the process of deriving dispersion equations, we introduce three second-order quantities,

$$\eta_1(y + nD) = (y + nD)\alpha_1 L_1 e^{\alpha_1 (y+nD)}$$
$$+\alpha(y + nD)\alpha_1 L_2 e^{-\alpha_1 (y+nD)} + L_3 e^{3\alpha_1 (y+nD)} + L_4 e^{-3\alpha_1 (y+nD)} \quad (4\text{-}9a)$$

$$\eta_2(y + nD) = [(y + nD)\alpha_1 L_1' + S]e^{\alpha_1 (y+nD)} + \alpha[-(y + nD)\alpha_1 L_2$$
$$+T]e^{-\alpha_1 (y+nD)} + L_3' e^{3\alpha_1 (y+nD)} + L_4' e^{-3\alpha_1 (y+nD)} \quad (4\text{-}9b)$$

and

$$\theta(y + nD) = \frac{i\alpha_1}{\varepsilon_1 \omega^2 k}[A_2 e^{\alpha_1 (y+nD)} + B_2 e^{-\alpha_1 (y+nD)} + C_2 e^{3\alpha_1 (y+nD)} + D_2 e^{-3\alpha_1 (y+nD)}] \quad (4\text{-}9c)$$

Thus the nonlinear magnetic field can be rewritten as

$$\vec{H} = A_x\{[e^{\alpha_1 (y+nD)} + \alpha' e^{-\alpha_1 (y+nD)} + \eta_1(y + nD) f_n)]\vec{e}_x$$
$$-\frac{ik}{\alpha_1}[e^{\alpha_1 (y+nD)} - \alpha' e^{-\alpha_1 (y+nD)} + \eta_2(y + nD) f_n]\vec{e}_y\} e^{-\beta nD} e^{i(kx - \omega t)} \quad (4\text{-}10a)$$

and the third-order magnetization is equal to

$$m_y^{(3)} = -\frac{ik}{\alpha_1} A_x \theta(y+nD) f_n e^{i(kx-\omega t)} e^{-n\beta D} \quad \text{(4-10b)}$$

The two formulae will be applied for solving the dispersion equations of the nonlinear surface and bulk polaritons from the boundary conditions satisfied by the wave fields.
Seeking the dispersion relations of AF polaritons should begin from the boundary conditions of the magnetic field H_x and magnetic induction field B_y continuous at the interfaces and surface ($y=-nD, -nD-d_1$ and 0). The results (4-3a) and (4-3b) related to the nth AF layer, as well as the solutions in the vacuum $\vec{H} = \vec{A}_0 e^{-\alpha_0 y} e^{i(kx-\omega t)}$ and in the nth NM layer $\vec{H} = [\vec{C} e^{\alpha_2(y+jD+d_1)} + \vec{D} e^{-\alpha_2(y+jD+d_1)}] e^{-\beta jD} e^{i(kx-\omega t)}$ will be used to determine the dispersion relations. In the following several paragraphs, we shall calculate the dispersion relations of the surface and bulk modes, respectively.

3. Bulk dispersion equation

For the bulk polaritons, there are 6 amplitude coefficients in the wave solutions, $A_x, \alpha', C_x, C_y, D_x$ and D_y. The magnetic induction .. in AF layers and $B_y = H_y$ in NM layers. The boundary conditions of B_y and H_x continuous at the interfaces ($y=-nD$ and $-nD-d_1$) imply four equations, and $\nabla \cdot \vec{H} = 0$ in a NM layer leads to two additional relations $C_y = -ikC_x / \alpha_2$ and $D_y = ikD_x / \alpha_2$. Thus we have

$$A_x[1+\alpha'+\eta_1(0) f_n] = (C_x e^{-\alpha_2 d_2} + D_x e^{\alpha_2 d_2}) e^{iQD} \quad \text{(4-11a)}$$

$$\frac{A_x}{\alpha_1}[\mu(1-\alpha'+\eta_2(0) f_n) + \theta(0) f_n] = \frac{1}{\alpha_2}(C_x e^{-\alpha_2 d_2} - D_x e^{\alpha_2 d_2}) e^{iQD} \quad \text{(4-11b)}$$

$$A_x[e^{-\alpha_1 d_1} + \alpha' e^{\alpha_1 d_1} + \eta_1(-d_1) f_n] = C_x + D_x \quad \text{(4-11c)}$$

$$\frac{A_x}{\alpha_1}[\mu(e^{-\alpha_1 d_1} - \alpha' e^{\alpha_1 d_1} + \eta_2(-d_1) f_n) + \theta(-d_1) f_n] = \frac{1}{\alpha_2}(C_x - D_x) \quad \text{(4-11d)}$$

From these four equations, we find the dispersion relation of the nonlinear bulk polaritons,

$$\cos(QD) - \cosh(\alpha_1 d_1)\cosh(\alpha_2 d_2) - \frac{\alpha_1^2 + \alpha_2^2 \mu^2}{2\alpha_1 \alpha_2 \mu} \sinh(\alpha_1 d_1)\sinh(\alpha_2 d_2) = \frac{1}{4} N \quad \text{(4-12)}$$

with the nonlinear factor N described by

$$\begin{aligned} N = {} & \eta_1(0)[-e^{-iQD} + \cosh(\alpha_2 d_2) e^{\alpha_1 d_1} + (\alpha_1 / \mu\alpha_2)\sinh(\alpha_2 d_2) e^{\alpha_1 d_1}] \\ & + [\eta_2(0) + \theta(0)/\mu][-e^{-iQD} + \cosh(\alpha_2 d_2) e^{\alpha_1 d_1} + (\alpha_2 \mu / \alpha_1)\sinh(\alpha_2 d_2) e^{\alpha_1 d_1}] \\ & + \eta_1(-d_1)[-e^{\alpha_1 d_1} e^{iQD} + \cosh(\alpha_2 d_2) - (\alpha_1 / \mu\alpha_2)\sinh(\alpha_2 d_2)] + [\eta_2(-d_1) \\ & + \theta(-d_1)/\mu][-e^{\alpha_1 d_1} e^{iQD} + \cosh(\alpha_2 d_2) - (\alpha_2 \mu / \alpha_1)\sinh(\alpha_2 d_2)] \end{aligned} \quad \text{(4-13)}$$

Due to the nonlinear interaction, the nonlinear term $N/4$ appears in the dispersion equation of the polaritons and is directly proportional to Δ. This term is a second-order quantity and makes a small correct to the dispersion properties of the linear bulk polaritons. Generally speaking, this nonlinear dispersion equation is a complex relation. However in some special circumstances it may be a real one. Here we illustrate it with an example. If $Q=0$, the bulk wave moves along the x-axis and the dispersion equation is a real equation for real α_1. For such a dispersion equation, ω has a real solution, otherwise the solution of ω is a complex number with the real part ω^{NL}, so-called the nonlinear mode frequency, and the imaginary part $\Delta\tau$, the attenuation or gain coefficient. In addition, it is very interesting that the unreciprocity of the bulk modes, $\omega(\bar{k}) \neq \omega(-\bar{k})$ with $\bar{k}=(k,Q,0)$, is seen, due to the existence of $\exp(-iQD)$ in the nonlinear term $N/4$ as a function of QD with the period 2π.

4. Surface dispersion relations

For the surface modes, note $f_n = \exp(-2n\beta D)$ and take the transformation $iQ \to \beta$ in equations (4-10), we can find

$$\cosh(\beta D) - \cosh(\alpha_1 d_1)\cosh(\alpha_2 d_2) - \frac{\alpha_1^2 + \alpha_2^2 \mu^2}{2\alpha_1 \alpha_2 \mu}\sinh(\alpha_1 d_1)\sinh(\alpha_2 d_2) = \frac{1}{4}N' e^{-2\beta n D} \tag{4-14}$$

in which N' can be obtained directly from Eq.(4-13) with the same transformation. This nonlinear term is directly proportional to the multiply of Δ and $\exp(-2n\beta D)$, so in the same condition the nonlinearity makes larger contribution to the bulk modes than the surface modes. We can use the linear expression of $\exp(-\beta D)$ to reduce the nonlinear term on the right-hand of Eq.(4-14), but have to derive its nonlinear expression to describe $\cosh(\beta D)$ on the left-hand, since its nonlinear part may has the same numerical order as that of $N' \exp(-2n\beta D)/4$. So we need another equation to determine it. Applying the boundary conditions at the surface, $y=0$ and $n=0$, we can find

$$\alpha_1[1+\alpha' + \eta_1(0)] = -\{\mu[1-\alpha' + \eta_2(0)] + \theta(0)\}\alpha_0 \tag{4-15}$$

Combining this with Eqs.(4-11a-c), the equation determining β is found,

$$\begin{aligned} e^{\beta D} = \{&(1+\alpha' + \eta_1(0) f_n)\cosh(\alpha_2 \mathrm{d}_2) + \alpha_2 \mu[1-\alpha' + (\eta_2(0) + \theta(0)/\mu) f_\mathrm{n}] \\ &\sinh(\alpha_2 \mathrm{d}_2)/\alpha_1\} / [\mathrm{e}^{-\alpha_1 d_1} + \alpha' e^{\alpha_1 d_1} + \eta(-d_1) f_n] \end{aligned} \tag{4-16}$$

with

$$\alpha' = \frac{1}{\alpha_0 \mu - \alpha_1}\{\alpha_0 \mu + \alpha_1 + \alpha_1 \eta_1(0) + \alpha_0[\mu \eta_2(0) + \theta(0)]\} \tag{4-17}$$

$f_n = \exp(-2n\beta D)$ in Eq.(4-16) also can be considered as an linear quantity since it always appears in the multiply of it and Δ. We also should note that there is a series of nonlinear surface eigen-modes as n can be any integer value equal to or larger than 1. Actually the nonlinear contribution decreases rapidly as n is increased, so only for small n, the nonlinear effect is important. In addition, increasing Δ and decreasing n have a similar effect in numerical calculation.

Because the nonlinear terms in Eqs.(4-12) and (4-14) all contain $\chi^{(3)}_{ijkl}$ directly proportional to $1/(\omega_r^2-\omega^2)^4$, the nonlinear effects may be too strong for us to use the third-order approximation for the nonlinear magnetization when ω is near to ω_r. In this situation we will take a smaller value of Δ to assure of the availability of this approximation.

We take the FeF_2/ZnF_2 superlattice as an example for numerical calculations, the physical parameters of FeF_2 are given in table 1. While the relative dielectric constant of ZnF_2 are $\varepsilon_2=8.0$. We apply the SL period $D=1.9\times10^{-2}cm$, and take $n=1$ for the surface modes. The nonlinear factor $\Delta=|A/(4\pi M_0)|^2$ is the relative strength of the wave field. The nonlinear shift in frequency is defined as $\Delta\omega=(\omega^{NL}-\omega)/\omega_r$, where the nonlinear frequency ω^{NL} and attenuation or gain coefficient $\Delta\tau$ as the real and imagine parts of the frequency solution from the nonlinear dispersion equations both are solved numerically. ω is determined by the linear dispersion relations.

	H_a	H_e	$4\pi M_0$	ε	γ
FeF_2	197kG	533kG	7.04 kG	5.5	1.97×10^{10} rad s^{-1} kG
MnF_2	7.87kG	550kG	5.65 kG	5.5	1.97×10^{10} rad s^{-1} kG

Table 1. Physical parameters for FeF_2 and MnF_2.

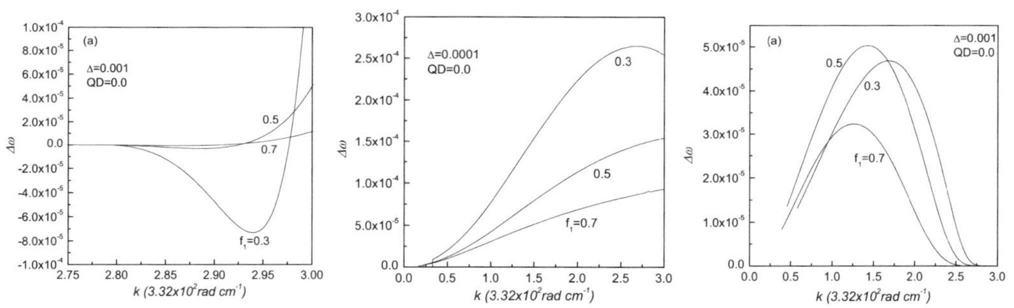

Fig. 6. Nonlinear shift in frequency(a) in the bottom band(b) in the middle band and (c) in the top band. After Wang & Li 2005.

We illustrate the nonlinear shift in frequency as function of the component of wave vector k in the three bulk-mode bands seperately in Fig. 6. is offered to illustrate the bottom band. (a) and ((b) show for $Q=0$ and π/D respectively. As shown in Fig.6(a), in the bottom band, when $f_1=0.3 and 0.5$, the nonlinear shift is downward or negative in the region of smaller k, but becomes positive from negative with the increase of k. For a SL with thicker AF layers, for example $f_1=0.7$, the shift is always positive. For the top bulk band, $\Delta\omega$ always is positive and possesses its maximum. In the middle band, Fig. 6(b) shows the positive frequency shift that increases basically with k. In terms of the shape of a band, the second-order derivative of linear frequency with respect to k, $\partial^2\omega/\partial k^2$ for a mode in it can be roughly estimated to be positive or negative. According to the Lighthill criterion $\Delta\omega\cdot\partial^2\omega/\partial k^2\le0$ for the existence of solitons(Lighthill,1965). One confirms from the figures that $\partial^2\omega/\partial k^2>0$ for modes in the top band, $\partial^2\omega/\partial k^2<0$ in the bottom band, but $\partial^2\omega/\partial k^2<0$ or $\partial^2\omega/\partial k^2>0$ in the middle band, depending on k. The soliton solution may be found since the Lighthill criterion can be fulfilled in the two bands. In the middle bulk band, the mode attenuation is vanishing, the nonlinearity is very evident and the nonlinear shift is positive.

We examine the surface magnetic polariton in the case of nonlinearity, which is shown in Fig.7. Similar to those in the middle bulk band, the surface-mode frequency also is very closed to ω_r, as a result, the nonlinear effect also is stronger. The attenuation $\Delta\tau=0$ as the dispersion equations are real. The shift $\Delta\omega$ is negative for $f_1\geq 0.3$, but positive for $f_1=0.1$. For $f_1=0.2$, it is positive and increases with k in the range of small k, but negative in the range of large k and its absolute value decreases as k is increased. Although there can be a series of surface eigen-modes in the nonlinear situation, the obvious nonlinear effect can be seen only for $n=1$, so that we present only the corresponding mode. One should note that the Lighthill criterion is satisfied for $f_1=0.1$ and 0.2, as a result, the surface soliton may form from the surface magnetic polariton.

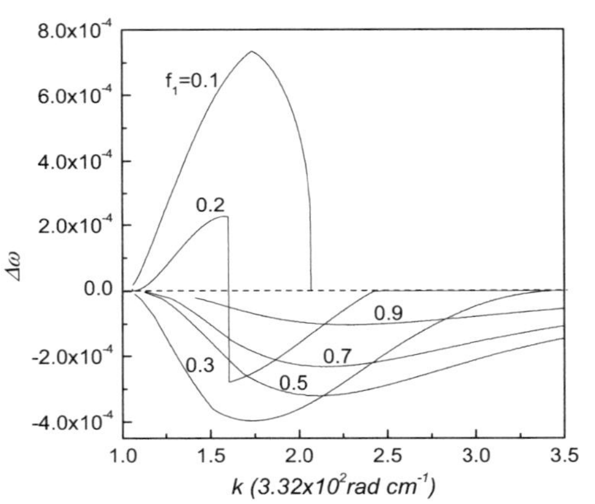

Fig. 7. Nonlinear frequency shift $\Delta\omega$ of the surface mode versus k for $\Delta=1.0\times10^{-4}$ and various values of f_1. After Wang & Li 2005.

4.2 Nonlinear infrared ransmission through and reflection off AF films

Finally, we discuss nonlinear transmission through the AF film. We assume that the media above and below the nonlinear AF film are both linear, but the film is nonlinear. Our geometry is shown in Fig. 8, where the anisotropy axis (the z axis) is parallel to the film surfaces and normal to the incident plane (the x-y plane). A linearly polarized radiation (TE wave) is obliquely incident on the upper surface.

Because we have known the nonlinear wave solution in the AF film and those above and below the film, to solve nonlinear transmission and reflection is a simple algebraical precess. Thus we directly present the finall results, the nonlinear refection and transmission coefficients

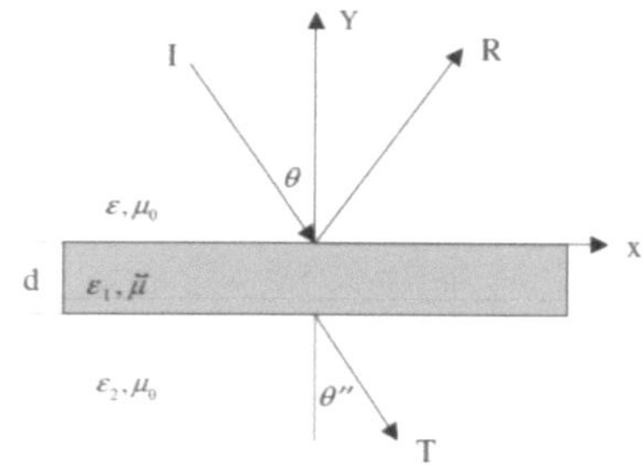

Fig. 8. Geometry and coordinate system for nonlinear reflection and transmission of an AF film with thickness d. R the reflection off and T the transmission through the film.

$$r = \frac{E_R}{E_I} = \frac{\mu_1\left(k_y - k_y''\right)\cos k_y' d + \left(k_y' - \mu_1^2 k_y k_y''/k_y'\right) i \sin k_y' d - NL_+}{\mu_1\left(k_y + k_y''\right)\cos k_y' d - \left(k_y' + \mu_1^2 k_y k_y''/k_y'\right) i \sin k_y' d - NL_-} \tag{4-18a}$$

$$t = \frac{E_T}{E_I} = \frac{\mu_1 k_y\left[2 - \eta_1(0) - \theta(0)\right] e^{ik_y'' d}}{\mu_1\left(k_y + k_y''\right)\cos k_y' d - \left(k_y' + \mu_1^2 k_y k_y''/k_y'\right) i \sin k_y' d - NL_-} \tag{4-18b}$$

in which the nonlinear terms $NL_\pm$ are shown with

$$\begin{aligned} NL_\pm = \frac{1}{2}\{&\pm\left(k_y - \mu_1 k_y''\right)\left[\cos\left(k_y d\right) \pm i\mu_1 k_y' \sin\left(k_y d\right)/k_y\right]\eta_1(-d)e^{ik_y d} \\ &+\left(\mp k_y + \mu_1 k_y'\right)\left[\cos\left(k_y d\right) - i\mu_1 k_y'' \sin\left(k_y d\right)/k_y\right]\eta_1(0) \\ &+\left(k_y - \mu_1 k_y''\right)\left[\pm i \sin\left(k_y d\right) + \mu_1 k_y' \cos\left(k_y d\right)/k_y\right]e^{ik_y d}\theta(-d) \\ &+\left(\pm k_y - \mu_1 k_y'\right)\left[i \sin\left(k_y d\right) - \mu_1 k'' \cos\left(k_y d\right)/k_y\right]\theta(0)\} \end{aligned} \tag{4-19}$$

Finally the reflectivity and transmissivity are defined as $R = |r|^2$ and $T = |t|^2$ (Klingshirn, C. F. ,1997). Here we should discuss a special situation. In the situation $(k_x = 0)$, from the expressions of L_1 to L_4 and L_1' to L_4', we find $\eta_1(y) = \theta(y) = 0$. It is quite obvious that one finds no nonlinear effects on the reflection and transmission in the case of normal incidence. For $\varepsilon > \varepsilon_2$, k_y'' becomes imaginary as the incident angle θ exceeds a special value, then the transmission vanishes. The nonlinear effect can be seen only from the reflection. Due to the complicated expressions for the reflection and transmission coefficients, more properties of R and T can be obtained only by numerical calculation of Eq. (4-18).

We take a FeF_2 film as an example for numerical calculations. with the physical parameters given in Table 1. The film thickness is fixed at $d = 30.0\mu m$ and the incident wave intensity $S_I = \sqrt{\varepsilon_0/\mu_0}\, E_I^2/2$, implicitly included in the nonlinear coefficients, is fixed at $S_I = 4.7 MWcm^{-2}$, corresponding to a magnetic amplitude of 16G in the incident wave. In the figures for numerical results, we use dotted lines to show linear results and solid lines to show nonlinear results. We shall discuss transmission and reflection of the AF film put in a vacuum. The transmission and reflection versus frequency ω are illustrated in Fig.9 for the incident angle $\theta = 30°$ and are shown in Fig.10 versus incident angle for $\omega/2\pi C = 52.8 cm^{-1}$

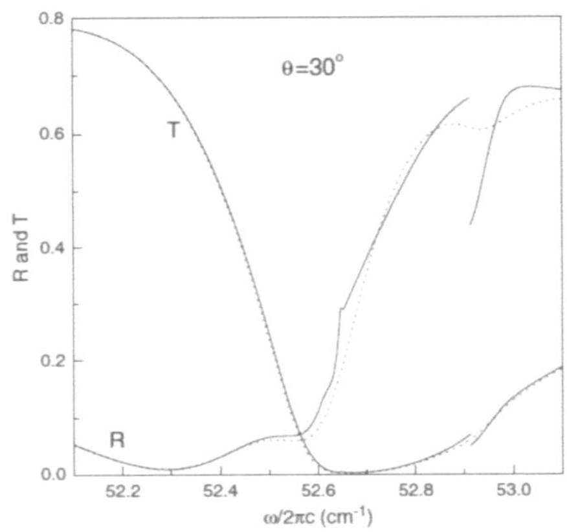

Fig. 9. Reflectivity and transmissivity versus frequency for a fixed angle of incidence of $30°$. After Bai, et. al. 2007.

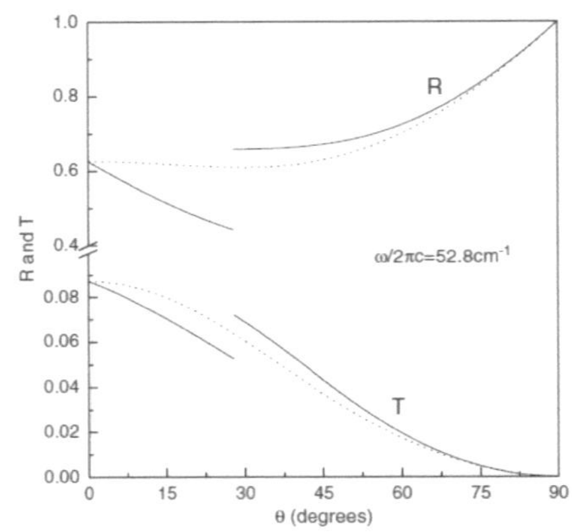

Fig. 10. Reflectivity and transmissivity versus angle of incidence for the frequency fixed at 52.8 cm^{-1}. After Bai, et. al. 2007.

First, the nonlinear modification is more evident in reflection for frequencies higher than ω_r. We see a very obvious discontinuity on the nonlinear R and T curves at $\omega/2\pi C = 52.9cm^{-1}$, corresponding to the smallest value of μ_1 whose real part changes in sign as the frequency moves cross this point. This causes the jump and obvious nonlinear modification, as the wave magnetic field is intense in the vicinity of this point. Secondly, R and T versus θ for a fixed frequency are shown in Fig.10. Here the discontinuity is also seen since the magnetic amplitude and the nonlinear terms vary with the wave vector $\vec{k}'$. It is more interesting that when the incident angle $\theta \le 27.5°$ the reflection and transmission are both lower than the linear ones, implying that the absorption is reinforced. However, in the range of $\theta \ge 27.5°$ they both are higher than the linear ones, and as a result the absorption is evidently restrained. The nonlinear influence disappears for normal incidence. we see the discontinuities on the reflection and transmission curves and the nonlinear effect is very obvious in the regions near to the jump points. The discontinuities are related to the bi-stable states. The nonlinear interaction also play an important role in decreasing or increasing the absorption in the AF film.

5. Second harmonic generation in antiferromagnetic films

In this section, the most fundamental nonlinear effect, second harmonic generation (SHG) of an AF film between two dielectrics (Zhou & Wang, 2008) and in one-dimensional photonic crystals (Zhou, et. al., 2009) have been analyzed based on the second-harmonic tensor elements obtained in section 2. We know from the expression of SH magnetization that if $H_0 = 0$ the SH magnetization is vanishing, as a result the SHG is absent. So the external magnetic field is necessary for the SHG. We take such an AF structure as example to describe the SHG theory, where the AF film is put two different dielectrics. In the coordinate system selected in Fig.11, the AF anisotropic field and dc magnetic field both parallel to the z-axis and the x-y plane as the incident plane. I is the incident wave, R the reflection wave and T the transmission wave, related to incident angle θ, reflection angle $-\theta$ and transmission angle θ', respectively. If a subscript s is added to the above quantities, they represent the corresponding quantities of second harmonic (SH) waves. The pump wave in the film is not indicated in this figure. The dielectric constants and magnetic permeabilities are shown in corresponding spaces.

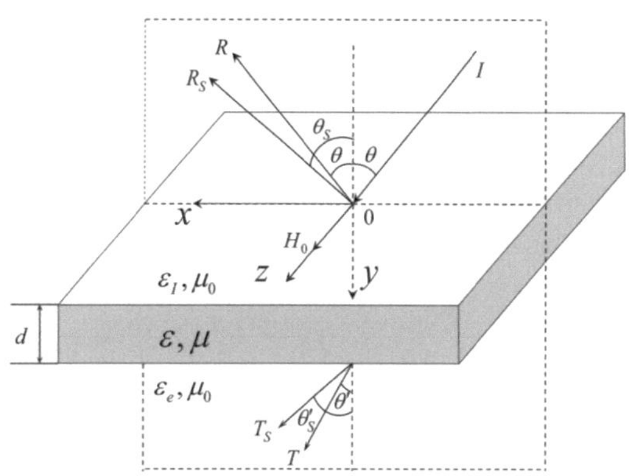

Fig. 11. Geometry and coordinate system.

Although we have obtained all elements of the SH susceptibility in section 2, but only two will be used in this geometry. It is because that a plane EMW of incidence can be decomposed into two waves, or a TE wave with the electric field normal to the incident plane and a TM wave with the magnetic field transverse to this plane. Due to no coupling between magnetic moments in the film and the TM wave (Lim, 2002, 2006; Wang & Li, 2005; Bai, et. al., 2007), the incident TM wave does not excite the linear and SH magnetizations, so can be ignored. Thus we take the TE wave as the incident wave I which produces the TE pump wave $\vec{H} = (H_x, H_y, 0)$ in the film. In this case, only one component of the SH magnetization can be found easily

$$m_z^{(2)}(\omega_s) = \chi_{xx}^{(2)}(\omega_s)(H_x H_x + H_y H_y) \tag{5-1}$$

with $\omega_s = 2\omega$ and the susceptibility elements

$$\chi_{xx}^{(2)}(2\omega) = \chi_{yy}^{(2)}(2\omega) = \frac{i\omega_m[\chi_{xx}^{(1)}\omega_0(\omega_r'^2 + \omega^2 - \omega_0^2) - \chi_{xy}^{(1)}\omega(\omega_r'^2 - \omega^2 + \omega_0^2)]}{M_0[\omega_r'^2 - (\omega - \omega_0)^2][\omega_r'^2 - (\omega + \omega_0)^2]} \tag{5-2}$$

The SHG magnetization arises as a source term in the harmonic wave equation and is excited by the pump wave, and in turn the pumping wave is induced by the incident wave. When the energy-flux density of the excited SH wave is much less than that of the incident wave, the assumption that the depletion of pump waves can be neglected(Shen, 1984) is commonly accepted. This assumption allows us to solve the pump wave in the film within the linear electromagnetic theory or with the linear optical method.

Based on the above assumption, to solve the pump wave is a linear problem. The method is well-known and just one usual optical process, so we give a simpler description for solving the pump wave in the film. Because the pump wave is a TE wave, we take its electric field to be

$$\vec{E} = \vec{e}_z[A_+\exp(ik_y y) + A_-\exp(-ik_y y)]\exp(ik_x x - i\omega t) \tag{5-3}$$

where A_+ and A_- show the amplitudes of the forward and backward waves in the film, respectively. The electric fields above and below the film are

$$\vec{E}_a = \vec{e}_z[E_0\exp(ik_{0y}y) + R_0\exp(-ik_{0y}y)]\exp(ik_{0x}x - i\omega t) \tag{5-4}$$

$$\vec{E}_b = \vec{e}_z T_0\exp(ik'_{0y}y)\exp(ik'_{0x}x - i\omega t) \tag{5-5}$$

The corresponding magnetic fields in different spaces are written as

$$\begin{aligned}\vec{H} = \frac{\exp(ik_x x - i\omega t)}{\omega\mu_0\mu_v}\{&\vec{e}_x k_y[(1+\delta)A_+\exp(ik_y y) + (1-\delta)A_-\exp(-ik_y y)] \\ &+ \vec{e}_y k_x[(\delta'-1)A_+\exp(ik_y y) - (\delta'+1)A_-\exp(-ik_y y)]\}\end{aligned} \tag{5-6a}$$

$$\begin{aligned}\vec{H}_a = \frac{\exp(ik_{0x}x - i\omega t)}{\omega\mu_0}\{&\vec{e}_x k_{0y}[E_0\exp(ik_{0y}y) - R_0\exp(-ik_{0y}y)] \\ &- \vec{e}_y k_{0x}[E_0\exp(ik_{0y}y) + R_0\exp(-ik_{0y}y)]\}\end{aligned} \tag{5-6b}$$

$$\vec{H}_b = \frac{T_0}{\omega\mu_0}(\vec{e}_x k'_{0y} - \vec{e}_y k'_{0x})\exp(ik'_{0y}y + ik'_{0x}x - i\omega t) \tag{5-6c}$$

where is $\mu_v = (\mu_1^2 - \mu_2^2)/\mu_1^2$ and μ_0 the vacuum magnetic permeability. $k_{0y} = k_0\cos\theta$ and $k_0 = \varepsilon_1^{1/2}\omega/c$ is the wave number in the above space, and $k'_{0y} = k'_0\cos\theta'$ and $k'_0 = \varepsilon_2^{1/2}\omega/c$ is the wave number below the film, but $k_y = [\varepsilon\mu_v(\omega/c)^2 - k_x^2]^{1/2}$. Here c is the vacuum velocity of light. $\delta = k_x\mu_2 / \mu_1 k_y$ and $\delta' = \mu_2 k_y / \mu_1 k_x$. The boundary conditions of the fields at the surfaces first require that $k_{0x} = k'_{0x} = k_x = k_0\sin\theta$, and these wave-number components should be real since we assume that dielectric constants and magnetic permeabilities in nonmagnetic media all are real values. In addition, using the boundary conditions, we also find the pump-field amplitudes $A_\pm = E_0 f_\pm$ with E_0 the electric amplitude of I, and

$$f_{\pm} = \frac{[\Delta(1 \mp \delta) \pm 1]\exp(-ik_y d)}{\cos k_y d(\Delta + \Delta') + i\delta(\Delta - \Delta')\sin k_y d - i[1 + \Delta'\Delta(1-\delta^2)]\sin k_y d} \tag{5-7}$$

where d is the film thickness, $\Delta = k_y / \mu_v k_{0y}$ and $\Delta' = k_y / \mu_v k'_{0y}$. The wave amplitudes R_0 and T_0 of R and T are not necessary for seeking the SHG, so they are given up here.To solve the output amplitudes of SHG, R_s and T_s, we should look for the solution of the SH wave equation in the film. In fact, there are three component equations, but only one contains a source term and this equation is

$$-(\frac{\partial^2}{\partial x^2} + \frac{\partial^2}{\partial y^2})H_{sz}(\omega_s) - \varepsilon(\omega_s / c)^2 H_{sz}(\omega_s) = \varepsilon(\omega_s / c)^2 m_z^{(2)}(\omega_s) \tag{5-8}$$

The other two are homogeneous and do not contain the field component $H_{sz}(\omega_s)$. In addition, the other SH components cannot emerge voluntarily without source terms, so it is evident that the SH wave is a TM wave. Because the SH magnetization and pump field in the film both have been given, to find the solution of equation (5-8) is easy. Let

$$\begin{aligned} H_{sz}(\omega_s) = [&A_s \exp(ik_{sy}y) + B_s \exp(-ik_{sy}y) + a\exp(2ik_y y) \\ &+ b\exp(-2ik_y y) + c]\exp(2ik_x x - i\omega_s t) \end{aligned} \tag{5-9}$$

with $k_{sy} = [\varepsilon(\omega_s / c)^2 - 4k_x^2]^{1/2}$. Substituting SH solution (5-9), expression (5-1) and solution (5-6a) into equation (5-8), we find the nonlinear amplitudes

$$a = E_0^2 \frac{\varepsilon_0 \chi_{zxx}^{(2)}(\omega_s) f_+^2}{\mu_0(\mu_v - 1)(\omega\mu_v / c)^2}[k_y^2(1+\delta)^2 + k_x^2(1-\delta')^2] \tag{5-10a}$$

$$b = E_0^2 \frac{\varepsilon_0 \chi_{zxx}^{(2)}(\omega_s) f_-^2}{\mu_0(\mu_v - 1)(\omega\mu_v / c)^2}[k_y^2(1-\delta)^2 + k_x^2(1+\delta')^2] \tag{5-10b}$$

$$c = -E_0^2 \frac{2\varepsilon_0 \varepsilon(\omega_s / c)^2 \chi_{zxx}^{(2)}(\omega_s) f_+ f_-}{\mu_0[4k_x^2 - \varepsilon(\omega_s / c)^2](\omega\mu_v / c)^2}[k_y^2(1-\delta^2) + k_x^2(\delta'^2 - 1)] \tag{5-10c}$$

Solution (5-9) shows that the SH wave in the film also propagates in the incident plane and it will radiate out from the film. We use

$$H_{sz}^a = R_s \exp[i(k_{sx}x - k_s y - \omega_s t)] \tag{5-11a}$$

to indicate the magnetic field of SH wave generated above the film and

$$H_{sz}^b = T_s \exp[i(k_{sx}x + k'_s y - \omega_s t) \tag{5-11b}$$

to represent the SH field below, with k_s and k'_s determined by $k_s^2 + k_{sx}^2 = \varepsilon_1(\omega_s / c)^2$ and $k_s'^2 + k_{sx}^2 = \varepsilon_2(\omega_s / c)^2$. The SH electric field in different spaces are found from to be

$$\vec{E}_s^a = \frac{R_s \exp[i(2k_x x - k_s y - \omega_s t)]}{\omega_s \varepsilon_0 \varepsilon_1}[k_s \vec{e}_x + k_{sx} \vec{e}_y] \tag{5-12a}$$

$$\vec{E}_s = -\frac{\exp[i(2k_x x - \omega_s t)]}{\omega_s \varepsilon_0 \varepsilon}\{\vec{e}_x[k_{sy}(A_s \exp(ik_{sy}y) - B_s \exp(-ik_{sy}y) + 2k_y a \exp(2ik_y y) \\ - 2k_y b \exp(-2ik_y y)] - 2k_x \vec{e}_y [a \exp(2ik_y y) + b \exp(-2ik_y y) + c]\} \quad (5\text{-}12b)$$

$$\vec{E}_s^b = \frac{T_s}{\omega_s \varepsilon_0 \varepsilon_2}(-k'_s \vec{e}_x + k_{sx}\vec{e}_y)\exp[i(2k_x x + k'_s y - \omega_s t)] \quad (5\text{-}12c)$$

Considering the boundary conditions of these fields continuous at the surfaces, there must be $k_{sx} = k'_{sx} = 2k_x$ and the these wave-number components all are real quantities, meaning the propagation angles of the SH outputs from the film

$$\theta_s = -\theta \quad (5\text{-}13a)$$

$$\theta'_s = arc\sin(\sqrt{\varepsilon_1 / \varepsilon_2}\sin\theta) \quad (5\text{-}13b)$$

It is proven that the SH wave outputs R_s and T_s have the same propagation direction as reflection wave *R* and transmission wave *T*, respectively.
Finally we solve the amplitudes of the output SH wave. The continuity conditions of H_{sz} and E_{sx} at the interfaces lead to

$$R_s = A_s + B_s + a + b + c \quad (5\text{-}14a)$$

$$R_s = \frac{\varepsilon_1}{\varepsilon k_s}[k_{sy}(-A_s + B_s) + 2k_y(-a + b)] \quad (5\text{-}14b)$$

$$T_s \exp(ik'_s d) = A_s \exp(ik_{sy}d) + B_s \exp(-ik_{sy}d) + a\exp(2ik_y d) + b\exp(-2ik_y d) + c \quad (5\text{-}14c)$$

$$T_s \exp(ik'_s d) = \frac{\varepsilon_2}{\varepsilon k'_s}\{k_{sy}[A_s \exp(ik_{sy}d) - B_s \exp(-ik_{sy}d)] \\ +2k_y[a\exp(2ik_y d) - b\exp(-2ik_y d)]\} \quad (5\text{-}14d)$$

After eliminating A_s and B_s from the above equations, we find the magnetic field-amplitudes of the output SH waves,

$$R_s = \frac{1}{S}\{[(\Delta_2 - \Delta_0)\cos k_{sy}d + i(\Delta_2\Delta_0 - 1)\sin k_{sy}d + \exp(2ik_y d)(-\Delta_2 + \Delta_0)]a \\ +[(\Delta_2 + \Delta_0)\cos k_{sy}d - i(\Delta_2\Delta_0 + 1)\sin k_{sy}d \\ -(\Delta_2 + \Delta_0)\exp(-2ik_y d)]b + [\Delta_2(\cos k_s d - 1) - i\sin k_s d]c\} \quad (5\text{-}15a)$$

$$T_s = \frac{\exp(-ik'_s d)}{S}\{[(\Delta_1 + \Delta_0)(e^{2ik_y d}\cos k_{sy}d - 1) - i(1 + \Delta_0\Delta_1)\exp(2ik_y d) \\ \sin k_{sy}d]a + [(\Delta_1 - \Delta_0)(\exp(-2ik_y d)\cos k_{sy}d - 1) + i(\Delta_0\Delta_1 - 1) \\ \exp(-2ik_y d)\sin k_{sy}d]b + [\Delta_1(\cos k_{sy}d - 1) - i\sin k_{sy}d]c\} \quad (5\text{-}15b)$$

where

$$S = [(\Delta_2 + \Delta_1)\cos k_{sy}d - i(1 + \Delta_2\Delta_1)\sin k_{sy}d] \quad \text{(5-15c)}$$

$\Delta_0 = 2k_y / k_{sy}$, $\Delta_1 = \varepsilon k_s / k_{sy}\varepsilon_1$ and $\Delta_2 = k_s'\varepsilon / (k_{sy}\varepsilon_2)$. We see from the expressions of *a*, *b* and *c* that SH amplitudes R_s and T_s are directly proportional to E_0^2, the square of electric amplitude of incidence wave. According to the definition of electromagnetic energy-flux density, $S_I = (\varepsilon_0\varepsilon_1 / \mu_0)^{1/2}|E_0|^2 / 2$ is the incident density, but the SH output densities are expressed as $S_R = (\mu_0 / \varepsilon_0\varepsilon_1)^{1/2}|R_s|^2 / 2$ and $S_T = (\mu_0 / \varepsilon_0\varepsilon_2)^{1/2}|T_s|^2 / 2$. We can conclude that the output densities are directly proportional to the square of the input (incident) density, or say the conversion efficiency $\alpha = S_{R,T} / S_I$ is directly proportional to the input density. For a fixed incident density, if the SH outputs are intense, the conversion efficiency must be high. Then, we are going to seek for the cases or conditions in which the SH outputs are intense.

The numerical calculations are based on three examples, a single MnF_2 film, SiO_2/MnF_2/air and ZnF_2/MnF_2/air, in which the MnF_2 film is antiferromagnetic. The relative dielectric constants are 1.0 for air, 2.3 for SiO_2 and 8.0 for ZnF_2. The relative magnetic permeabilities of these media are 1.0. There are two resonance frequencies in the dc field of $1.0kG$, $\omega_1/2\pi c = 9.76cm^{-1}$ and $\omega_2/2\pi c = 9.83cm^{-1}$. We take the AF damping coefficient $\tau = 0.002$ and the film thickness $d = 255\mu m$. The incident density is fixed at $S_I = 1.0kW / cm^2$, which is much less than that in the previous papers (Almeida & Mills, 1987; Kahn, et. al., 1988; Costa, et. al., 1993; Wang & Li, 2005; Bai, et. al., 2007).

We first illustrate the output densities of a single film versus frequency ω and incident angle θ with Fig.12 (a) for S_R and (b) for S_T. Evidently in terms of their respective maxima, S_R is weaker than S_T by about ten times. Their maxima both are situated at the second resonant frequency ω_2 and correspond to the situation of normal incidence. The figure of S_R is more complicated than that of S_T since additional weaker peaks of S_R are seen at large incident angles.

Next we discuss the SH outputs of SiO_2/MnF_2/air shown in Fig.13. Incident wave *I* and reflective wave *R* are in the SiO_2 medium and transmission wave *T* in air. The maximum peak of S_R is between the two resonant frequencies and in the region of $\theta > \theta_c' = 41.3^{\circ}$. For the given parameters, this angle just satisfies $\sin\theta_c = \sqrt{\varepsilon_2 / \varepsilon_1}$ and is related to $k_{0y}' = 0$, so it can be called a critical angle. When $\theta > \theta_c'$, k_{0y}' is an imaginary number and transmission *T* vanishes. For $\theta < \theta_c'$, S_R is very weak and numerically similar to that of the single film. However, the maximum of S_T is about four times as large as that of S_R, and S_T decreases rapidly as the incident angle or frequency moves away from θ_c' or the resonant frequency region. We find that the maxima of S_R and S_T are in intensity higher than those shown in Fig.12 by about 40 and 13 times, respectively.

Finally we discuss the SH outputs of ZnF_2/MnF_2/air, with the dielectric constant of ZnF_2 larger than that of SiO_2. The spectrum of S_R is the most complicated and interesting, as shown in Fig.14 (a). First we see two special angles of incidence. The first angle has the same definition as θ_c' in the last paragraph and is equal to 20.1°. The second defined as θ_c corresponds to $k_y = 0$ and is equal to 55°. For $\theta > \theta_c$, k_y becomes an imaginary number

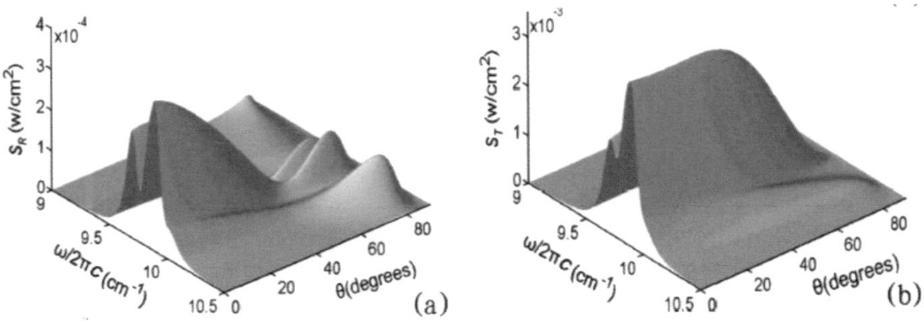

Fig. 12. SH outputs of a single AF film (MnF_2 film), S_R and S_T versus the incident angle and frequency. After Zhou & Wang, 2008.

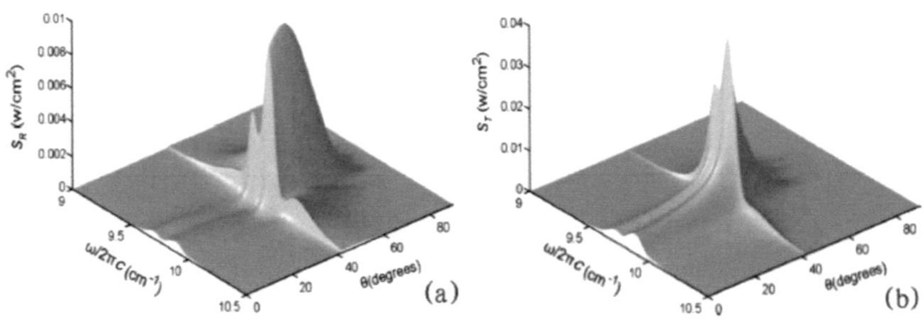

Fig. 13. SH outputs of SiO_2/MnF_2/air, S_R and S_T versus the incident angle and frequency. After Zhou & Wang, 2008.

and the incident wave I is completely reflected, so the SH wave is not excited. On this point, Fig.13(a) is completely different from Fig.12(a). More peaks of S_R appear between the two critical angles, but the highest peak stands between the two resonant frequencies and is near to θ_c. Outside of the region between θ_c and θ_c', we almost cannot see S_R. For S_T, the pattern is more simple, as shown in Fig.14 (b). Only one main peak is seen clearly, which arises at θ_c' and occupies a wider frequency range. Different from Fig.13, the maxima in Fig.14(a) and Fig.14(b) are about equal. Comparing Fig.14 with Fig.12, we find that the maximums of S_R and S_T are larger than those shown in Fig.12 by about 240 times and 20 times, respectively.

For the SH output peaks in Fig.13 and Fig.14, we present the explanations as follows. The pump wave in the film is composed of two parts, the forward and backward waves corresponding to the signs + and $-$ in Eq.(5-3), respectively. The transmission (T) vanishes and the forward wave is completely reflected from the bottom surface of the film as k_{0y}' is equal to zero or an imaginary number. In this situation, the backward wave as the reflection wave is the most intense and equal in intensity to the forward wave. The interference of the two waves at the bottom surface makes the pump wave enlarged, and further leads to the appearance of the T_s-peak in the vicinity of the critical angle θ_c'. The intensity of R_s, however, depends on that of the pump wave at the upper surface. When the phase difference between the forward and backward waves satisfies $\phi = \pm 2k\pi$ (k is an integer) at the surface, the interference results in the peaks of R_s. Thus the interference effect in the film plays an important role in the enhancement of the SHG.

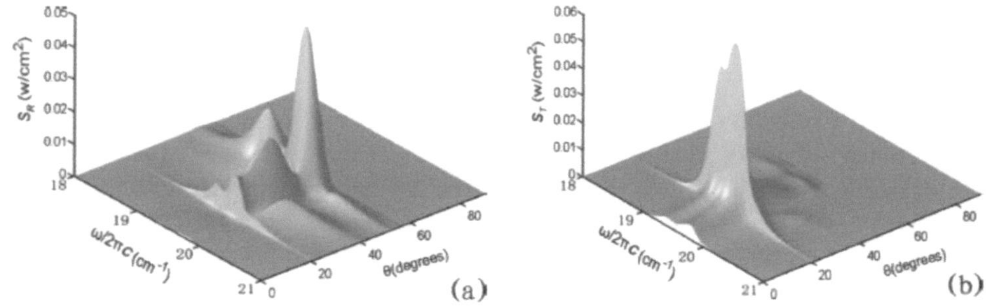

Fig. 14. SH outputs of ZnF_2/ MnF_2/air, S_R and S_T versus the incident angle and frequency. After Zhou & Wang, 2008.

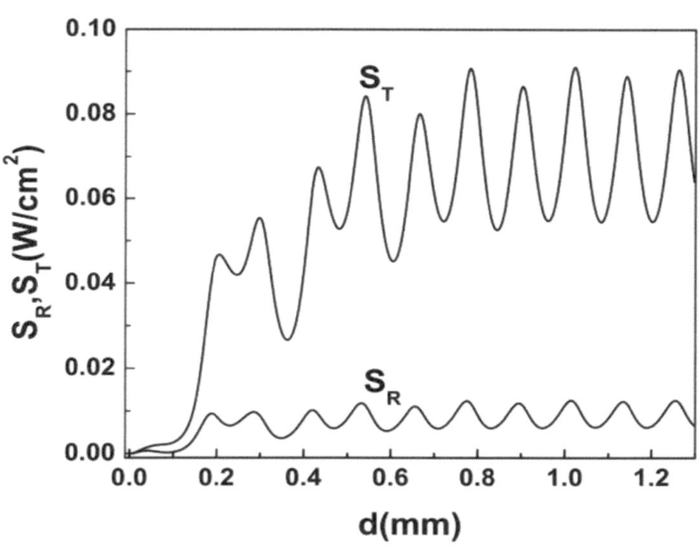

Fig. 15. SH outputs of SiO_2/MnF_2/air. S_R and S_T versus the film thickness for $\omega = 9.84cm^{-1}$ and $\theta = 41.3^\circ$. After Zhou & Wang, 2008.

It is also interesting for us to examine the SH outputs versus the film thickness. We take the SiO_2/MnF_2/air as an example and show the result in Fig.15. We think that the SH fringes result from the change of optical thickness of the film, and the SH outputs reache their individual saturation values about at $d = 800\mu m$, 0.09 W/cm^2, and 0.012 W/cm^2. If we enhance the incident wave density to 10.0kW/cm^2, the two output densities are increased by 100 times, to 9.0W/cm^2 and 1.2W/cm^2, or if we focus S_I on a smaller area, higher SH outputs are also obtained, so it is not difficult to observe the SH outputs.
If we put this AF film into one-dimension Photonic crystals (PCs), the SHG has a higher efficiency(Zhou, et. al., 2009). It is because that when some AF films as defect layers are introduced into a one-dimension PC, the defect modes may appear in the band gaps. Thus electromagnetic radiations corresponding to the defect modes can enter the PC and be greatly localized in the AF films. This localization effect has been applied to the SHG from a traditional nonlinear film embedded in one-dimension photonic crystals(Ren, et. Al., 2004 ; Si, et. al., 2001 ; Zhu, et.al., 2008, Wang, F., et. al. 2006), where a giant enhancement of the SHG was found.

6. Summary

In this chacter, we first presented various-order nonlinear magnetizations and magnetic susceptibilities of antiferromagnets within the perturbation theory in a special geometry, where the external magnetic field is pointed along the anisotropy axis. As a base of the nonlinear subject, linear magnetic polariton theory of AF systems were introduced, including the effective-medium method and transfer-matrix-method. Here nonlinear propagation of electromagnetic waves in the AF systems was composed of three subjects, nonlinear polaritons, nonlinear transmition and reflection, and second-harmonic generation. For each subject, we presented a theoretical method and gave main results. However, magnetically optical nonlinearity is a great field. For AF systems, due to their infrared and millimeter resonant-frequency feature, they may possess great potential applications in infrared and THz technology fields. Many subjects parallel to the those in the traditional nonlinear optics have not been discussed up to now. So the magnetically nonlinear optics is a opening field. We also hope that more experimental and theoretical works can appear in future.

7. Acknowledgment

This work is financially supported by the National Natural Scienc Foundation of China with grant no.11074061 and the Natural Science Foundation of Heilongjiang Province with grant no.ZD200913.

8. References

Almeida, N. S. & Mills, D. L.(1987); Nonlinear Infrared Response of Antiferromagnets. *Phys. Rev. B*. Vol.36, (1987), pp.2015-2023.

Almeida, N. S. & Mills, D. L.(1988). Effective-medium Theory of Long-wavelength Spin Waves in Magnetic Superlattices, *Phys. Rev. B*, Vol.38, (1988), pp.6698–6710.

Almeida, N. S. & Tilley, D. R.(1990). Surface Poaritons on Antiferromagnetic Supperlattices, *Solid State Commun.*, Vol.73, (1990), pp.23-27.

Bai, J.; Zhou, S.; Liu, F. L. & Wang, X. Z.(2007). Nonlinear Infrared Transmission Through and Reflection Off Antiferromagnetic Films. *J. Phys.: Condens. Matters*, Vol.19, (2007), pp.046217-046227.

Balakrishnan,R. ; Bishop, A. R. & Dandoloff, R.(1992). Geometric Phase in the Classical Continuous Antiferromagnetic Heisenberg Spin Chain, *Phys. Rev.Lett.* Vol.64, (1990), pp.2107-2110; Anholonomy of a Moving Space Curve and Applications to Classical Magnetic Chains,*Phys. Rev. B* Vol. 47, (1992), pp.3108-3117.

Balakrishnan, R. & Blumenfeld, R.(1997). On the Twist Excitations in a Classical Anisotropic Antiferromagnetic Chain, *Phys. Lett. A*, Vol.237, (1997), pp.69-72.

Barnas, J.(1988). Spin Waves in Superlattices. I General Dispersion Equations for Exchange Magnetostatic and Retarded Modes, *J. Phys. C: Solid state Phys*, Vol. 21,(1988) pp. 1021-1036.

Boardman, A. & Egan, P. (1986). *Surface Wave in Plasmas and Solids*, Vukovic, S. (Ed.), 3, World Publ., Singapore.

Born, M.; Wolf, E.(1964), *Principles of optics: electromagnetic theory of propagation, interference and diffraction of light.* Oxford, Pergamon Press.

Camley, R. E. & Mills, D. L.(1982). Surface-Polaritons on Uniaxial Antiferromagnets, *Physical Review B*, Vol.26 No.3, (1982), pp. 1280-1287.

Camley, R. E.; Cottam, M. G. & Tilley, D. R.(1992). Surface-Polaritons in Antiferromagnetic Superlattices with Ordering Perpendicular To the Surface, *Solid State Communications*, Vol.81 No.7, (February 1992), pp. 571-574.

Cao, S. & Caillé, A.(1982). Polaritons Guides Dans Une Lame Antiferromagnetique, *Solid State Commun*. Vol. 43, No.6, (August, 1982), pp.411-413.

Costa, B. V. ; Gourea M. E. & Pires, A. S. T.(1993). Soliton Behavior in an Antiferromagnetic Chain, *Phys. Rev.B* Vol. 47, (1993), pp.5059-5062.

Daniel, M. & Bishop, A. R.(1992). Nonlinear Excitations in the Classical Continuum Antiferromagnetic Heisenberg Spin Chain, *Phys. Lett. A*, Vol.162, (1992), pp.162-166.

Daniel, M. & Amuda, R.(1994). On the Spin Excitations in the Classical Continuum Heisenberg Antiferromagnetic Spin Systems, *Phys. Lett. A*, Vol.191, (1994), pp.46-56.

Dumelow, T. &Tilley, D. R.(1993). Optical Properties of Semiconductor Superlattices in the Far Infrared, *J. Opt. Soc. Amer. A*, Vol.10, (1993), pp.633-645.

Elmzughi, F.G.; Constantinou, N. C. & Tilley, D.R.(1995a). The Effective-medium Theory of Magnetoplasma Superlattices, *J. Phys.: Condens. Matter*, Vol. 7, (1995), pp.315-326.

Elmzughi, F. G.; Constantinou, N. C. & Tilley, D. R.(1995b). Theory of Electromagnetic Modes of a Magnetic Superlattice in a Transverse Magnetic Field: An Effective-Medium Approach, *Phys. Rev. B*, Vol.51, (1995), pp.11515-11520.

Fiebig, M.; Frohlich, D.; Krichevtsov, B. B. & Pisarev, R. V.(1994). Second Harmonic Generation and Magnetic-Dipole-Electric-Dipole Interference in Antiferromagnetic Cr_2O_3, *Phys. Rev. Lett.* , Vol. 73, (1994), pp.2127-2130.

Fiebig, M.; Frohlich, D.; Lottermoser, T.; Pisarev, R. V. & Weber, H. J.(2001). Second Harmonic Generation in the Centrosymmetric Antiferromagnet NiO, *Phys. Rev. Lett.* Vol.87, (2001), pp.137202.

Fiebig, M.; Pavlov, F V. V. & Pisarev, R. V. (2005). Second-Harmonic Generation as a Tool for Studying Electronic and Magnetic Structures of Crystals: Review, *J. Opt. Soc. Am. B*, Vol.22, (2005), pp.96-118.

Jensen,M. R. F.; Parker,T. J.; Abraha, K. & Tilley, D. R.(1995). Experimental Observation of Magnetic Surface Polaritons in FeF_2 by Attenuated Total Reflection, *Phys. Rev. Lett.* Vol,75, (1995),pp.3756–3759.

Kahn,L.; Almeida, N. S. & Mills, D. L.(1988). Nonlinear Optical Response of Superlattices: Multistability and Soliton Trains, *Phys. Rev. B*, Vol.37, (1988), pp.8072-8081.

Klingshirn, C. F. (1997), Chapter3, In: *Semiconductor Optics*, Springer, Berlin.

Lighthill, M. J.(1965). Contributions to the Theory of Waves in Nonlinear Dispersive Systems, *J. Inst. Math. Appl.* Vol.1, (1965),pp.269-306.

Lim, S. C.; Osman, J. & Tilley, D. R.(2000). Calculations of Nonlinear Magnetic Susceptibility Tensors for a Uniaxial Antiferromagnet. *J. Phys. D, Applied Physics*, Vol.33, (2000), pp.2899-2910.

Lim, S. C.(2002). Magnetic Second-harmonic-generation of an Antiferromagnetic Film, *J. Opt. Soc. Am. B*, Vol.19, (2002), pp.1401-1410.

Lim, S. C.(2006). Second Harmonic Generation of Magnetic and Dielectric Multilayers, *J. Phys.: Condens. Matter*, Vol.18, (2006), pp.4329-4343.

Morrish, A. H. (2001). *The Physical Principles of Magnetism*, Wiley-IEEE Press, ISBN 978-0-7803-6029-7.

Oliveros, M. C.; Almeida, N. S.; Tilley, D. R.; Thomas, J. & Camley, R. E.(1992). Magnetostatic Modes and Polaritons in Antiferromagnetic Nonmagnetic Superlattices. *Journal of Physics-Condensed Matter*, Vol.4, No.44, (November 1992), pp. 8497-8510.

Raj, N. & Tilley, D. R.(1987). Polariton and Effective-medium Theory of Magnetic Superlattices, *Phys. Rev. B*, Vol.36, (1987), pp.7003–7007.

Raj, N. & Tilley, D. R.(1989), The Electrodynamics of Superlattices, Chapter 7 of The Dielectric Function of Condensed Systems. Elsevier, Amsterdam.

Ren, F. F. ; Li, R. ; Chen, C. ; Wang, H. T. ; Qiu, J. ; Si, J. & Hirao, K.(2004). Giant Enhancement of Second Harmonic Generation in a Finite Photonic Crystal with a Single Defect and Dual-localized Modes, *Phys. Rev. B*, Vol.70, (2004), pp. 245109 (4 pages).

Stamps, R. L. & Camley, R. E.(1996). Spin Waves in Antiferromagnetic Thin Films and Multilayers: Surface and Interface Exchange and Entire-Cell Effective-Medium Theory. *Physical Review B*, Vol.54, No. 21, (December 1996), pp. 15200-15209.

Shen, Y. R.(1984), *The Princinples of Nonliear Optics*, (Wiley), pp. 86-107.

Si, B ; Jiang, Z. M. & Wang, X.(2001). Defective Photonic Crystals with Greatly Enhanced Second-harmonic Generation, *Opt. Lett.* Vol. 26, (2001), pp.1194-1196.

Song, Y. L.; Ta, J. X..; Li, H. & Wang, X. Z.(2009). Presence of Left-handness and Negative Refraction in Antiferromagnetic/ionic-crystal Multilayered Film, *J. Appl. Phys.* Vol.106, (2009) pp. 063119.

Ta, J. X.; Song, Y. L. & Wang, X. Z. (2010), Magneto-phonon Polaritons of Antiferromagnetic/ion-crystal Superlattices, J. Appl. Phys. Vol.108, (2010) pp.013520 (4 pages).

Vukovic, S.; Gavrilin, S.N. & Nikito, S.A.(1992). Bistability of Electromagnetic Waves in an Easy-Axis Antiferromagnet Subjected to a Static Magnetic Field. *Phys. Solid State.*, Vol.34, (1992), pp.1826-1828.

Wang, F. ; Zhu, S. N. ; Li, K. F. & Cheah, K. W.(2006). Third-harmonic Generation in a One-dimension Photonic-crystal-based Amorphous Nanocavity, *Appl. Phys. Lett.* Vol.88, (2006), pp.071102 (3 pages).

Wang, Q. & Awai, I.(1998). Frequency Characteristics of the Magnetic Spatial Solitons on the Surface of an Antiferromagnet. *J. Appl. Phys.* Vol.83, (1998), pp.382-387.

Wang,Q.; Wu,Z.; Li, S. & Wang, L.(2000). Nonlinear Behavior of Magnetic Surface Waves on the Interface between Ferromagnet and Antiferromagnet. *J. Appl. Phys.*, Vol.87, (2000), pp.1908-1913.

Wang, J. J. ; Zhou, X. F.; Wan, W. L. & Wang, X. Z, Transmission by Antiferromagnetic-Nonmagnetic Multilayers, *J. Phys.: Condens Matter*, Vol. 11,(1999) pp. 2697-2705.

Wang, X. Z. & Tilley D. R.(1987). Retarded Modes of a Lateral Antiferromagnetic /nonmagnetic Superlattice, *Phys. Rev. B*, Vol.52, No. 18, (November 1987), pp.13353–13357.

Wang, X. Z. & Fu, S. F.(2004). Dispersion Properties of Nonlinear Bulk Polaritons in Uniaxial Antiferromagnetic/nonmagnetic Superlattices *J. Magn. Magn. Mater.* Vol.271, (2004), pp.334-347.

Wang, X. Z. & Li, H.(2005). Nonlinear Polaritons in Antiferromagnetic /Nonmagnetic Superlattices. *Phy. Rev. B*, Vol.72, (2005), pp.054403-054412.

Wright, E.; Stegeman, G. (1992). *Nonlinear planar waveguide. Anisotropic & nonlinear opt. waveguide*, Elsevier Science Publisher B. V., pp.117.

Zhou, S.; Li, H.; Fu, S. F. & Wang, X. Z.(2009). Second Harmonic Gerneration from an Antiferromagnetic Film in One-dimensional Photonic Crystals. *Phys. Rev. B*, Vol.80, (2009), pp.205409 (12 pages).

Zhou, S.; Wang, X. Z.(2008). A Method of Enhancing Second-Harmonic Generation of Antiferromagnetic Film. *Journal of the Optical Society of America B*, Vol.25, (2008), pp.1639~1644.

Zhou, S.(2010). Magnetically optical nonlinearity of antiferromagnetic/dielectric systems, Doctorial thesis, Ch.5 (Harbin University of Science and Technology,2010).

Zhu, N. & Cao, S.(1987). Magnetic Polaritons in Antiferromagnetic/ nonmagnetic Multilayers, Physics Letters A, Vol. 124, No.9, (October, 1987), pp. 515-522.

Zhu, Q. ; Wang, D. & Zhang, Y.(2008). Design of Defective Nonlinear Photonic Crystals for Multiple Wavelengths' Second Harmonic Generation, *J. Opt. A: Pure Appl. Opt.* Vol.10, (2008), pp.025201 (4 pages).

4

Quasi-planar Chiral Materials for Microwave Frequencies

Ismael Barba[1], A.C.L. Cabeceira[1], A.J. García-Collado[2], G.J. Molina-Cuberos[2], J. Margineda[2] and J. Represa[1]
[1]*University of Valladolid*
[2]*University of Murcia*
Spain

1. Introduction

The growing development in the new communication technologies requests devices to perform new features or to improve the old ones. The trend is to develop new artificial materials reproducing well-known properties already present in other frequency ranges (such as optics) or materials with properties inexistent in the nature. Among the first kind, artificial chiral media, based on the random inclusion of metallic particles with chiral symmetry into a host medium are worth to mention (Fig. 1). Nevertheless, the fabrication techniques up-to-date are quite expensive and produce samples not easy to be tailored and with imperfections, such as intrinsic anisotropy and non-homogeneity (non-uniform density and orientation of inclusions), as well as heavy losses.

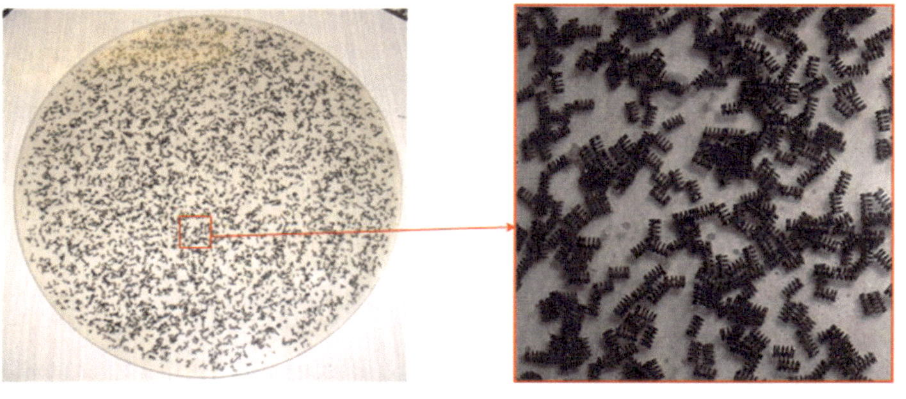

Fig. 1. Helix-based artificial chiral material. The sample is a 30 cm diameter disk fabricated by dispersion of six-turn stainless-steel helices in an epoxy resin with a low curing temperature. The helices are 2 mm height and 1.2 mm outer diameter.

During the last years, alternative methods, based on a periodic distribution of planar or quasi-planar chiral particles, have been proposed. This alternative presents the possibility of using conventional printed-circuit fabrication techniques to manufacture the structure. At the same time, the use of via holes provides additional flexibility to select the type of

inclusions from helices to cranks or even pseudo-chiral inclusions such as Ω's. As a consequence, the realization of the bulk material, staggering printed circuit plates, gives rise to axial anisotropy.
In this chapter, we are going to present some of the research performed during the last years in the field of chiral materials implementation by means of quasi-planar technologies. Section 2 presents an introduction on chiral materials, as well as (2.2) different approaches in order to implement them: traditional (random) distribution and new, periodic distributions. In the last case, we present different alternative (planar) implementations, finishing with our own (quasi-planar) proposal.
Section 3 shows the two complementary analysis techniques we have employed: numerical analysis, as well as experimental measures, both in free and guided propagation, with a previous fabrication of the samples. Finally, we present (section 4) the results we have obtained (rotation angle of polarization).

2. Chiral materials

Chiral materials are characterized by asymmetric microstructures in such a way that those structures and their mirror images are not superimposable. As a consequence, right- and left-hand circularly polarized waves propagate through the material with different phase velocities and, in case the medium is lossy, absorption rates. Electromagnetic waves in chiral media show the following interesting behavior (Lindell et al., 1994):
1. Optical (electromagnetic) rotatory dispersion (ORD), causing a rotation of polarization;
2. Circular dichroism (CD): due to the different absorption coefficients of a right- and left-handed circularly polarized wave, the nature of field polarization is modified, making linear polarization of a wave to change into elliptical polarization.

These properties have drawn considerable attention to chiral media and may open new potential applications in microwave and millimeter-wave technology: antennas and arrays (Lakhtakia et al., 1988; Viitanen et al., 1998), twist polarizers (Lindell et al., 1992), antireflection coatings (Varadan et al., 1987; Kopyt 2010), etc. It has been also proposed as a way to achieve negative refraction index (Pendry, 2004; Tretyakov et al., 2005). Also, many papers on the analysis of free and guided electromagnetic wave propagation through chiral media have been published, both in time (González-García et al., 1998; Demir et al., 2005; Pereda et al., 2006) and frequency (Xu et al., 1995; Alú et al., 2003; Pitarch et al., 2007; Gómez et al., 2010) domain. For this reason, during the last years, there has been an extensive research on new designs that enhance the above-mentioned properties, as we will see in the next sections of this chapter.

2.1 Constitutive relationships

In contrast to isotropic materials, characterized by their permittivity and permeability, bi-isotropic materials show a cross coupling between electric and magnetic fields, their constitutive relations being:

$$\vec{D} = \varepsilon\vec{E} + \eta\vec{H}$$
$$\vec{B} = \zeta\vec{E} + \mu\vec{H} \tag{1}$$

where the four scalars ε, μ, η, ζ are function of frequency ω. When the following condition holds:

$$\zeta = -\eta = \frac{j\kappa}{c_0} \tag{2}$$

c_0 being light speed in vacuum, the medium is said to be "chiral". The parameter κ is the "chirality" or "Pasteur" parameter (Lindell et al., 1994). In the frequency domain, this leads to the following constitutive relationships:

$$\begin{aligned} \vec{D}(\omega) &= \varepsilon \vec{E}(\omega) - \frac{j\kappa(\omega)}{c_0}\vec{H}(\omega) \\ \vec{B}(\omega) &= \mu \vec{H}(\omega) + \frac{j\kappa(\omega)}{c_0}\vec{E}(\omega) \end{aligned} \tag{3}$$

The real part of the chirality parameter is related with the rotation angle of the polarization plane (ORD) in a distance d by means of the following expression:

$$\theta = 2d\omega \frac{\kappa'}{c_0} \tag{4}$$

Considering electromagnetic field propagation through a homogeneous chiral medium, it is convenient to introduce new field variables, $\vec{E}_{\pm}$ and $\vec{H}_{\pm}$ ("wavefield vectors"), being the following linear combinations of the electric and magnetic fields:

$$\vec{E}_{\pm} = \frac{1}{2}\left(\vec{E} \mp jZ\vec{H}\right), \vec{H}_{\pm} = \frac{1}{2}\left(\vec{H} \pm \frac{j}{Z}\vec{E}\right) \tag{5}$$

where Z is the wave impedance of the medium, $Z = \sqrt{\mu/\varepsilon}$. Actually, the two wavefields $\{\vec{E}_+, \vec{H}_+\}$ and $\{\vec{E}_-, \vec{H}_-\}$ are plane right-circularly and left-circularly polarized waves, respectively. The advantage of introducing these new vectors is that they satisfy the Maxwell equations in an equivalent isotropic medium, so we may use well-known solutions for fields in simple isotropic medium to obtain solutions for wave propagation through chiral media (Lindell et al., 1994). These wavefield vectors will "see" equivalent simple isotropic media with the equivalent parameters:

$$\varepsilon_{\pm} = \varepsilon\left(1 \pm \frac{\kappa}{\sqrt{\varepsilon\mu}}\right), \mu_{\pm} = \mu\left(1 \pm \frac{\kappa}{\sqrt{\varepsilon\mu}}\right) \tag{6}$$

It is clear that, if κ is high enough, one of the wavefield vectors correspond to a backward wave (Tretyakov et al., 2005). That means that, for one of the two possible circularly polarized waves, travelling through a highly chiral material, this one behaves as a left-handed (Veselago) metamaterial.

2.2 Chiral implementations

2.2.1 Random distributions

Traditionally, artificial chiral media at microwave frequencies are fabricated by embedding conducting helices into a host, as shown in Fig 1. The dimensions of these helices determine the bandwidth where the optical activity takes place (Lindman, 1920; Tretyakov et al., 2005). Nevertheless, chirality is a geometrical aspect, therefore helices are not the only possibility, so other type of inclusions, like metal cranks (Molina-Cuberos et al., 2009; Cloete et al., 2001) have been proposed also; an example may be seen in Fig. 2.

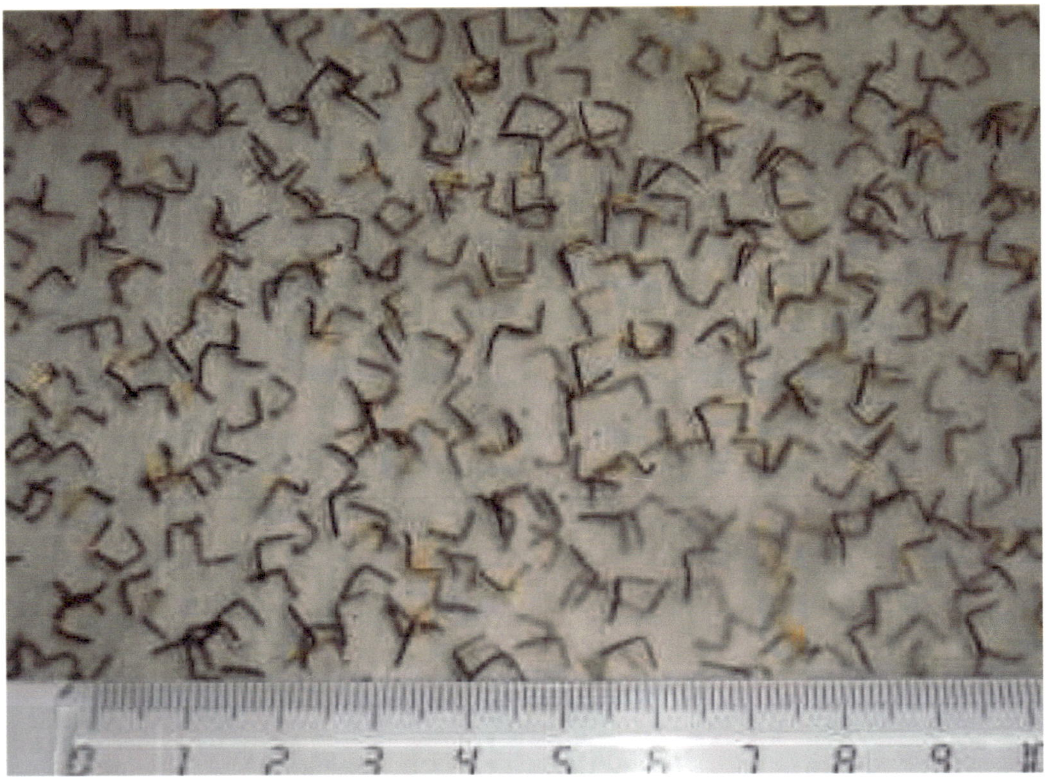

Fig. 2. Crank-based artificial chiral material. Chiral elements were produced from a 0.4 mm diameter and 12.6 mm length copper wire by bending in three segments by two 90 angles, all with the same handedness. The elements were dispersed in an epoxy resin with a low curing temperature (Molina-Cuberos et al., 2009).

In any case, it is necessary to be careful with the fabrication procedure to assure isotropy and homogeneity. The inclusions must be randomly oriented with no special direction. If the particles are placed in an aligned configuration, the result is a macroscopically bianisotropic material, leading to matrix coefficients for the constitutive parameters (Lindell et al., 1994). Also, a random distribution trends to present local density variations and accidental alignments (see detail in Fig. 1), which causes spatial variations of the constitutive relationships. At the same time, this procedure involves other drawbacks like high cost and difficulty in cutting and molding the material.

In the case of random distribution of cranks, the problems associated to the lack of homogeneity are enhanced. For the same total wire length cranks are bigger than helices, which makes the number density of cranks to be lower than the one using helices and increases the inhomogeneity. Molina-Cuberos et al. (2009) found fluctuations of the transmitted wave depending on the sample position and orientation with respect to the antenna. Therefore, several measurements and a mean value of the rotation angle were carried out.

2.2.2 Periodical distributions

The problems associated to the lack of homogeneity in chiral media based on random distribution of particles as helices or cranks can be reduced or even eliminated by designing periodical lattices. By an adequate distribution of metallic cranks is possible to build chiral media with homogeneous, isotropic and reciprocal behavior at microwave range (García-Collado et al., 2010), Fig. 3 shows an example of such medium.

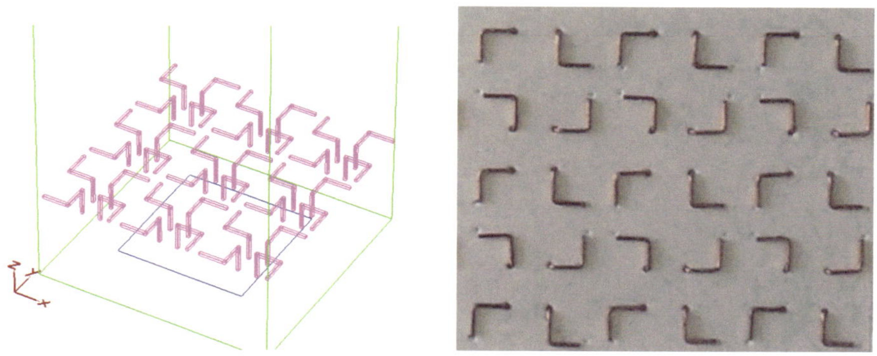

Fig. 3. Detailed view of a periodical lattice of cranks with the same handedness produced from 0.68 mm diameter and 15 mm length copper wire by bending in three segments by two 90 degrees angles. One of the segments is introduced perpendicularly into the host medium, polyurethane foam with a relative permittivity close to one. The backside of the medium is completely free of metal. Right: photograph of the lattice (García-Collado et al., 2010). Left: MEFIsTo™ model in which we may see the geometry of the cranks

For these reasons, alternative methods of manufacturing chiral materials have been proposed in recent years: Pendry et al. (Pendry et al.; 2004) proposed a periodical distribution of twisted Swiss-rolls, Kopyt et al. (Kopyt et al.; 2010) a distribution of chiral honeycombs. Nevertheless, most of the alternatives rely on planar and quasi-planar technologies, like Printed Circuit Board (PCB) technology, or even integrated circuit technology, for THz and optical materials. They provide a low-cost technique, which allows a high flexibility in the design of the elementary cell.

Planar technologies make use of two-dimensional elements, in order to obtain media with chiral response. The general concept of chirality, from a geometrical point of view, can be defined in a plane geometry (two dimensions): a structure is considered to be chiral in a plane if it cannot be brought into congruence with its mirror image, unless it is lifted from the plane (Le Guennec, 2000a, 2000b). In this case, it is possible to design a 2D-chiral medium consisting on flat elements possessing no line of symmetry in the plane, and which allows the use of planar technology to manufacture it.

However, electromagnetic activity (electromagnetic rotatory dispersion and circular dichroism) is a phenomenon that takes place in the three dimensional space. Some authors have tried to find electromagnetic activity in thus 2D structures: Papakostas et al. (2003) found a rotation of the polarization plane of a wave incident on a 2D-chiral planar structure like showed in Fig. 4, remarking its apparently nonreciprocal nature: when observed from the back side instead of the front, the sense of the twist is reversed, suggesting then a nonreciprocal polarization rotation similar to that observed in the Faraday effect. That interpretation of this result has opened a discussion on the possibility of a violation of reciprocity and time reversal symmetry (Schwanecke et al., 2003). Kuwata-Gonokami et al.

(2005) concluded that such structures are actually chiral in 3D (taking into account air-metal and substrate-metal interfaces), and their electromagnetic activity must arise from this three dimensional nature.

Fig. 4. Planar chiral structure made by arrays of gammadions arranged in two-dimensional square gratins (Papakostas et al., 2003).

Other groups have achieved three-dimensional chirality by means of multilayered structures of plane elements. The elements may be 2D-chiral (Rogacheva et al., 2006; Plum et al., 2007, 2009) or even non chiral (Zhou et al., 2009): in both cases, the 3D-chirality is obtained by means of a twist between layers (Fig. 5). These structures resulted to give an extremely strong rotation, as well as a negative index of refraction for one of the circularly polarized waves.

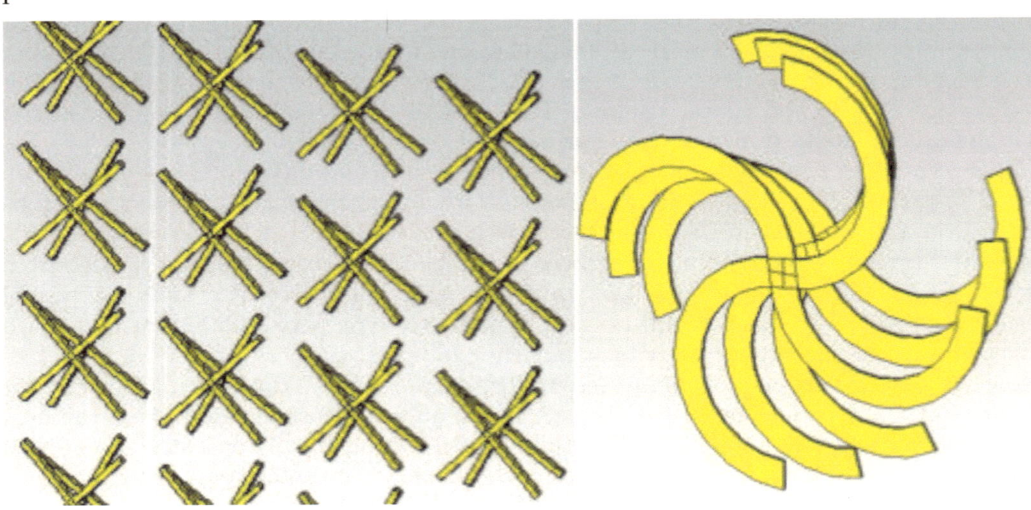

Fig. 5. Left: schematic representation of a chiral "cross-wire" simple like the one studied by Zhou et al. (2009). Right: schematic representation of a unit cell of a chiral structure, constructed from planar metal rosettes separated by a dielectric slab (Rogacheva et al., 2006; Plum et al., 2007, 2009)

Finally, it is possible also to construct 3D chiral samples using of quasi-planar technology: in this case, three dimensional PCB technology is employed, involving two-sided boards plus the use of via holes to connect both sides of the board. Such approximation was proposed by Marqués et al. (2007) and also by the authors of this chapter (Molina-Cuberos et al., 2009; Barba et al., 2009).

Fig. 6. Photograph of a structure similar to the shown in Fig. 3, but manufactured by means of Printed Circuit technology.

In our research, we have designed different chiral distributions of "molecules" and implemented them by different means: one (Fig. 3) is made by using metal cranks introduced into a polyurethane tablet; the second one (Fig. 6) is, as mentioned, made using PCB technology. We have designed and analyzed different distributions; some of them have been implemented and their behavior measured experimentally, while other ones have been modeled using numerical techniques. More details may be read in the following sections.

3. Analysis

We have worked, first, with the numerical analysis of the designed materials, which allows the study of their electromagnetic behavior at high frequency, previous to the effective construction of the same ones. We have used two different commercially available software in time domain:

a. MEFiSTo™, based on TLM method.
b. CST Studio Suite™ 2009, based on the finite integration technique (FIT).

Both methods are complete tools to solve electromagnetic problems in 3D, allowing the graphic visualization of the electromagnetic field propagation and its interaction with materials and boundaries during the simulation. The principal advantage of simulating in the time domain is that it most closely resembles the real world. In our case, it allows to obtain a very broadband data with a single simulation run with much less memory requirements than required in frequency-domain methods.

The experimental set-up used is based on a previous one for permittivity and permeability measurements at X-band (8.2 - 12.4 GHz) (Muñoz et al., 1998), and adapted to measure electromagnetic activity (Molina-Cuberos et al., 2009; García-Collado et al., 2010). Fig. 7

shows a diagram of the experimental set-up, where the incident wave is linearly polarized in the vertical direction.

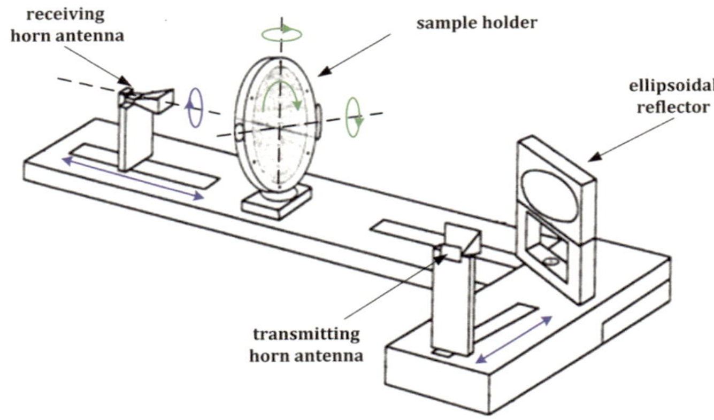

Fig. 7. Schematic diagram of the free space setup for the experimental determination of the rotation angle and the three constitutive parameters of isotropic chiral material in the X-Band (not to scale).

The transmitting and receiving antennas are 10-dB-gain rectangular horns. An incident beam is focused by an ellipsoidal concave mirror (30 cm x 26 cm), which produces a roughly circular focal area of about 6 cm in diameter, which is lower than sample size, so that diffraction problems are avoided with relatively small samples. The transmitting antenna is placed at one of the mirror foci (35 cm) and the sample at the other one. The sample holder is midway between the mirror and the receiving antenna and is able to rotate around the two axes perpendicular to the direction of propagation. The receiving antenna, located at 35 cm from the sample, can rotate about the longitudinal axis, which allows the measurement of the scattering parameters (S parameters) corresponding to any polar transmission. The interested reader is referred to Muñoz et al. (1998), Gómez et al. (2008) and García-Collado et al., (2010) for a detailed description of the measurement setup and technique. Here, we briefly present the measurement process:

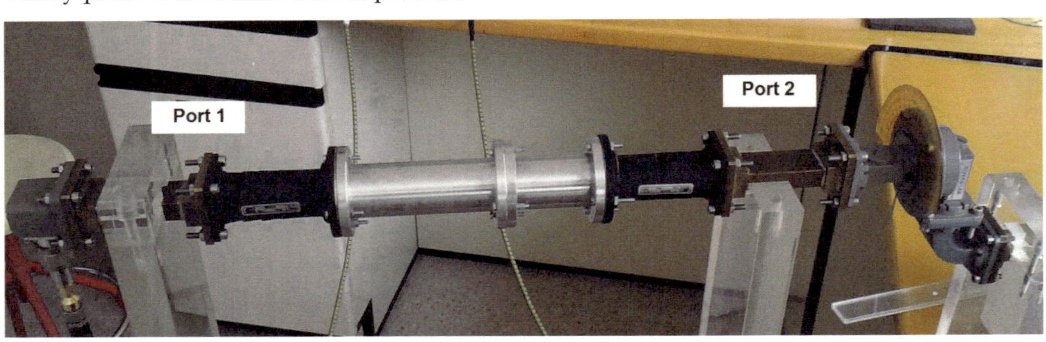

Fig. 8. Waveguide setup for the experimental determination of the rotation angle produced by a chiral material in X-band. A cylindrical sample is located in the circular waveguide and fed through port 1. The transmitted wave is measured in port 2 by a rotating connection, which allows determining the component of the electrical field parallel to the TE_{10} mode of the rectangular wave.

First a two-port "through-reflect-line"' (TRL) calibration is performed at the two waveguide terminals of the network analyzer, PNA-L N5230A, where the antennas are connected. Then, a time domain (TD) transform is used to filter out mismatches from the antennas, edge diffraction effects, and unwanted multiple antenna-mirror-sample reflections or reflections from other parts of the system by means of the ``gating'' TD option of the network analyzer. The rotation angle of the transmitted polarization ellipse is defined as the difference between the polarization direction of the incident wave and the direction of the major axis of the transmitted elliptically polarized wave. Rotation can be determined by looking for the minimum value of the transmitted wave or by measuring the transmission coefficient for co- and cross-polarization, S_{21CO} and S_{21CR} (Balanis, 1989):

$$\begin{aligned} OA &= \left[\frac{1}{2}\left(S_{21CO}^2 + S_{21CR}^2 + \left(S_{21CO}^4 + S_{21CR}^4 + 2S_{21CO}^2 S_{21CR}^2 \cos(2\phi)\right)^{1/2}\right)\right]^{1/2} \\ OB &= \left[\frac{1}{2}\left(S_{21CO}^2 + S_{21CR}^2 - \left(S_{21CO}^4 + S_{21CR}^4 + 2S_{21CO}^2 S_{21CR}^2 \cos(2\phi)\right)^{1/2}\right)\right]^{1/2} \\ \tau &= \frac{\pi}{2} - \frac{1}{2}\tan^{-1}\left(\frac{2S_{21CO}S_{21CR}}{S_{21CO} - S_{21CR}}\cos(2\phi)\right) \end{aligned} \quad (7)$$

where OA and OB are the major and minor axes, respectively, ϕ the phase difference between S_{21CO} and S_{21CR}, and τ the tilt of the ellipse, relative to the incident wave.
In principle, the precise angle of rotation cannot be determined by this measurement alone, there is an uncertainty of $2n\pi$, where n is an integer. To determine the angle uniquely, we make use of measurements far away from the resonance range, where it is expected, and found that the rotation angle goes to zero. Once the scattering coefficients are known, it is also possible to retrieve the constitutive parameters (ε, μ, κ) of the sample.

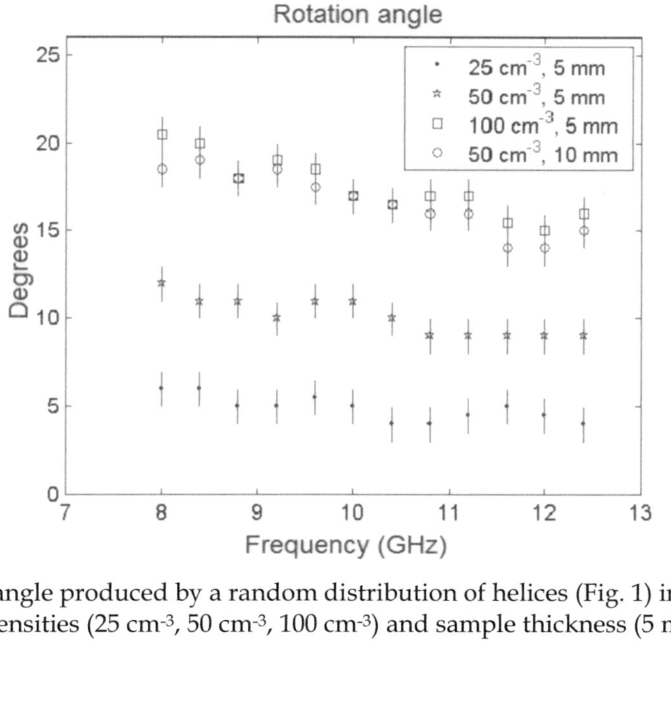

Fig. 9. Rotation angle produced by a random distribution of helices (Fig. 1) in a host, for different helix densities (25 cm^{-3}, 50 cm^{-3}, 100 cm^{-3}) and sample thickness (5 mm, 10 mm).

Several analyses of the chiral effects, by making use of waveguide setup, have been also developed; see for example Brewitt-Taylor et al. (1999). In order to test the effect of metallic cranks in waveguide, some samples were initially designed to produce chiral isotropic materials with a resonance frequency at X-band and placed into a section of circular waveguide. Fig. 8 shows the experimental set-up. The sample is excited in a rectangular waveguide and fed to the circular waveguide through a rectangular-circular waveguide transition, the dominant mode in the rectangular waveguide is TE_{10}, and the polarization is perpendicular to the resistive film of the transition which absorbs any cross-polarized field. The dominant mode is TE_{11}, in the empty circular waveguide, and $HE_{\pm 11}$ in the chirowaveguide. After the sample, a section of circular waveguide, which can rotate around the longitudinal axis, is connected to a rectangular guide through a transition. The rotation angle of the transmitted wave is obtained by measuring the minimum value of the transmitted wave; we have found that this procedure is more accurate than the determination of the maximum on the transmitted wave.

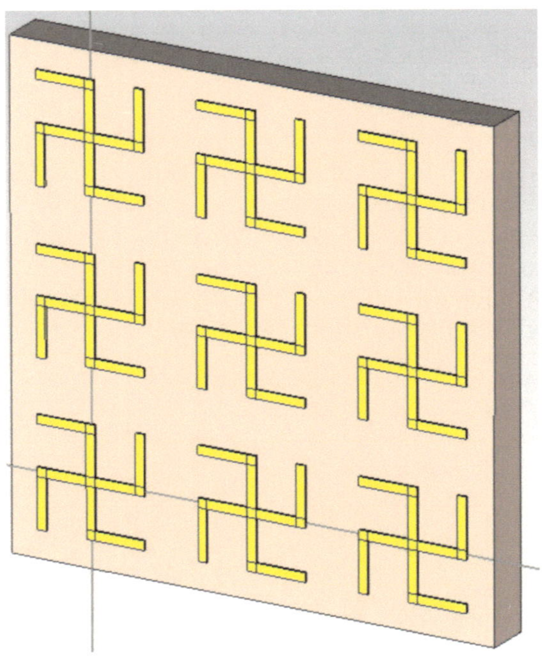

Fig. 10. Schematic illustration of an array of gammadions, chiral in 2D. Each gammadion is assumed to be of copper, and occupies a square of 6x6 mm. There is a separation of 3 mm between each gammadion. The board is 2.5 mm thick.

4. Results

4.1 Random distributions (helices)

Fig. 9 shows the rotation angle produced by a distribution of helices in a host medium for densities ranging from 0 cm^{-3} to 100 cm^{-3}, the error bars showing the uncertainties in the angle determination. Chiral elements are six-turn stainless-steel helices that are 2 mm long and 1.2 mm in outer diameter. The elements were dispersed in an epoxy resin with a low curing temperature. We observe that the rotation angle decreases when the frequency increases, which means that the resonance frequency is below the measurement range. As it can be expected, the rotation angle increases with the number density of inclusions and with the sample width, following a nearly linear relation. Similar behavior has been found in other experiments with helices (Brewitt-Taylor et al., 1999) or cranks (Molina-Cuberos et al., 2005).

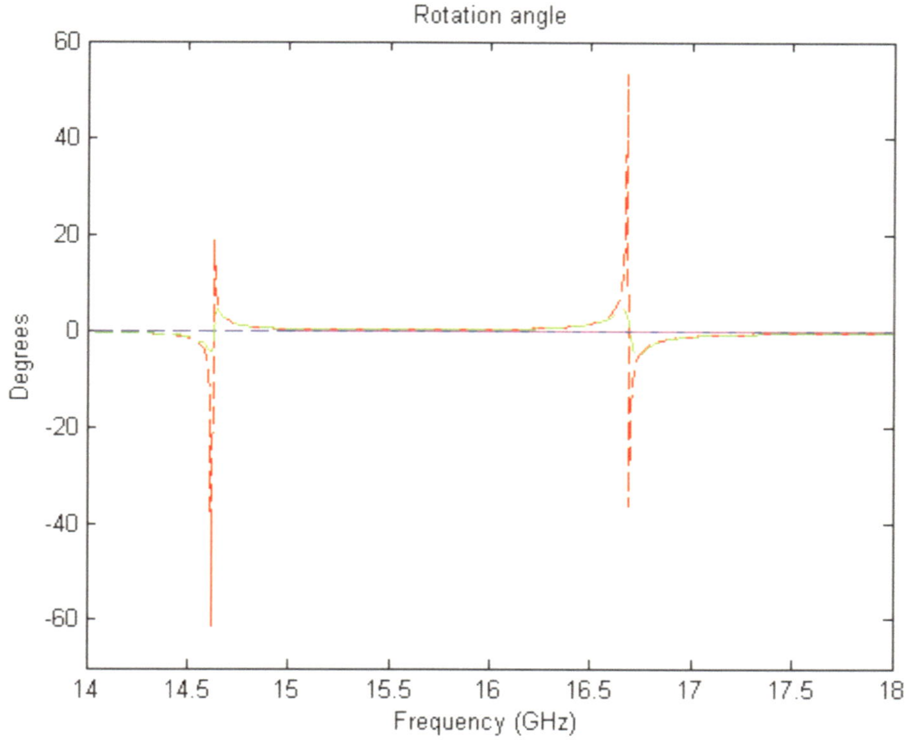

Fig. 11. Rotation of the polarization plane for a plane wave normally incident over a planar array of gammadions (Fig. 10), and for different supporting boards: free space (magenta), FR4 (blue), unlossy CER-10 (green) and lossy CER-10 (red). The result is the same in front and back incidence.

4.2 Planar distributions

We have modeled, using CST Studio Suite™ 2009, the rotation of the polarization plane, for a plane wave normally incident over a plane structure, similar, at a different scale, to the one studied by Papakostas et al. (2003). Our structure is also an array of gammadions (Fig. 10) that, in this case, presents resonance in the microwave band. The rotation has been determined assuming different properties of the board that supports the array: first assuming it has the same properties as vacuum, second, a typical material on PC Boards (FR4, $\varepsilon_r \approx 4.3$) and, finally, a high permittivity material, like Taconic CER-10 ($\varepsilon_r \approx 10$), all present in CST Studio Suite™ 2009 library. The results are shown in Fig. 11. In the first case (vacuum), the structure is symmetrical in a normal axis, so it is not chiral in 3D (the specular image is coincident with the result of a rotation around a longitudinal axis), so there is no electromagnetic activity (no rotation). When taking into account the effect of the board, the structure becomes 3D chiral. In this case, we observe electromagnetic activity, which increases when the properties of the board (permittivity or losses) are higher, i.e., when there is more difference with free space.

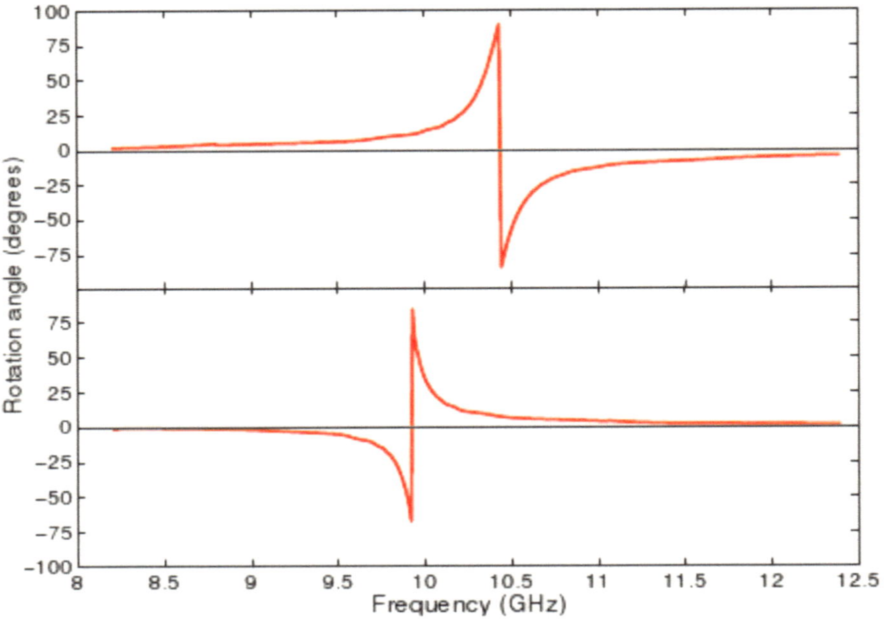

Fig. 12. Two examples of the rotation angle produced by a periodical lattice of metallic cranks formed by three equal size segments (5 mm) cranks for left-handed cranks with a separation of 6.9 mm (up) and right-handed cranks with a separation of 9.1 mm (down). [Reprinted from García-Collado et al. (2010) © 2010 IEEE]

4.3 Quasi-planar distributions (cranks)

Fig. 12 shows two examples of the rotation angle produced by periodical lattices of cranks as the one represented in Fig. 3. Both plots correspond to the cranks with the same total length, 15 mm, and different handedness and separation. It can be observed that the sign of the rotation produced by a periodical lattice of cranks depends on the handedness of the elements, as it has been observed in chiral composites formed by randomly oriented elements. In a periodical lattice, the distance of the elements also affects to the characteristic frequencies. In this case, the resonance frequency decreases from 10.4 GHz (up) to 9.8 GHz when the crank separation distance changes from 6.9 mm to 9.1 mm. We do not observe any non-reciprocal effect, i.e. the rotation angle is the same if the wave is incident in the opposite direction.

These results are compared with other ones, obtained by means of time-domain modeling of the same structure, using MeFisTo-3D. In this case, the four cranks of each gammadion are separated 6 mm, while there are 4 mm of distance between two consecutive gammadions. The results are showed in Fig. 13, showing a good agreement between both measures.

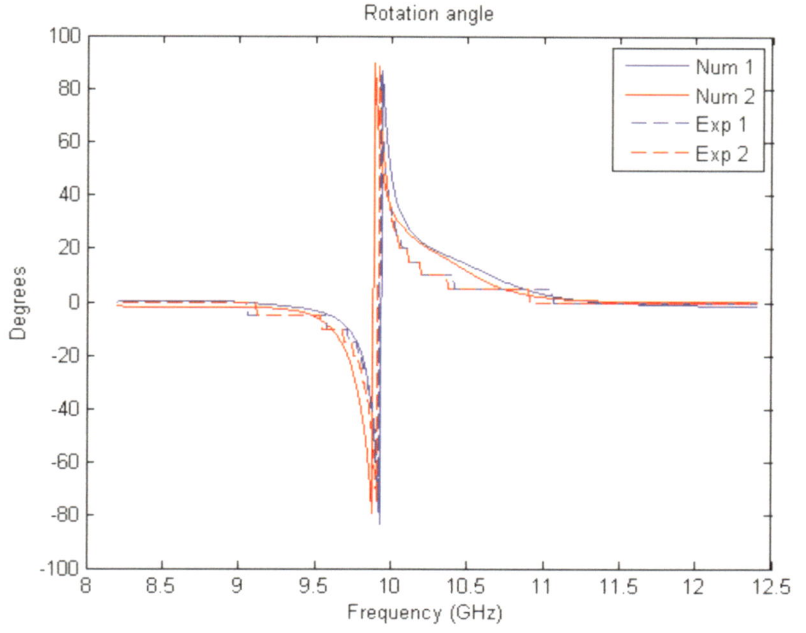

Fig. 13. Rotation of the polarization angle for a plane wave normally incident over a quasi-planar periodic array of right-handed cranks as shown in Figs. 3 and 6: numerical (Num) and experimental (Exp) results. 1 and 2 represent the two possible directions of the propagation wave (incident from front and back side, respectively).

Finally, we propose a different distribution of cranks (Fig. 14). In this case, there is a higher concentration of cranks in the same surface, so it is expected to obtain a higher gyrotropy too. That distribution is also geometrically reciprocal.

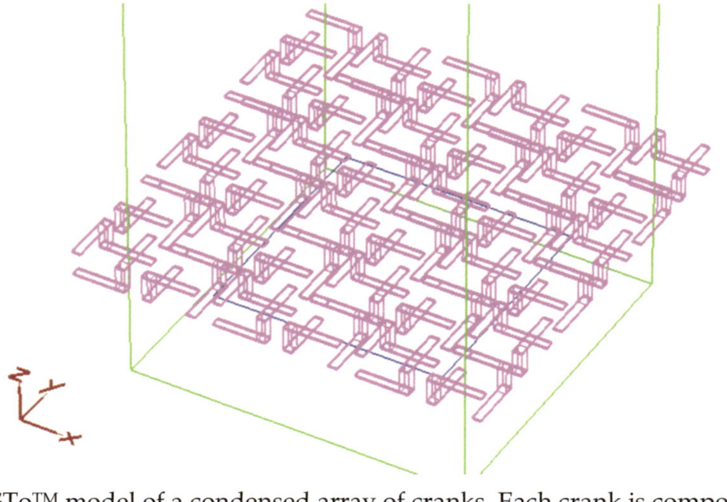

Fig. 14. MEFiSTo™ model of a condensed array of cranks. Each crank is composed by two arms, 3mm long, one in each side of the board (1.5 mm of thickness), plus a via connecting both.

The electromagnetic behavior of such distribution has been modeled using MEFiSTo™: we have obtained the rotation of the polarization plane after a normal transmission through that array. The angle of rotation does not depend on the initial polarization of the incident wave (that is, the medium behaves like a biisotropic one, at least in a transversal axis), and it is the same in the two directions of propagation (reciprocal). The result is shown in Fig. 15.

It is worth to mention the couple of discontinuities between -90° and 90° that may be observed in the figure. Such discontinuities are common to most of the distributions we have studied: when we see only one of them (Fig. 12 and Fig. 13) it is caused by the limitations in broadband that suffer our experimental bank. At the same time, other authors (Zhou et al., 2009) find a similar behavior in frequency, being usually assumed to correspond to resonance frequencies. We believe this behavior does not correspond to a real jump in the rotation frequency, but it is a consequence of the measurement procedure, in which the result is normalized between -90° and 90°. If we normalize between 0 y 180° the result in Fig. 15 would be as shown in Fig. 16.

More important: if we study the propagation through several layers of our material, we may draw the rotation angle like in Fig. 17. There, it is demonstrated that the response is lineal (the rotation angle is proportional to the width of the material (number of layers) and, then, the resonance frequency does not depend on the number of layers.

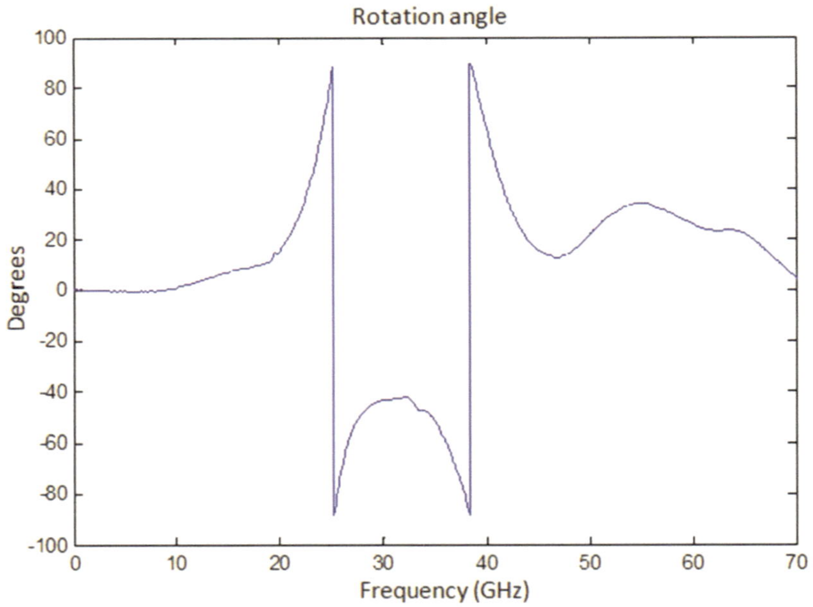

Fig. 15. Rotation of the polarization plane for a plane wave normally incident over a condensed array of cranks (Fig. 14), normalizing between -90° and 90°

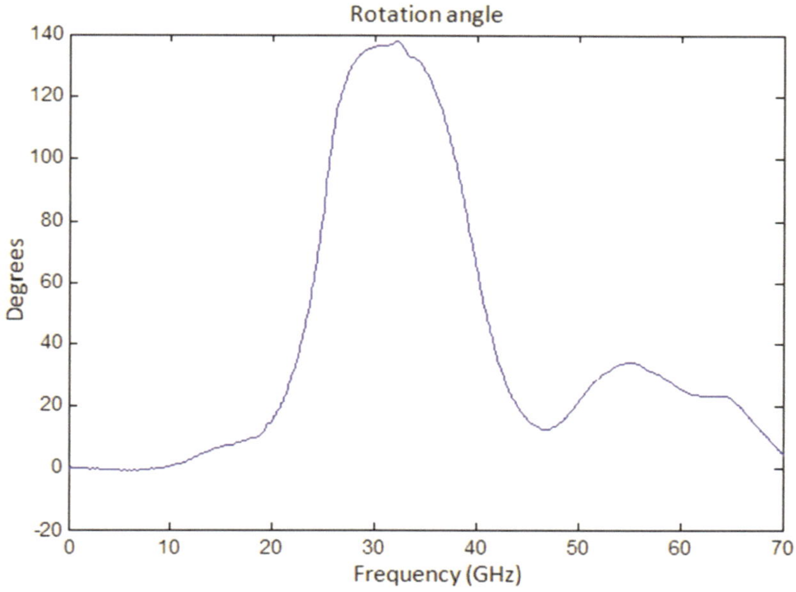

Fig. 16. Rotation of the polarization plane for a plane wave normally incident over a condensed array of cranks like represented in Fig. 14, normalizing between 0° and 180°

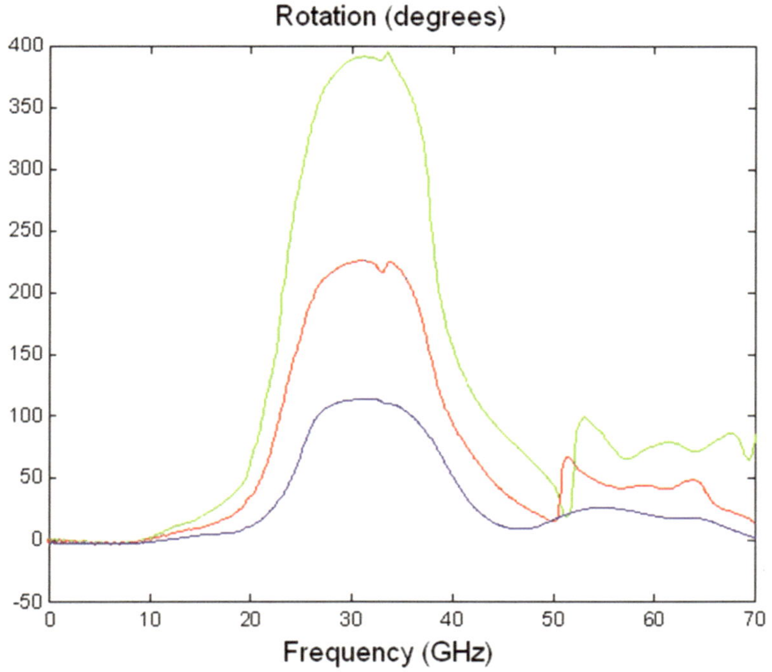

Fig. 17. Rotation of the polarization angle for a wave linearly polarized, incident over a condensed distribution of cranks like shown in Fig. 14, for one (blue line), two (red) or three (green) parallel boards. [Reprinted from Barba et al. (2009) © 2009 IEEE]

The chiral material for waveguide experiments was built as described in section 3. However, there are some inherent restrictions in the design due to the limited size of the sample. The radius of the waveguide is similar, in magnitude, to the one of the crank, which strongly limits the number of elements that can be placed on a one-layer distribution, without contact among the elements. Fig. 18 shows two examples produced by four metallic cranks in a foam host medium (left) and eight cranks (right). We have experimentally observed, as it could be deduced by considering symmetry reasons, that other distributions of cranks do not present an isotropic behavior.

In order to analyze the response of a single cell, we have measured the rotation angle after a transmission through a group of four cranks, making use of the waveguide setup described in section 3.Fig. 19 shows the rotation angle for cranks formed by equal-size segments, with a total length L ranging from 13.5 mm to 18 mm (Fig. 18). For example, for L = 15 mm, a clear resonance frequency is observed at f_0= 10.08 GHz, the angle is negative below f_0 and positive above f_0. It can be also observed that resonance frequency decreases when the length of the cranks increases, which is in agreement with similar observations found in composites formed by randomly oriented helices (Busse et al., 1999) or cranks (Molina-Cuberos et al., 2009). The experimental resonance frequencies are 8.24 GHz, 9.04 GHz, 10.1 GHz and 11.7 GHz, very close to a relation $\lambda = 2L$.

We have previously checked that the rotation angle does not depend on the relative orientation between cranks and incident wave, i.e. the sample presents an isotropic and homogeneous behavior. This fact does not occur in other configurations with odd number of cranks or with less symmetry properties. In the last case, the observed gyrotropy is a non-chiral effect and other electromagnetic effects, if any, hide the rotation due to chirality. In general we have found isotropic behavior when the sample presents symmetry under 45 degrees rotation, although other rotation symmetries are not ruled out.

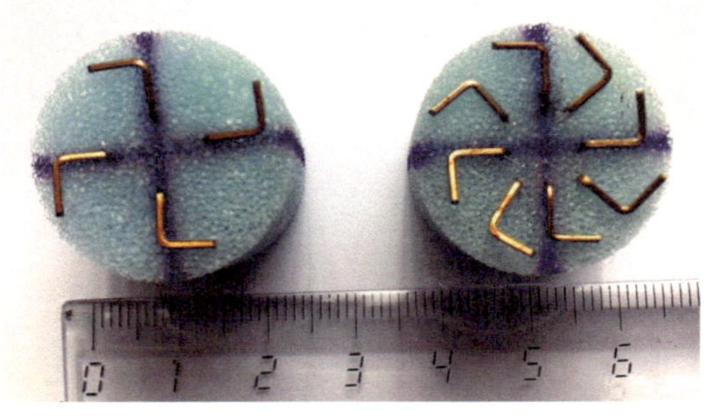

Fig. 18. Cylindrical samples used for the experimental determination of chiral effect by using a waveguide setup.

5. Conclusion

We have studied different periodical distributions, planar and quasi-planar, which show chiral behavior. We have observed that even when using a planar distribution, its electromagnetic activity comes from its 3D geometry. The rotation will be stronger, then, if we enhance this 3D characteristic. Two possibilities have been studied: some researchers prefer to use multilayered distributions of planar geometries, with a twist between adjacent layers, while we prefer to use two face metallization, with vias connecting both faces of every board: that may present the advantage of obtaining similar electromagnetic activity, combined with thinner structures. The results we have obtained, both using numerical time-domain modeling and experimental measurements seem to support our claim

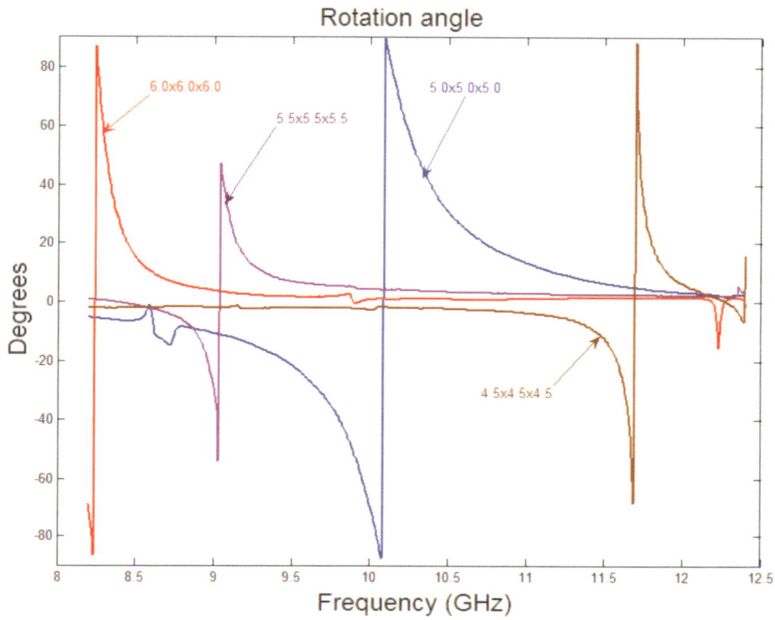

Fig. 19. Rotation angle produced by the samples composed by four cranks in foam (Fig. 18), as a function of the size of the cranks.

6. References

Alú, A.; Bilotti, F. & Vegni, L. (2003), Generalized transmission line equations for bianisotropic materials, *IEEE Transactions on Microwave Theory and Techniques*. Vol. 51, No. 11 (November 2003), pp. 3134–3141. ISSN 0018-9480.

Barba, I; Cabeceira, A.C.L.; Gómez, A. & Represa, J. (2009), Chiral Media Based on Printed Circuit Board Technology: A Numerical Time-Domain Approach, *IEEE Transactions on Magnetics*. Vol. 45, No. 3 (March 2009), pp. 1170-1173. ISSN 0018-9464.

Bahr, A.J. & Clausing, K.R. (1994). An approximate model for artificial chiral material, *IEEE Transactions on Antennas and Propagation*. Vol. 42, No. 12 (December 1994), pp. 1592-1599, ISSN 0018-926X.

Balanis, C.A. (1989). *Advanced Engineering Electromagnetics*, John Wiley & Sons, ISBN 0-471-62194-3, New York, NY, USA.

Brewitt-Taylor, C.R.; Lederer, P.G.; Smith F.C. & Haq S. (1999). Measurements and prediction of helix-loaded chiral composites, *IEEE Transactions on Antennas and Propagation*. Vol. 47, No. 4 (April 1999), pp. 692-700, ISSN 0018-926X.

Cloete, J.H.; Bingle, M. & Davidson, D.B. (2001). The Role of Chirality in Syntehtic Microwave Absorbers, *International Journal of Electronics and Communications*, Vol. 55, No. 4 (April 2001), pp. 233-239, ISSN 1434-8411.

Condon, E.U. (1937). Theories of optical rotatory power, *Reviews of Modern Physics,* Vol. 9, (October 1937), pp. 432-457, ISSN 0064-6861.

Demir, V.; Elsherbeni, A.Z. & Arvas, E (2005) FDTD formulation for dispersive chiral media using the Z transform method, *IEEE Transactions on Antennas and Propagation,* Vol. 53, No. 10 (October 2005), pp. 3374–3384, ISSN 0018-926X

García-Collado A.J.; Molina-Cuberos, G.J.; Margineda, J.; Núñez, M.J. & Martín, E. (2010). Isotropic and homogeneous behavior of chiral media based on periodical inclusions of cranks, *IEEE Microwaves and Wireless Components Letters,* Vol. 20, No 3, pp. 176-177, (March 2010), ISSN 1531-1309.

Gómez A.; Lakhtakia, A.; Margineda, J.; Molina-Cuberos, G.J; Núñez, M.J.; Saiz Ipiña, J.A, & Vegas A. (2008). Full-Wave hybrid technique for 3-D isotropic-chiral-material discontinuities in rectangular waveguides: Theory and Experiment, *IEEE Transactions on Microwave Theory and Techniques,* Vol. 56, No. 12 (December 2008), pp. 2815-2824, ISSN 0018-9480.

Gómez, A.; Lakhtakia, A.; Vegas, A. & Solano, M.A. (2010) Hybrid technique for analyzing metallic waveguides containing isotropic chiral materials, *IET Microwaves, Antennas & Propagation,* Vol. 4, No. 3 (March 2010), pp. 305-315, ISSN 1751-8725.

González-García, S.; Villó-Pérez, I.; Gómez-Martín, R. & García-Olmedo, B. (1998) Extension of Berenger's PML for bi-isotropic media, *IEEE Microwave and Guided Wave Letters,* Vol. 8, No. 9 (September 1998), pp. 297–299, ISSN 1051-8207

Kopyt, P.; Damian, R.; Celuch, M. & Ciobanu, R. (2010) Dielectric properties of chiral honeycombs - Modelling and experiment, *Composites Science and Technology,* Vol. 70, No. 7 (July 2010), pp. 1080-1088, ISSN 0266-3538.

Kuwata-Gonokami, M.; Saito, N.; Ino Y.; Kauranen, M.; Jefimovs, K.; Vallius, T.; Turunen, J. & Svirko, Y. (2005). Giant Optical Activity in Quasi-Two-Dimensional Planar Nanostructures, *Physical Review Letters,* Vol. 95, No. 22 (November 2005), pp. 227401, ISSN 0031-9007.

Lakhtakia, A.; Varadan, V.V. & Varadan, V.K. (1988) Radiation by a straight thin-wire antenna embedded in an isotropic chiral medium, *IEEE Transactions on Electromagnetic Compatibility,* Vol. 30, No. 1 (February 1988), pp. 84–87, ISSN 0018-9375.

Le Guennec, P. (2000a) Two-dimensional theory of chirality. I. Absolute chirality, *Journal of Mathematical Physics,* Vol. 41, No. 9 (September 2000), pp. 5954-5985, ISSN 0022-2488.

Le Guennec, P. (2000a) Two-dimensional theory of chirality. I. Relative chirality and the chirality of complex fields, *Journal of Mathematical Physics,* Vol. 41, No. 9 (September 2000), pp. 5986-6006, ISSN 0022-2488.

Lindell, I.V.; Tretyakov, S.A. & Oksanen, M.I. (1992) Conductor-backed Tellegen slab as twist polarizer, *Electronic Letters,* Vol. 28, No. 3 (30th January 1992), pp. 281–282, ISSN 0013-5194.

Lindell, I.V.; Sihvola, A.H.; Tretyakov, S.A. & Viitanen, A.J. (1994). *Electromagnetic Waves on Chiral and Bi-Isotropic Media,* Artech House, ISBN 0-89006-684-1, Norwood, MA, USA.

Lindman, K.F. (1920) Uber eine durch ein isotropes system von spiralformigen resonatoren erzeugte rotationspolarisation der elektromagnetischen wellen, *Annalen der Physik,* Vol. 368. No. 23 (May 1920), pp. 621–644, ISSN 1521-3889.

Marqués, R.; Jelinek, J. & Mesa, F. (2007). Negative refraction from balanced quasi-planar chiral inclusions, *Microwave and Optical technology Letters*, Vol. 49, No. 10 (October 2007), pp. 2606-2609, ISSN 0895-2477.

Molina-Cuberos, G.J.; García-Collado A.J.; Margineda, J.; Núñez, M.J. & Martín, E. (2009). Electromagnetic Activity of Chiral Media Based on Crank Inclusions, *IEEE Microwave and Wireless Components Letters*. Vol. 19, No. 5 (May 2009), pp. 278-280. ISSN 1531–1309.

Muñoz, J.; Rojo, M. Parreño, A. & Margineda J. (1998). Automatic measurement of permittivity and permeability at microwave frequencies using normal and oblique free-wave incidence with focused beam. *IEEE Transactions on Instrumentation and Measurement*, Vol 47, No. 4 (August 1998), pp. 886-892, ISSN 0018-9456.

Papakostas, A.; Potts, A.; Bagnall, D.M.; Prosvirnin, S.L.; Coles, H.J. & Zheludev, N.I. (2003). Optical Manifestations of Planar Chirality, *Physical Review Letters*, Vol. 90, No. 10 (March 2003), pp. 107404, ISSN 0031-9007.

Pendry, J.B. (2004). A Chiral Route to Negative Refraction, *Science*, Vol. 306, No. 5700 (November 2004), pp. 1353-1355, ISSN 0036-8075.

Pereda, J.A.; Grande, A.; González, O. & Vegas, A. (2006). FDTD Modeling of Chiral Media by Using the Mobius Transformation Technique, *IEEE Antennas and Wireless Propagation Techniques*, Vol. 5, No. 1 (December 2006), pp. 327–330, ISSN 1536-1225.

Pitarch, J.; Catalá-Civera, J.; Peñaranda-Foiz, F. & Solano, M.A. (2007). Efficient modal analysis of bianisotropic waveguides by the Coupled Mode Method, *IEEE Transactions on Microwave Theory and Techniques*, Vol. 55, No. 1 (January 2007), ISSN 0018-9480.

Plum, E.; Fedotov, V.A.; Schwanecke, A.S.; Zheludev, N.I. & Chen, Y. (2007). Giant optical gyrotropy due to electromagnetic coupling, *Applied Physics Letters*, Vol. 90, No. 22 (May 2007), pp. 223113, ISSN 0003-6951.

Plum, E.; Zhou, J.; Dong J.; Fedotov, V.A.; Koschny, T.; Soukoulis, C.M. & Zheludev, N.I. (2009). Metamaterial with negative index due to chirality, *Physical Review B*, Vol. 79, No. 3 (January 2009), pp. 035407, ISSN 1098-0121.

Rogacheva, A. V.; Fedotov, V.A.; Schwanecke, A.S. & Zheludev, N.I. (2006). Giant Gyrotropy due to Electromagnetic-Field Coupling in a Bilayered Chiral Structure, *Physical Review Letters*, Vol. 97, No. 17 (October 2006), pp. 177401, ISSN 0031-9007.

Schwanecke, A.S.; Krasavin, A.; Bagnall, D.M.; Potts, A.; Zayats, A.V. & Zheludev, N.I. (2003). Broken Time Reversal of Light Interaction with Planar Chiral Nanostructures, *Physical Review Letters*, Vol. 91, No. 24 (December 2003), pp. 247404, ISSN 0031-9007.

Tretyakov, S.A.; Sihvola, A.H. & Jylhä, L. (2005). Backward-wave regime and negative refraction in chiral composites, *Photonics and Nanostructures: Fundamentals and Applications*, Vol. 3, No. 2-3 (December 2005), pp. 107–115, ISSN 1569-4410.

Varadan, V.K.; Varadan, V.V. & Lakhtakia, A. (1987). On the possibility of designing antireflection coatings using chiral composites, *Journal of Wave Matter Interaction*, Vol. 2 , No. 1 (January 1987), pp. 71–81, ISSN 0887-0586.

Viitanen, A.J. & Lindell, I.V. (1998). Chiral slab polarization transformer for aperture antennas, *IEEE Transactions on Antennas and Propagation*, Vol. 46 , No. 9 (September 1998), pp. 1395–1397, ISSN 0018-926X.

Xu, Y. & Bosisio, R.G. (21995) An efficient method for study of general bi-anisotropic waveguides, *IEEE Transactions on Microwave Theory and Techniques*, Vol. 43, No. 4 (April 1995), pp. 873–879, ISSN 0018-9480.

Zhou, J.; Dong, J.; Wang, W.; Koschny, T.; Kafesaki, M. & Soukoulis, C.M. (2009). Negative refractive index due to chirality, *Physical Review B*, Vol. 79, No. 12 (March 2009), pp. 121104, ISSN 1098-0121.

5

Electromagnetic Waves in Contaminated Soils

Arvin Farid[1], Akram N. Alshawabkeh[2] and Carey M. Rappaport[2]
[1]Boise State University,
[2]Northeastern University
USA

1. Introduction

Soil is a complex, potentially heterogeneous, lossy, and dispersive medium. Modeling the propagation and scattering of electromagnetic (EM) waves in soil is, hence, more challenging than in air or in other less complex media. This chapter will explain fundamentals of the numerical modeling of EM wave propagation and scattering in soil through solving Maxwell's equations using a finite difference time domain (FDTD) method. The chapter will explain how: (i) the lossy and dispersive soil medium (in both dry and fully water-saturated conditions), (ii) a fourth phase (anomaly), (iii) two different types of transmitting antennae (a monopole and a dipole), and (iv) required absorbing boundary conditions can numerically be modeled. This is described through two examples that simulate the detection of DNAPL (dense nonaqueous-phase liquid) contamination in soil using Cross-well radar (CWR). CWR —otherwise known as cross-borehole GPR (ground penetrating radar)—modality was selected to eliminate the need for simulation of the roughness of the soil-air interface. The two examples demonstrate the scattering effect of a dielectric anomaly (representing a DNAPL pool) on the EM wave propagation through soil.
The objective behind selecting these two examples is twofold: (i) explanation of the details and challenges of numerical modeling of EM wave propagation and scattering through soil for an actual problem (in this case, DNAPL detection), and (ii) demonstration of the feasibility of using EM waves for this actual detection problem.
In addition, the results of an experimental simulation of Example 1 (the case with the monopole antenna) will be analyzed and discussed. The results of the corresponding numerically simulated example will then be compared and validated against the above-mentioned experimental results. A conclusion section will close the chapter.
Before explaining the numerical modeling and its challenges, some background about DNAPLs and their detection technologies, including the CWR method, is explained in the following section.

2. Background

Cross-well radar (CWR), otherwise known as cross-borehole ground-penetrating radar (cross-borehole GPR) is a minimally invasive method that uses high frequency electromagnetic (EM) waves transmitted and received by antennae in the subsurface to image objects of contrasting dielectric properties. In order to assess the feasibility of using

CWR for detection of dense nonaqueous-phase liquid (DNAPL) pools in deeper layers, this technique should be numerically simulated and evaluated. CWR is more appropriate for deep investigations. For near-surface sensing such as bridge-deck health monitoring, methods such as GPR are more appropriate (Belli et al., 2009 and 2009a). GPR studies require addressing the scattering due to the rough soil-air interface in the forward model as well as through the inversion process (Firoozabadi et al., 2007).This chapter describes a numerical modeling approach to Maxwell's equations using a finite difference time domain (FDTD) solution with both monopole and dipole antennae to simulate the scattering and propagation of EM waves in DNAPL-contaminated media. An FDTD code, originally developed for detection of mines using two-dimensional (2D) surface-reflection ground-penetrating radar (GPR) through non-dispersive media, was revised and upgraded for three-dimensional (3D) cross-borehole wave propagation in heterogeneous soils. The three dimensional FDTD code was enhanced to accommodate dispersive media and was used to model EM wave propagation in heterogeneous soils. This chapter describes the effect of the radiation patterns of two different antennae on propagation and scattering of EM waves through the soil subsurface and its potential for detection of DNAPL pools. In order to evaluate the feasibility of using the CWR method to detect DNAPL pools, illustrative examples with and without the presence of the DNAPL pool were analyzed. The results show considerable diagnostic potential to detect contaminated zones with DNAPLs using EM waves through CWR.

DNAPLs are separate-phase hydrocarbon liquids denser than water, such as chlorinated solvents (tetrachloroethene (PCE) and trichloroethylene (TCE)), wood preservatives, coal tar wastes, and pesticides. DNAPLs may not usually be found as a free phase in soil cores or accumulated in monitoring wells. Based on this lack of observable evidence of DNAPL pools, investigators may conclude that no DNAPL is present, when it may be present in substantial quantities at residual saturation as large as 70% to 80% of the total porosity (ITRC, 2000).

There are varieties of invasive techniques to detect DNAPLs such as direct push probe techniques (e.g., direct soil sampling or indirect sampling such as negative ion sensor) and use of in situ tracers (e.g., PITT, or partitioning interwell tracer test), excavating test pits, and groundwater profiling. Most invasive techniques just provide point-sources of information and may help the DNAPL pool to spread through the substantial number of required boreholes and excavated trenches, while noninvasive geophysical techniques noninvasively or minimally invasively detect DNAPLs. They can use different types of waves (e.g., electromagnetic, acoustic, etc.) to indirectly reconstruct images to characterize or detect anomalies as DNAPLs within heterogeneous media. These methods are minimally invasive techniques that discriminate the contrast between local physical properties of the background and target to produce images of the subsurface.

Cross-well P-wave transmission at 90 kHz was used in a sand pack before and after introducing NAPLs by Geller et al. (2000). The results indicated that small NAPL saturations may be more easily detected with amplitude than with travel time data, but the relationships between the amplitude changes and NAPL saturation may be more complex than those for velocity.

Cross-hole complex resistivity was also applied to a contaminated vadose zone by Grimm and Olhoeft (2004) to predict the general distribution of DNAPLs at parts per thousand concentrations, specifically widespread near-surface contamination and in the vadose zone immediately underneath the source.

Complex resistivity (CR) is a technique that so far has shown a potential in detecting DNAPL pools (Blackhawk Geoservices Inc., 2008).

Smith-Rose (1993 & 1935) is among the earliest scientists who studied the dielectric permittivity of different soils. Some physical properties of a typical DNAPL (i.e., PCE) are compared with those of different soils, water, and air in Table 1.
As seen, dielectric permittivity and effective (= DC + alternating field) electrical conductivity of PCE are lower than those of water and relatively similar to air or some dry soils. Thus, a PCE-contaminated region of the saturated zone should provide a contrast with a similar magnitude of the unsaturated-saturated zone interface (water table). The values of dielectric permittivity for water and soil are dispersive (depend on frequency), but most DNAPLs are nonpolar molecules with minimal frequency dependence and therefore with almost no dispersive characteristics. In frequencies greater than 10 MHz and less than 1 GHz, the dielectric permittivity is controlled by the polarization of individual water molecules and is therefore dependent on moisture content (Binley et al., 2001). Dielectric permittivity in the lower MHz range of frequency (< 10 MHz) depends on particle shape and mineralogy, electrolyte type and concentration, particle orientation, and soil electrolyte interaction (Rinaldi & Francisca, 1999). In the upper MHz range of frequency (10 to 1000 MHz), the real part of dielectric permittivity is affected by the polar contribution of bound and free water molecules (Hoekstra & Doyle, 1971; Hoekstra & Delaney, 1974; Selig & Mansukhani, 1975; Dobson et al., 1985; Hallikainen et al., 1985; and Arulanandan, 1964). Sachs and Spiegler (1964) presented a model for dielectric permittivity of soil mixtures at different frequencies. This model is based on an equivalent circuit for conductive particles in electrolytes and fitted in the dielectric plane by adjustment of three parameters. Arulanandan and Smith (1973) explained some aspects of dielectric dispersion based on the Sachs and Spiegler (1964) model in a frequency range from 10^6 to 10^8 Hz. Another model that explains the dispersion of the soil relative dielectric permittivity, otherwise known as dielectric constant, was presented by Thevanayagham (1995), which considers the effect of particle orientation relative to the electric field. Weedon and Rappaport (1997) and Rappaport et al. (1999) presented a model to predict the dispersive nature of the dielectric permittivity and electrical conductivity of soils, using a rational Z-transform approximation function of the conductivity. This technique uses the values of dielectric permittivity and electrical conductivity of the bulk mixture, and does not deal with the components and their volumetric content. It is basically a technique to convert the bulk values for the matrix to a form applicable to the FDTD method to take the dispersive nature of the resultant mixture into account.

Material	Density ρ (g/cm3)	Real Part of Dielectric Constant* (ε_r)		**** Effective Electrical Conductivity σ_e (1/Ω.m)	
		f = 100 MHz	f = 1.5 GHz	f = 100 MHz	f = 1.5 GHz
PCE	1.62	2.28	2.28	2.63×10^{-6}	1.52×10^{-4}
Water**	1.00	78	77.14	2.2×10^{-3}	0.509
Dry Sand	1.30	2.55	2.55	1.8×10^{-4}	2.1×10^{-3}
Soil***	1.60	6.912	6.50	0.1329	0.1

* Dielectric constant is also known as relative dielectric permittivity
** Pure water
*** 20% moist Bosnian clay loam
**** Effective electrical conductivity (σ_e) = Static or DC conductivity (σ_s) + Alternating field conductivity (σ_a)

Table 1. Physical Properties of PCE, Water and Soil at 25°C (Degrees Celsius) (Brewster & Annan, 1994), (Von Hippel, 1953), (Hipp, 1974), (Weedon & Rappaport, 1997), (Rappaport et al., 1999).

The loss (attenuation) in soil is a function of a variety of factors such as soil type and mineralogy, moisture content, electrical conductivity, and frequency. Higher frequency EM waves attenuate faster. In other words, soil acts as a low-pass filter. However, higher frequency EM waves provide higher image resolution. There is a tradeoff between the penetration depth and image resolution. Therefore, feasibility of the CWR to detect an average size DNAPL pool using proper frequency range is the first step to detection, which is the main goal of this chapter.

Soil, water, and DNAPL mixed at DNAPL saturations of 0% to 50% (i.e., water saturation of 50% to 100% below water table; i.e., no air) have been evaluated by Ajo-Franklin et al. (2004) to have a dielectric constant of 9 (50% water saturation soil with 50% DNAPL saturation) to 24 (fully (100%) water-saturated soil with 0% DNAPL saturation). These values are different from the dielectric constant of pure DNAPL ($\varepsilon' \approx 2.3$) and vary based on porosity and degree of DNAPL saturation as the result of mixture.

There are some expected limitations to using CWR to detect DNAPLs in soil and groundwater such as: (a) use of CWR to detect DNAPLS may not perform as well in dry or partially saturated zones as in water saturated zones, due to the weaker contrast in electromagnetic properties of DNAPLs and dry soils compared to the one between DNAPLs and water saturated soils, (b) the concentration (or contaminant saturation) must be fairly high, which makes the detection of DNAPLs using CWR often limited to identifying DNAPL pool sources (high DNAPL saturation) and not plumes, and (c) the spacing between the wells strongly influences the effectiveness of CWR (as the separation between transmitting and receiving antennae increases, the radar wave amplitude attenuates, which creates greater difficulty in distinguishing the wave from background noises) (ITRC, 2000).

Therefore, in this chapter a simple case of DNAPL dominantly replacing water in the pores (DNAPL saturation of 80% to 100%) is simulated numerically. Based on Bruggeman-Hanai-Sen (BHS) model (Sen et al., 1981), due to low dielectric constant of soil grains (≈ 5, (Ajo-Franklin et al., 2004)), the bulk dielectric constant of the soil mixture within the DNAPL pool (close to 100% DNAPL saturation) values between 2.3 and 5. Hence, for comparison, acrylic was selected as the dielectric object of $\varepsilon' \approx 2.6$ (Weast, 1974) since it has been used in another experimental work by the authors (Farid et al., 2006). This strong contrast in dielectric constant between the dielectric constant of the acrylic (≈ 2.6) and water-saturated soils (≈ 15 - 25) makes detecting the scatterer in the saturated soil using radar-based geophysical methods feasible.

The U.S. Geological Survey used 500-MHz surface GPR among other methods to monitor the location and migration of the subsequent plume (Sneddon et al., 2000).

Bradford and Wu (2007) tested 3D multi-fold GPR on a small scale controlled DNAPL release to detect contaminated soil with demonstrated success.

While GPR is the least invasive radar-based method, it is not a practical one for deep investigations. Cross-Well Radar (CWR), otherwise known as cross-borehole GPR, may be a more effective method for deep contaminant detection. CWR is a method that uses EM wave transmission measurement across borehole antennae as opposed to reflection measurements used in surface-reflection imaging methods, such as GPR. CWR overcomes some of depth limitations of surface-reflection imaging GPR. CWR uses antennae that are lowered into sampling wells with cables. Radar waves are emitted from a transmitting antenna in one well and propagate through the ground to a receiving antenna in a second well. The subsurface geology and pore fluids scatter, refract, or reflect the waves. DNAPL pools also provide dielectric permittivity and electrical conductivity contrasts within relative moist soils, which in turn, cause refractions or reflections on EM waves. These scatterings and reflections make DNAPLs amenable to detection by radar. Other than the magnitude of the scattering field, phase, and hence, travel time can also be used for other techniques such as travel time tomography to reconstruct images of the background medium and anomalies within the medium.

Success of radar-based methods as well as any other detection technique depends on the strength of the contrast in properties of the background and scatterer (e.g., DNAPLs, moisture). Dielectric permittivity and electrical conductivity are the main parameters controlling such differences for radar-based techniques.

CWR relies on a one-way travel time whereas the surface-reflection imaging GPR relies on the 2-way travel time. Therefore, the effective antenna separation of CWR can be twice as far as the penetration depth of surface-reflection GPR, theoretically. GPR antennae are designed to be relatively impedance-matched to the ground, and the reflection coefficient at the air-soil interface is close to unity leading to the potential for strong internal reflection. However, there is still the disadvantage of reflections at the air-soil interface, which reduces the amount of received waves reflected by the underground objects. The performance of detection using CWR is highly dependent on seasonal conditions, in particular soil moisture (Daniels et al., 1992). Soil mineral composition and physicochemical properties also influence radar signals. For example, clays tend to attenuate these signals and therefore may limit the skin depth and effectiveness of the method (Anderson & Peltola, 1996). Considering these limiting factors, feasibility of the technique for DNAPL detection should be evaluated by simulating implementation of ultra-wideband CWR. Further understanding of the behavior of 3D EM waves in soil is necessary for the future implementation and feasibility evaluation of CWR for detecting DNAPLs. To achieve this goal, the detection technique using monopole and dipole antennae is modeled numerically in the time domain using an FDTD technique. A critical issue addressed in this research, by direct simulation of the antennae, is the wave interaction at the antenna/soil interface. The technique models the transmitting antenna but measures the electrical field at all grid points on any desired cross-sectional or depth slice (instead of modeling receiving antennae). This is equivalent to having hundreds of receiving antennae in soil, which is not practical. In practical techniques such as cross-well tomography, few antennae are installed and used alternatively as transmitters or receivers to collect data in a multiple-depth, multiple-location manner, and the outcome is used for inversion and image processing techniques (the authors are working on different aspects of cross-well tomography and results will be published in the future). Obviously, the simulation technique presented in this chapter is used only for preliminary evaluation of the feasibility of detecting DNAPL pools using CWR. The outcome can also be used as a forward modeling for future inversion techniques.

Besides, another goal of this chapter is to evaluate the feasibility of DNAPL detection and to investigate if there are points with strong signature by the object on the scattered field, not to discuss inverse modeling. Therefore, receiving antennae are not modeled. The results model no noise. Nevertheless, comparison between the values of the scattered field (by the target) with those of the incident field will provide helpful information that can lead to the critical noise/signal ratio, even without applying any noise. This is discussed in more detail in the following sections. Experimental works are being conducted to evaluate typical levels of noise, compare them to the levels derived from this technique, then apply the noise to the simulation and reevaluate the feasibility (Farid et al., 2006).

3. FTDT modeling of maxwell's equations

Finite Difference Time Domain (FDTD) is a relatively powerful and very popular method because of its simplicity and despite its instability problems. Therefore, a 3D FDTD code was developed and used to simulate the antenna performance, coupling, and interactions in soils. The FDTD technique was originally introduced by Yee (1966) and is based on time and spatial discretization of Maxwell's equations to obtain solutions for the EM field in the time domain. The technique is numerically implemented by continuously sampling the electromagnetic field over the wave propagation medium discretized into a grid. The differential form of Maxwell's equations in the time domain is as follows (Grant & Philips, 1990; Sheriff, 1989; and Balanis, 1989).

$$\nabla \times \underline{E} = -\frac{\partial \underline{B}}{\partial t} - \underline{J_m} \tag{1}$$

$$\nabla \times \underline{H} = \frac{\partial \underline{D}}{\partial t} + \underline{J_e} \tag{2}$$

$$\nabla \cdot \underline{D} = \rho_e \tag{3}$$

$$\nabla \cdot \underline{B} = \rho_m \tag{4}$$

along with the constitutive relations,

$$\underline{B} = \mu \underline{H} \tag{5}$$

$$\underline{D} = \varepsilon \underline{E} \tag{6}$$

$$\underline{J_e} = \sigma_e \underline{E} \tag{7}$$

$$\underline{J_m} = \sigma_m \underline{H} \tag{8}$$

where **E** is the electric field density, **H** is the magnetic field intensity, **D** is the electric displacement or electric flux density, **B** is the magnetic flux density, $\underline{J_e}$ is the impressed (source) electric current density, $\underline{J_m}$ is the impressed (source) magnetic conductive current density, σ_e is the electrical conductivity, σ_m is the magnetic resistivity, ε is the dielectric permittivity, μ is the magnetic permeability, ρ_e is the electric charge density, and ρ_m is the (monopole) magnetic charge density. All of the field parameters, **E**, **H**, **B**, **D**, $\underline{J_e}$, and $\underline{J_m}$ are assumed to be functions of position and time t, while material parameters μ, ε, and σ are functions of position and may be dispersive functions of frequency (and thus, through the Fourier Transform, also of time). Note that the usual multiplicative constitutive relations are written as convolutions for dispersive soil media. Equations (1) through (4) can be also written in integral forms. These four partial differential equations are discretized in a 3D rectangular grid, using a central-difference approximation of two consecutive values for the field components in both space and time. Due to the nature of electric and magnetic fields (e.g., coil loops), these two cannot be located at the same spatial location. In the discretization process, electric and magnetic field components are assigned to the edges of complementary interlocking cubical meshes. Thus, half-space indices are introduced for different field components. Besides, half time steps are introduced to perform the finite difference computation of **E** based on **H** and vice versa. Spatial and time steps are respectively represented by the lower indices (i, j, k) and the upper index (n). Some details of the simulation process are explained in the following. Fig. 1 helps to better understand this staggered time and space grid. As seen in the superscripts in the figure, there is ½ time step difference between **E** and **H**. Due to the central difference approximation technique in time, **H** is present at both the $(n + ½)^{th}$ and $(n - ½)^{th}$ time steps in the equations along with **E** only at the nth time step.

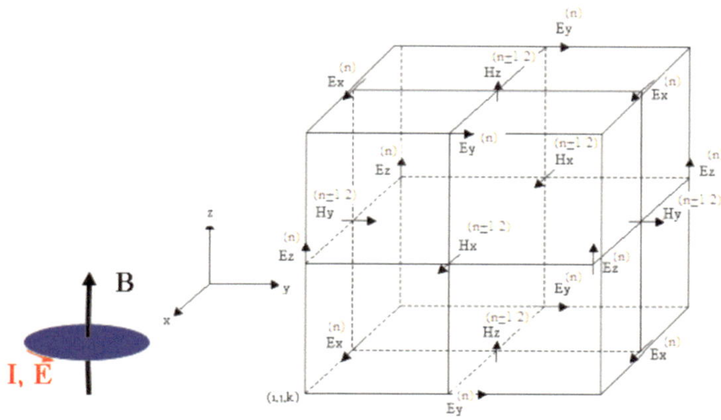

Fig. 1. 3D interlocking FDTD cell (Yee cell, 1966).

3.1 Advantages and limitations

The most advantageous characteristic of the FDTD method is its simplicity. Solving Maxwell's equations using FDTD is a simple iterative procedure. However, there are several limitations to the numerical implementation of the FDTD technique through the above-mentioned difference equations. These limitations have been and are subject of research. Some of these challenging limitations impose restrictions on the grid size and time step increments that affect stability and accuracy of the method (Kunz & Luebbers, 1993).

One fundamental restriction requires the longest side of the grid cell to be much shorter than the shortest wavelength of the wave within the cell. A very common restriction assumed in practice is $\lambda/10$, where λ is the shortest appreciable wavelength in the excitation signal. This constraint is imposed by both sampling limitation and grid dispersion errors (Kosmas, 2006).

The second restriction develops from the scale and geometry of the problem. The method uses a uniform grid to model small antennae along with a large soil medium with or without contrasting objects. Accordingly, the geometry imposes challenging limitations, especially in computation cost. For example, in the case of a DNAPL contaminated site, dimensions of the background in horizontal directions may be of the order of tens of meters by tens of meters in the XY-plane, while the antenna thickness may be of the order of millimeters to centimeters. This requires solving a large uniform grid with a small grid-cell size in the X and Y directions. At the same time, a small vertical (Z-direction) cell size is required to account for thin DNAPL pools. One way to solve this problem is by using a non-uniform grid, which adds more difficulties to the task of satisfying stability conditions (e.g., Courant condition).

The third limitation is the time-step restriction required to satisfy the Courant condition. The Courant condition restricts time increments to a range in which waves do not travel too far in each time increment, or:

$$\frac{1}{V_{p,\max}\Delta t} \geq \left(\sum_{m=1}^{N} \frac{1}{\Delta x_m{}^2} \right)^{1/2} \tag{9}$$

where N is the grid dimension, x_m is the grid cell size in the mth direction, and $V_{p,max}$ is the maximum wave phase velocity within the model. In a 3D case, Equation (9) will be written in the following form:

$$\frac{1}{V_{p,\max}\Delta t} \geq \left(\frac{1}{\Delta x^2} + \frac{1}{\Delta y^2} + \frac{1}{\Delta z^2} \right)^{1/2} \tag{10}$$

In the case of N-dimensional isotropic cells, Equation (9) can be simplified to $V_{p,\max}\Delta t / \delta \leq 1 / \sqrt{N}$ (e.g., in a three dimensional case, $N = 3$, and $\Delta x = \Delta y = \Delta z = \delta$), where the left hand side is called the Courant number. Different cases have demonstrated that using smaller values of Δt does not necessarily improve the results. However, smaller values for the Courant number may sometimes yield satisfactory results (Kosmas, 2006).

4. Modeling procedure

4.1 Modeling dispersive nature of soil

Soil is a complex medium to model because of its heterogeneous, lossy, and dispersive nature. Accurate computation of the behavior of soil over a wideband of frequency requires either several individual frequency domain calculations or a robust deconvolution of $\mathbf{E}(t)$ from $\mathbf{D}(t)$ in the time domain (Rappaport et al., 1999). The convolutional approach to modeling dispersion in soil approximates the frequency domain dispersive complex dielectric constant with rational functions of $j\omega$ and multiplies the constitutive relations by the denominator. Then, the results are inverse-Fourier-transformed. This model is called the Debye or Lorentz model (Kashiwa & Fukai, 1990; and Gandhi, 1993). Weedon and Rappaport (1997) and Rappaport and Winton (1997) simplified the problem by modeling the conductivity as a simple rational function of Z-transform. Modeling conductivity in terms of the Z-transform variable (Z^{-1} readily transforms to "time delays") simplifies the process of conversion of the generalized dispersive Ohm's Law ($\mathbf{J}(Z) = \boldsymbol{\sigma}(Z)\ \mathbf{E}(Z)$) to the time domain. To keep the simplicity of the FDTD method, Z-transform function of conductivity in the time domain is given as (Weedon & Rappaport, 1997; and Rappaport et al., 1999):

$$\sigma(Z) = \frac{\mathrm{J}(Z)}{\mathrm{E}(Z)} = \frac{b_0 + b_1 Z^{-1} + b_2 Z^{-2}}{1 + a_1 Z^{-1}} \tag{11}$$

Using this function, both real and imaginary components of conductivity depend on the sampling interval Δt and the coefficients of the rational Z-transform function (a_1, b_0, b_1, and

b_2). The real part of conductivity corresponds to measured conductivity, and the imaginary part corresponds to real dielectric permittivity. Ampere's law can be written in the Z-domain as follows.

$$\nabla \times \underline{\mathrm{H}(Z)} = \varepsilon_{av} \frac{1 - Z^{-1}}{\Delta t} \underline{\mathrm{E}(Z)} + \frac{(1 + Z^{-1})}{2} \sigma(Z) \underline{\mathrm{E}(Z)} \tag{12}$$

4.2 Absorbing boundary conditions

As in simulation of any diffusion application (e.g., heat transfer or water flow through porous media), wave propagation through infinite media is practically modeled in a finite grid. To model the infinity of the flow or wave propagation through the required infinite media by the finite number of grid cells, appropriate boundary conditions are required. The lattice termination absorbing boundary condition used in the FDTD code was based on the second-order one-way wave propagation equation of Mur (1981) and adjusted for lossy soils by Talbot and Rappaport (2000) using the rational function approximation of Equation (11).

4.3 Antenna and excitation

There are some factors that categorize the excitation, such as the source geometry, excitation signal type, and hard or soft source of propagation.

Hard and Soft Sources: The code uses soft sources for all different antenna types. A hard source specifies the total field at the excitation point. These types of sources are usually avoided as they cause undesirable reflections at the source point. Alternatively, a soft source specifies the additional field supplied above the existing background field at the source point. The soft source can be specified at points on a radiating aperture, while hard sources are used for current sources flowing along metal structures. In 2D cases, soft sources propagate well, but extensive testing of 3D FDTD cases has shown difficulties involved in propagation from a soft source unless the physical body of the antennae is modeled. This paper uses soft sources for both monopole and dipole antenna cases of the following sections.

Excitation Signal: A cosine-modulated Gaussian signal (Equation (13)) was used to excite both antennae because of its simplicity and frequency content

$$E^n(i,j,k) = E_0 \exp\left[-\left(\frac{t - t_0}{W} \right)^2 \right] \cos(2\pi f t + \phi) \tag{13}$$

where E_0 is the amplitude of the pulse, t is any arbitrary time instant, and t0 is the time instant corresponding to the peak of the Gaussian. Ideally, a Gaussian pulse has infinite duration. However, in numerical implementation it ought to be truncated. Thus, W is chosen as the duration of the pulse. Parameters t_0 and W can be chosen, but not absolutely arbitrarily. In this chapter, W was chosen based on the bandwidth used in a scaled experiment by the authors (Farid et al., 2006). Inappropriate choices of these parameters may introduce instability and noise into the FDTD computation. Starting and truncation of the Gaussian signal should be sufficiently smooth (i.e., the field value at the truncation time should be sufficiently small) to alleviate introduction of frequencies with spectrum levels above the one that a single or double precision calculation can tolerate. Large spectrum energy of the incident field may cause instability and noise, which in turn, results in computation corruption. This is true for any type of pulse. For the Gaussian type pulse, experience has shown that a choice of $t_0 \geq 4W$ ensures smooth truncation, and in turn, good performance (Kosmas, 2006).

4.4 Geometry of the source

To model the DNAPL detection and evaluate its feasibility, two different antenna types were modeled, a monopole and a dipole. Some fundamental parameters and facts about these simulations are explained in the following.

- The monopole antenna is simply a coaxial cable, schematically shown in Fig. 2(a) with its shield removed at a specific length from the tip (referred to as extended dielectric length). In a simple monopole antenna, the most efficient radiation occurs when the exposed dielectric and center conductor length is ¼ of the excitation wavelength. For the range of frequency of interest of this and other ongoing and further research (400-2200 MHz), and based on the dielectric permittivity of the dielectric part of the antennae (≈ 2.1), this length should be between 2.35 cm and 12.94 cm. Different types of antennae have different radiation patterns. The radiation pattern is a visual way of representing how an antenna distributes energy into the surrounding space. Different methods (parameters) can be used to demonstrate the wave propagation through space. One popular method is to use power as the key parameter. This method visualizes how and in which direction power is concentrated by the antenna to propagate away. This is possible by drawing contours of equal power. Other parameters, such as field magnitude or phase, can be used for visualization purposes. Contours of these parameters are drawn to visualize radiation patterns. A monopole antenna radiates waves in all different directions.
- A dipole antenna consists of two straight metallic or dielectric parts connected at the center to a feed line. A typical dipole antenna is shown in Fig. 2(b). This antenna type constitutes the main RF radiating and receiving element in various sophisticated types. Dipoles are inherently balanced antennae due to bilateral symmetry. The best dipole length is usually ½ wavelengths.

5. Evaluating feasibility of DNAPL detection

As mentioned before, two different antennae are simulated in the code to model the problem and evaluate the feasibility of the use of the CWR method to detect dielectric objects in soil. The first case models a small-scale modeled monopole antenna as a soft source of the wave within a fully water-saturated sandy soil medium (degree of water-saturation = $S_r = 1$) and the gravimetric moisture content = $w = 17\%$). The second case is the simulation of a dipole antenna in a larger size medium of the same fully saturated sandy soil ($S_r = 1$ and $w = 17\%$). This value of moisture content is selected to model a soil medium similar to the existing fully saturated sandy soil in a pilot-scale facility constructed at Northeastern University (referred to as SoilBED) for subsurface sensing and imaging experimentation. This 17% moisture content is the reference moisture content for much further experimental research by the authors (for more information refer to Farid et al., 2006) in the SoilBED facility. Hydrological modeling of DNAPLs and the dielectric properties converted through a generic petrophysical relationship in the simulation is desired but does not fit in the extent of this chapter. Details of the two above-mentioned simulated cases follow.

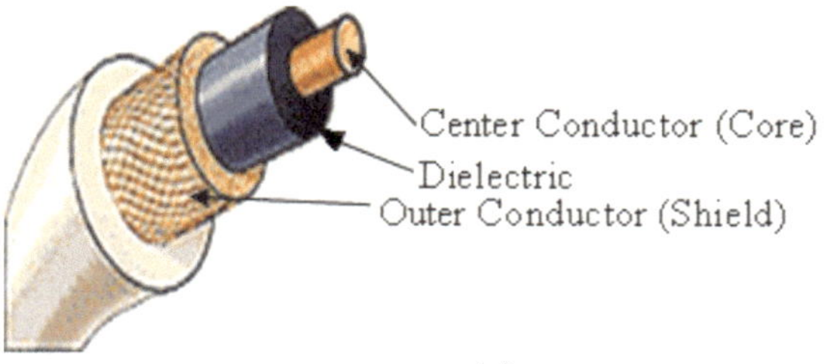

(a)

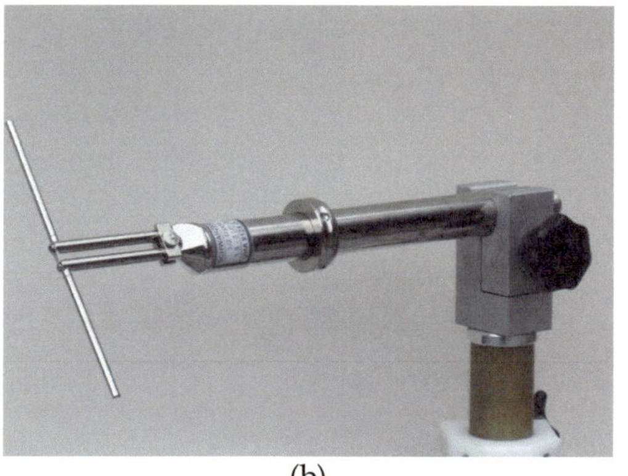

(b)

Fig. 2. Antennae: (a) Monopole antenna derived from a coaxial cable by removing a part of outer conductor, (b) a UHF–Half–Wave Dipole (Wikipedia, 2011).

5.1 Monopole antenna case (A)

The pilot-scale simulation of the SoilBED facility (Farid et al., 2006) is explained in this section. In the first case, a 5 mm-thick monopole antenna is modeled within a fully saturated sandy soil background. The size of the medium under study was selected to satisfy limitations of the FDTD code as well as the experiment. Table 2(a) summarizes details about the geometry and grid size of the soil medium. The simulation is driven by a cosine modulated Gaussian time pulse at a reasonably high frequency (1.5 GHz). To accommodate the simulation of the dispersive soil and stability of the FDTD code at this frequency, the time increment Δt = 2 psec was used. The dispersive properties of the soil for this choice of center frequency and time-step are modeled with the Z-transform function coefficient set ε_{Ave} , a_1, b_0, b_1, and b_2, given in Table 2(b).

To relate the results to the field site, the model can be scaled up in size while scaling down the frequency. To evaluate the feasibility of the DNAPL detection method using monopole antennae, wave propagation through the background soil and scattered EM wave propagation by a DNAPL pool were modeled and analyzed. The geometry details of the monopole transmitting antenna modeled in this case are tabulated in Table 2(c). The drive signal excites the top of the simulated coaxial cable feeding the monopole antenna in a conventional radial field pattern. The electric field components on all grid points of different cross-sectional and depth slices of the medium were computed and then visualized using MATLAB.

First, the background medium was analyzed. Then, a rectangular acrylic plate as a representative of a DNAPL pool was modeled within the soil medium. Fig. 3 schematically shows the simulated geometry (the monopole antenna, the DNAPL pool, and the soil medium). Details of the geometry of the DNAPL pool scatterer are listed in Table 2(d).

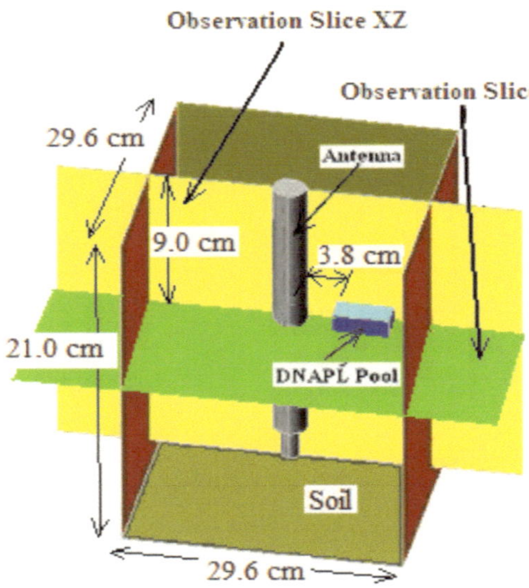

Fig. 3. SoilBED, antenna, observation slice and rectangular DNAPL pool (3 cm × 3 cm × 1 cm).

Geometry	Size
Simulated grid	149 × 149 × 29
Grid cell size	0.2 cm × 0.2 cm × 1 cm
Entire grid size	29.6 cm × 29.6 cm × 28 cm
Soil thickness	21 cm
Air thickness	7 cm

Table 2.a. Details of the simulated medium

Parameter	Value
ε_{Ave} (Dielectric Permittivity)	20.9
a_1	-0.8985
b_0	-34.3627
b_1	68.7577
b_2	-34.3945

* Due to solving the problem at $\Delta t = 2$ psec, the FDTD code is very sensitive, and all digits are necessary to satisfy the stability conditions

Table 2.b. Soil properties, used for the simulation of the fully saturated sandy soil at $f = 1.5$ GHz, $\Delta t = 2$ psec, and 17% gravimetric moisture content (w)

Antenna Details	Size
Antenna depth	120 mm
Perfectly conducting core wire thickness	1 mm
Extended dielectric length	20 mm
Extended dielectric thickness	3 mm
Perfectly conducting outer conductor (shield) thickness	3 mm
Frequency	1.5 GHz
Gaussian width	0.667 nsec
Gaussian peak	5 nsec

* The dielectric constant and effective electrical conductivity of the extended dielectric of the antenna are respectively assumed to be 2.1 and zero (Ω^{-1}).

Table 2.c. Geometry details of the simulated monopole antenna

DNAPL Pool Geometry	Size
Horizontal cross-section	3 cm × 3 cm × 1 cm
Depth	9 cm
Clear separation from the antenna	3.8 cm
Coordinate of the pool center*	6 cm, 0 cm, -2 cm

* With respect to the center of the grid

Table 2.d. Details of the DNAPL pool scatterer

To evaluate the wave propagation, the following observation slices were selected. Different components of electric field were computed and visualized on these slices.

- A cross-sectional (horizontal: XY-plane) slice, cutting through the antenna and DNAPL pool at the depth of 9 cm. Z and X components of the electric field (E_z and E_x) are shown on this slice in Fig. 4.

  Up to this point, only the three vector components of the electric field were visualized. Now, the power is depicted. The intensity of a rapidly varying field is often displayed on a dB scale, enabling the visualization of small amplitude levels. This scale is given by $20 \log_{10} | E / E_{max} |$. It is important to note that on the selected depth slice, E_y equals 0, and hence $\mathbf{E} = E_x \vec{i} + E_z \vec{k}$. In addition, since the time domain signals are all purely real, but may have positive or negative values, the dB scale is artificially augmented with positive values to indicate negative field values and better display the oscillating nature of the rapidly decaying wave. The sign of corresponding E_z governs the sign of the dB value. It should be stressed that 0 dB is the maximum field intensity, and positive dB values correspond to weaker signals with the opposite sign.
- A depth (vertical: XZ-plane) slice, passing through the antenna and DNAPL pool. This slice (XZ-plane) was chosen because the YZ-plane does not intersect the DNAPL pool. Due to symmetry, E_y is zero on this XZ slice, and hence $|\mathbf{E}|$ can be computed by only E_x and E_z. Results are shown in Fig. 5.

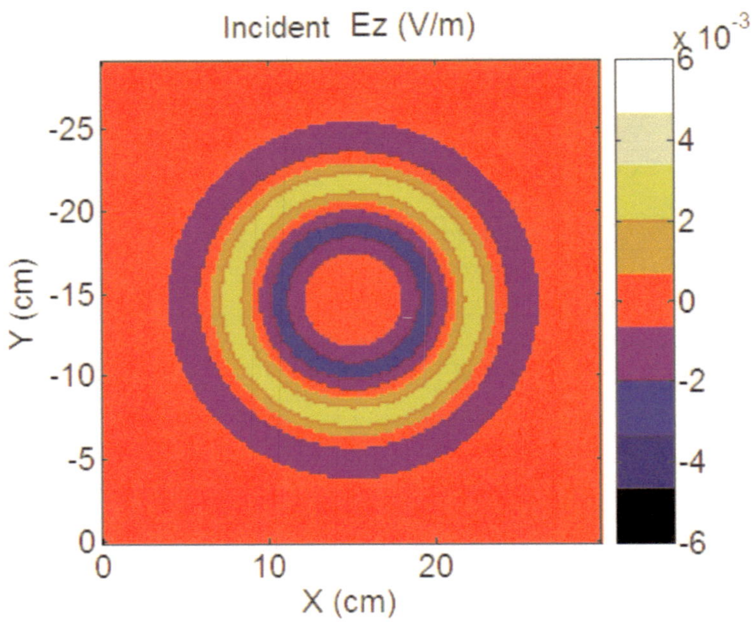
Incident Ez (V/m)
x 10^-3
6
4
2
0
-2
-4
-6
-25
-20
-15
-10
-5
0
Y (cm)
0
10
20
X (cm)

(a)

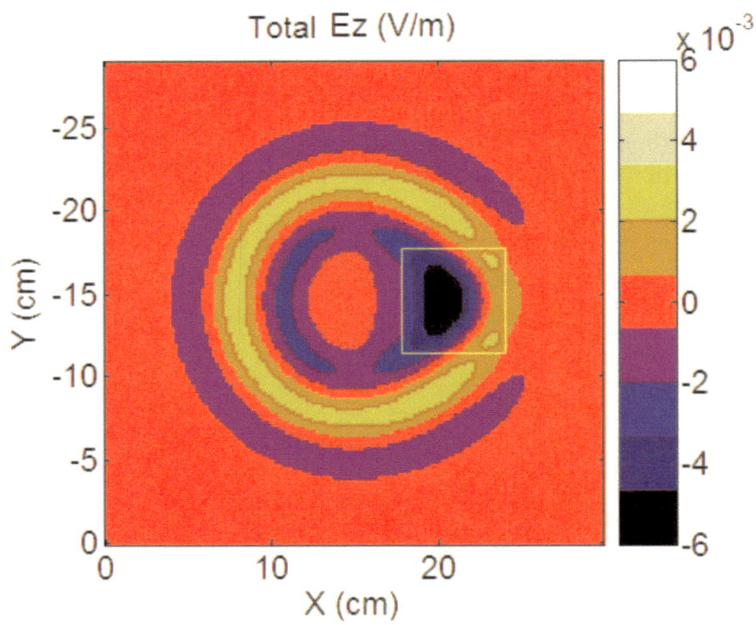
Total Ez (V/m)
x 10^-3
6
4
2
0
-2
-4
-6
-25
-20
-15
-10
-5
0
Y (cm)
0
10
20
X (cm)

(b)

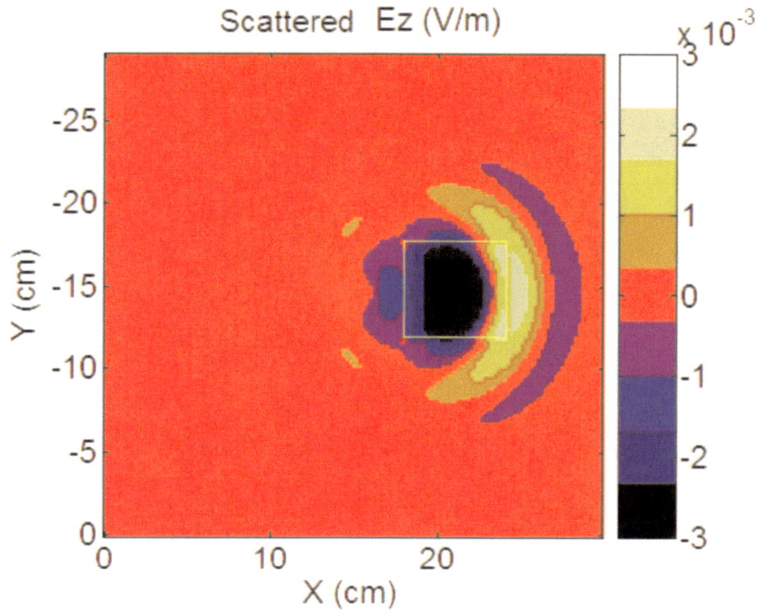
Scattered Ez (V/m)
x 10^-3
3
2
1
0
-1
-2
-3
-25
-20
-15
-10
-5
0
Y (cm)
0
10
20
X (cm)

(c)

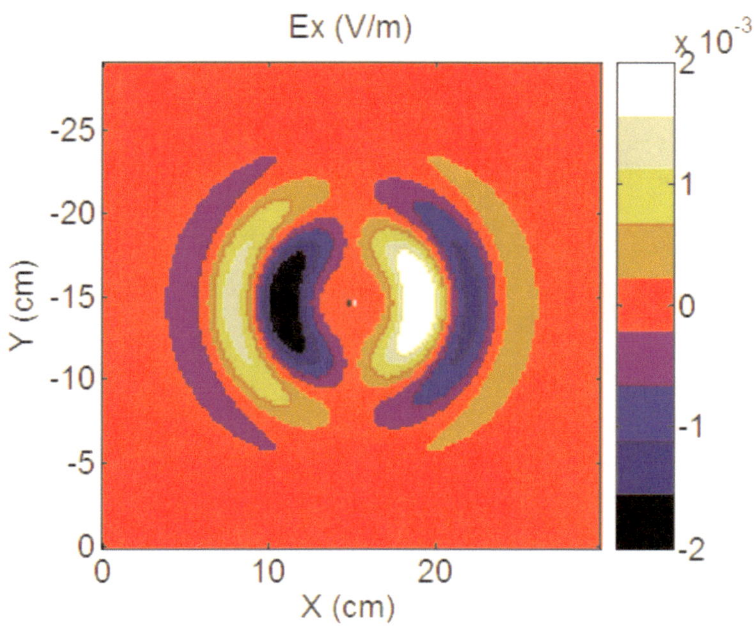
Ex (V/m)
x 10^-3
2
1
0
-1
-2
-25
-20
-15
-10
-5
0
Y (cm)
0
10
20
X (cm)

(d)

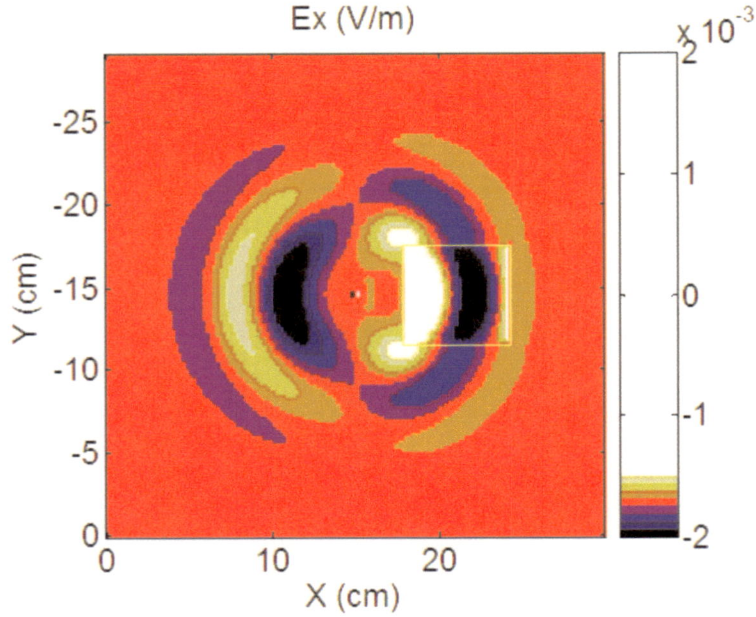

(e)

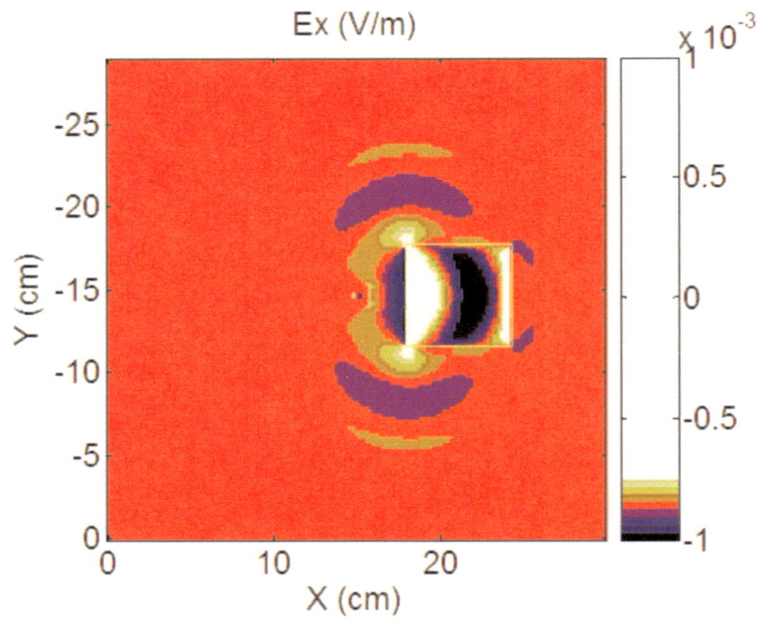

(f)

Fig. 4. Electric field simulated on the cross-sectional slice (XY-plane) at t = 3.6 nsec (the extent of the DNAPL pool is marked by a yellow box): Z-component of the electric field: a) Incident, b) Total, and c) Scattered; and X-component of the electric field: d) Incident, e) Total, and f) Scattered.

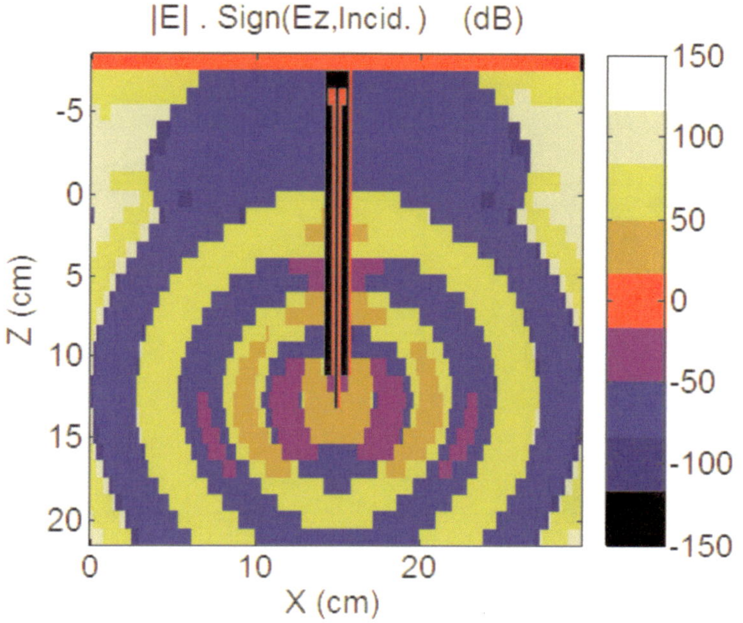
|E| . Sign(Ez,Incid.) (dB)
Z (cm)
X (cm)
-5
0
5
10
15
20
0
10
20
150
100
50
0
-50
-100
-150

(a)

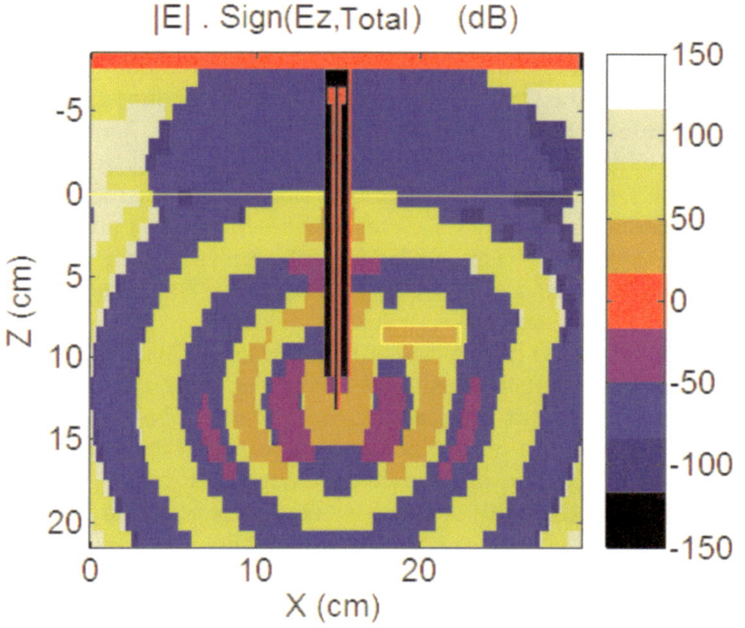
|E| . Sign(Ez,Total) (dB)
Z (cm)
X (cm)
-5
0
5
10
15
20
0
10
20
150
100
50
0
-50
-100
-150

(b)

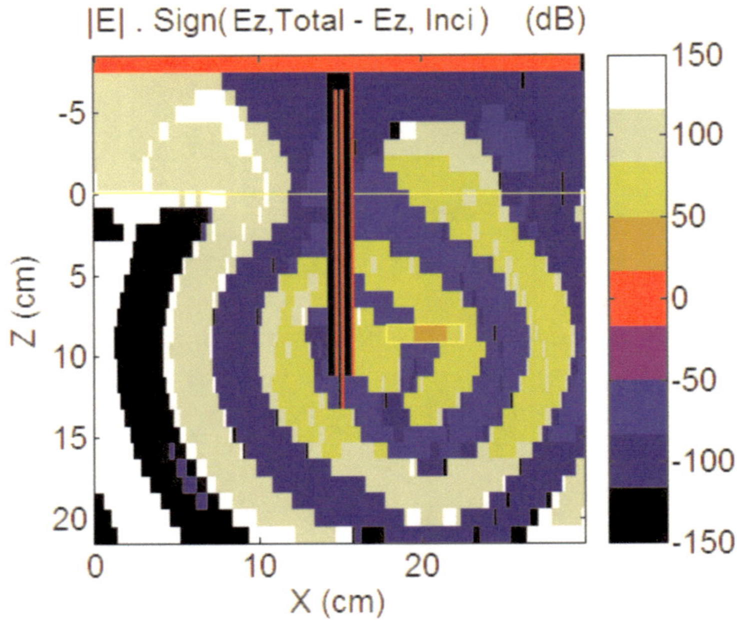

(c)

Fig. 5. Electric field [Sign($E_{z,Total}$ - $E_{z,Incident}$)] × 20 $\log_{10}$(|E| or |$E_x \vec{i}$ +$E_z \vec{k}$ |), on the depth slice (XZ-plane) at time t = 3.6 nsec (the extent of the DNAPL pool and soil-air interface are marked in yellow): a) Incident image, b) Total image, and c) Scattered image.

This case was initially analyzed without DNAPL contamination (incident field or background) and then with the DNAPL pool (total field). The scattered field by the DNAPL pool target can be computed by subtracting the two previous fields. Three figures are shown for each slice and for each electric field component and for: (i) "incident" (i.e., background, no target), (ii) "total" = background + DNAPL pool target as the scatterer, and (iii) "scattered" (i.e., signature of the target). All results shown in Fig. 4 are captured at t = 3.6 nsec. As seen, incident results of Figs. 4(a) and 4(d) are symmetric, while the total field results shown in Figs. 4(b) and 5(e) are not symmetric. The resulting scattered field information shown in Figs. 4(c) and 4(f) is asymmetric as well.

The incident, total, and scattered (target signature) fields are shown in Fig. 5. The monopole antenna was modeled as a Z-polarized antenna. Therefore, the Z-component of the electric field is the major component, but the scattered field by the DNAPL pool is also readily visible on the X and Y component plots. Since E_z dominates and the scattered field is visible on the Z-component (Fig. 5(c)), the scattered field shown on the dB plot will be clear as well. Further studies (that do not fit in this chapter) show weaker scattered Z-component in dry sandy soils. Different components can be experimentally measured using a receiving antenna with a different polarization (e.g., an X or Y polarized antenna, which is simply a monopole placed horizontally) than the Z-polarized (vertical) transmitting antenna. The scattered field is comparable to the incident field in this case. This potential can also be demonstrated in a different form as shown in Fig. 6.

This figure shows that there is a considerable magnitude and travel time difference between the total and incident fields received at a receiver located right above the DNAPL pool. The strong magnitude difference (more than 100%) and time difference (around 100 psec) between the two signals illustrate the potential of the cross-borehole GPR method to detect DNAPL pools. The early arrival of the total field is caused by the increase in the velocity of EM waves through the DNAPL pool due to its lower dielectric permittivity compared to the saturated soil. The increase in the magnitude of the total field is, on the other hand, caused by lower loss through the DNAPL pool due to its lower electrical conductivity. This illustrates a great potential for DNAPL detection using CWR in saturated soils, if the thickness and size of the pool is a reasonable fraction of the wavelength.

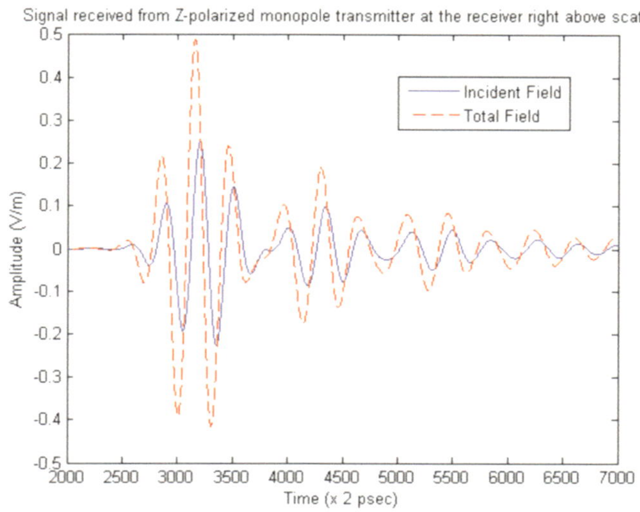

Fig. 6. Z-component of total and incident electric fields due to the monopole antenna, received at a receiver located right above the DNAPL pool.

5.2 Dipole antenna case (B)

The above-mentioned small size monopole case can be scaled up to a more realistic size contaminated site. However, scaling up the results may cause some problems that do not allow a simple and direct generalization from small numerical models to real size contaminated sites. For example, in a non-dispersive medium, linear enlargement of the size can be simply interpreted to a linear increase in the wavelength and decrease in the frequency. However, in a dispersive medium, any change in the frequency causes variations in the dielectric properties of the medium. This change in the dielectric constant causes variations in the wave velocity, which in turn adds nonlinearity to the scaling process from the simulated medium up to the real size.

Therefore, to evaluate the scaling issues in a dispersive medium and study the effect of different radiation patterns of different antennae, another case with a more realistic size of soil medium surrounding a dipole antenna was modeled. The dipole is also larger than the monopole, since the smallest object to be modeled (the antenna) controls the uniform grid size in X and Y directions and size limitations of the FDTD code. The details about the grid size and the geometry of the soil medium for this case are tabulated in Table 3(a).

To decrease the computation cost, a much larger grid cell (3 cm in X and Y directions, and 5 cm in Z direction) was modeled (Table 3(a)). To satisfy sampling limitations (grid size < $\lambda/10$) and study the scaling effect, the wavelength should be larger. Therefore, the frequency was selected to be 100 MHz (lower than 1.5 GHz in Case A). To satisfy the Courant's condition for the new grid size, the time increment was increased to Δt = 50 psec.

Geometry	Size
Simulated grid	149 × 149 × 69
Grid cell size	3 cm × 3 cm × 5 cm
Entire grid size	444 cm × 444 cm × 340 cm
Soil thickness	305 cm
Air thickness	35 cm

Table 3.a. Details of the simulated medium

The soil medium is exactly the same fully water-saturated sandy soil modeled in the previous case with 17% gravimetric moisture content. However, dielectric properties of the dispersive soil at the different frequency and time increments (f = 100 MHz, and Δt = 50 psec) are different. Therefore, the dielectric constant and coefficients (ε_{Ave}, a_1, b_0, b_1, and b_2) of the Z-transform function required to model the dispersive electrical conductivity of the soil were recomputed for the new frequency and time increment. The new soil parameters are listed in Table 3(b). A center-fed resistively tapered ½ wavelength dipole antenna is modeled as the transmitter. The particular details of the resistive dipole are avoided by modeling the antenna electromagnetically as simply a tapered half-wave surface current source residing on the exposed coaxial insulator (maximum at the center, the point where the feed line joins the elements, and zero at the ends of the elements). This type of antenna may be used in a PVC-lined borehole filled with water. Therefore, the model simulates the antenna surrounded by water. Obviously, to model the dispersive nature of water and maintain the symmetry and accuracy on the circular interface around the antenna, water is modeled using the same technique used to model lossy dispersive soils (Weedon & Rappaport, 1997). For the same reason, the dielectric portion is modeled using the same technique used for lossy dispersive soils, despite the non-lossy and non-dispersive nature of the dielectric material. The PVC casing was ignored during the simulation to simplify the

Parameter	Value
ε_{Ave} (Dielectric Permittivity)	14.9251
a_1	-0.8985
b_0	1.04948
b_1	-1.9896
b_2	0.94093

* Due to solving the problem at Δt = 50 psec, the FDTD code is very sensitive, and all digits are necessary to satisfy the stability conditions.

Table 3.b. Soil parameters, used for the simulation of the fully saturated sandy soil at f = 100 MHz, Δt = 50 psec, and 17% gravimetric moisture content

modeling and because the wall of the PVC is very thin compared to the wavelength (780 mm) of the EM wave. The dipole antenna is Z-polarized and the excitation signal is a 100 MHz cosine-modulated Gaussian pulse, progressively delayed along the antenna in the Z-direction (i.e., points along the Z-directed dipole are excited with a progressive phase delay proportionate to the traveling time of the current fed through the midpoint and along the dielectric portion of the dipole). Table 3(c) summarizes the details about the structure of the dipole antenna.

Antenna Details	Size
Antenna depth	1800 mm
Borehole diameter	240 mm
Perfectly conducting core wire thickness	22 mm
Extended dielectric length	500 mm
Extended dielectric thickness	64 mm
Perfectly conducting outer conductor (shield) thickness	43 mm
Depth of water-filled borehole	500 mm
Frequency	100 MHz
Gaussian width	10 nsec
Gaussian peak	75 nsec

* The dielectric constant and effective electrical conductivity of the extended dielectric are respectively assumed to be 2.1 and zero (Ω-1).

Table 3.c. Details of the simulated dipole antenna

First, the wave propagation through the soil background was analyzed. Then, a rectangular DNAPL pool was modeled within the soil medium. Fig. 7 schematically shows the geometry of this DNAPL pool. The details about the geometry of the DNAPL pool scatterer are listed in Table 3(d).

DNAPL Pool Geometry	Size
Horizontal area	45 cm × 45 cm × 15 cm
Depth	90 cm
Clear distance to the antenna	22.5 cm
Coordinate of the pool center*	90 cm, 0 cm, 45 cm

* With respect to the center of the grid

Table 3.d. Details about the DNAPL pool scatterer

Similar to Case A, the transmitting antenna is modeled in the code, but rather than modeling receiving antennae, the three different components of the electric and magnetic fields are computed at all grid points on the following depth and cross-sectional slices.

- A cross-sectional (horizontal: XY-plane) slice, cutting through the antenna and DNAPL pool at the depth of 90 cm. Z and X components of the electric field (E_z and E_x) are shown on this slice (Fig. 8).
- A depth (vertical: XZ-plane) slice, passing through the antenna and DNAPL pool. The magnitude of the power, derived from both E_x and E_z, is shown on this slice in Fig. 9 (E_y is zero on this slice due to symmetry).

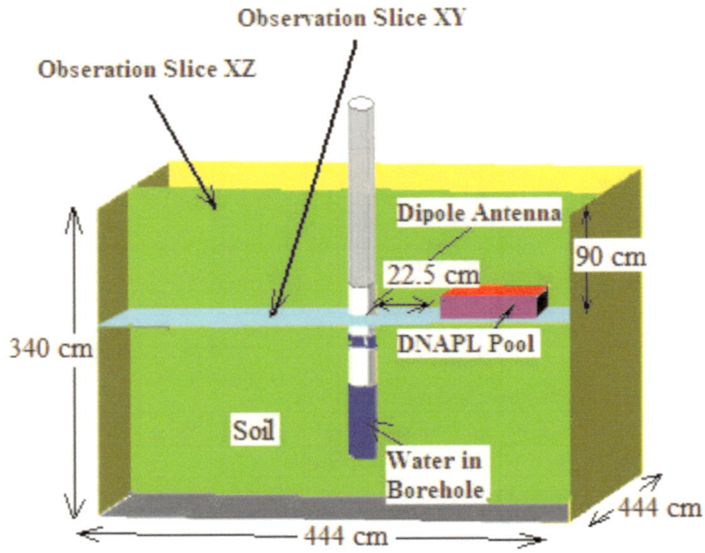

Fig. 7. Schematic representation of the borehole dipole antenna geometry and DNAPL pool (45 × 45 cm × 15 cm).

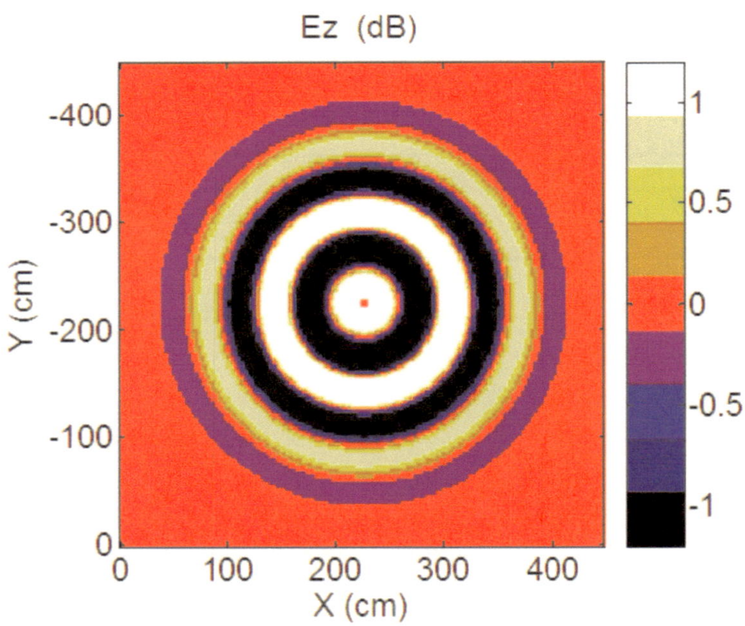

(a)

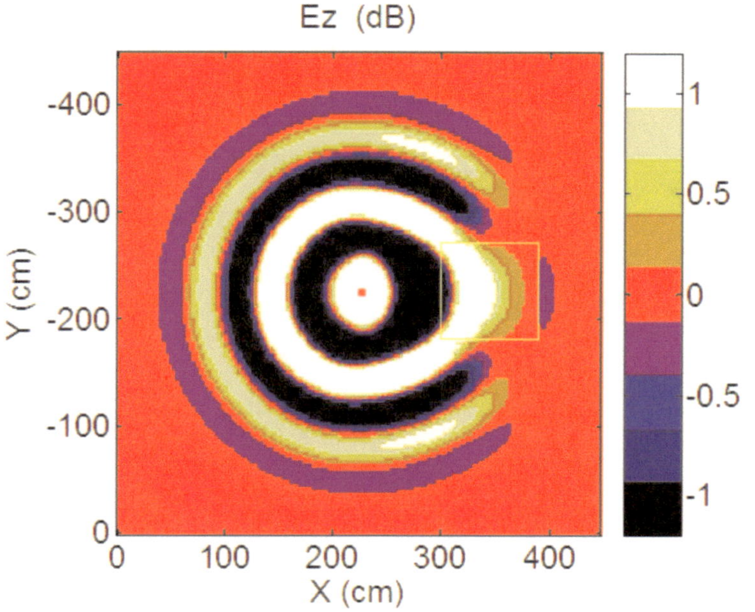
Ez (dB)
-400
-300
-200
-100
0
Y (cm)
0
100
200
300
400
X (cm)
1
0.5
0
-0.5
-1

(b)

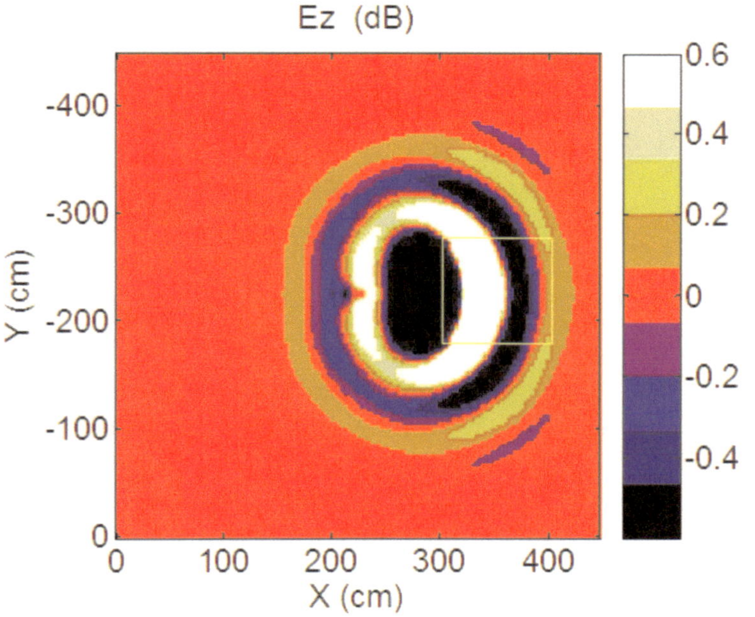
Ez (dB)
-400
-300
-200
-100
0
Y (cm)
0
100
200
300
400
X (cm)
0.6
0.4
0.2
0
-0.2
-0.4

(c)

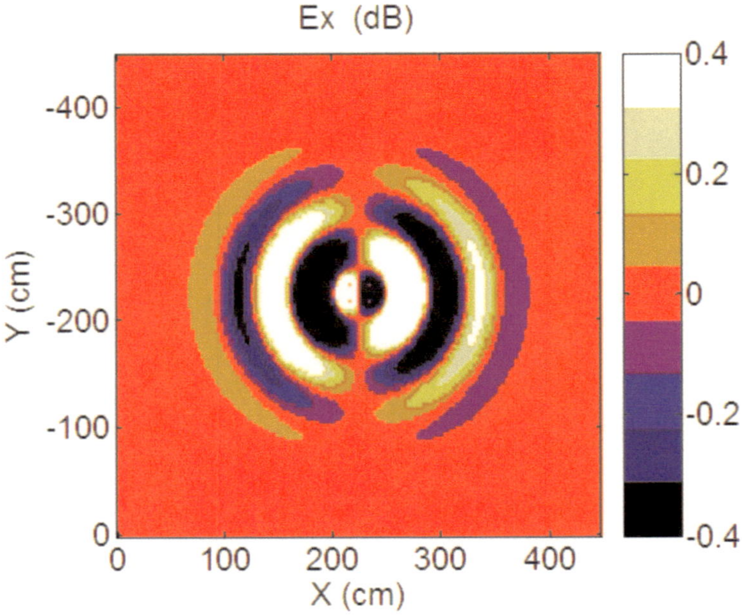
Ex (dB)
-400
-300
-200
-100
0
Y (cm)
0
100
200
300
400
X (cm)
0.4
0.2
0
-0.2
-0.4

(d)

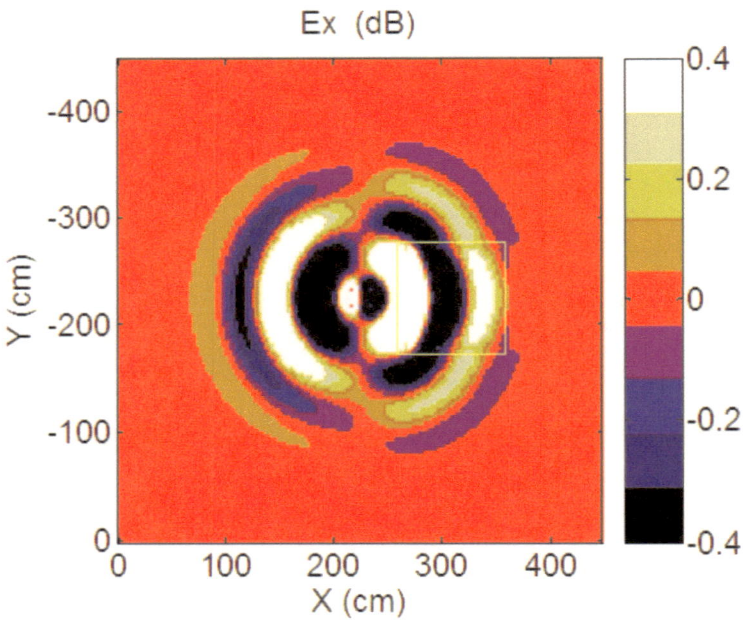
Ex (dB)
-400
-300
-200
-100
0
Y (cm)
0
100
200
300
400
X (cm)
0.4
0.2
0
-0.2
-0.4

(e)

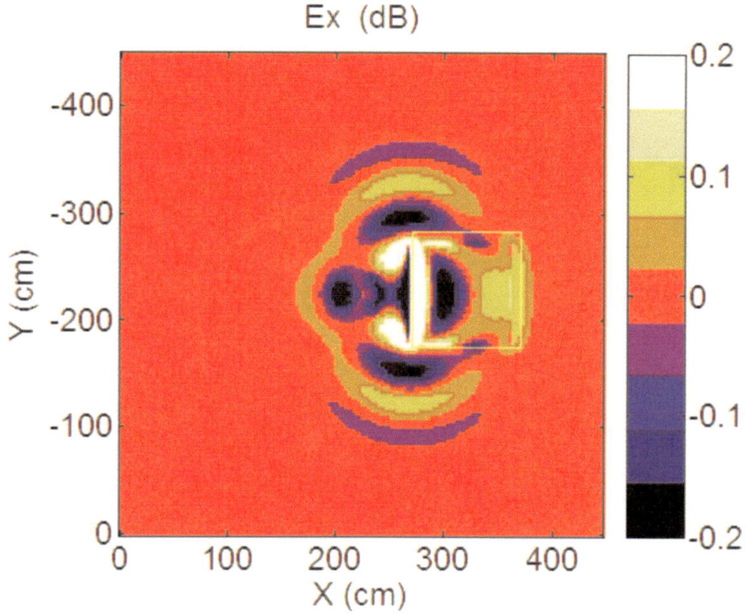

(f)

Fig. 8. Electric field simulated on the cross-sectional slice (XY-plane) at t = 90 nsec (the extent of the DNAPL pool is marked by a yellow box): Z-component of the electric field: a) Incident, b) Total, and c) Scattered X-component of the electric field: d) Incident, e) Total, and f) Scattered.

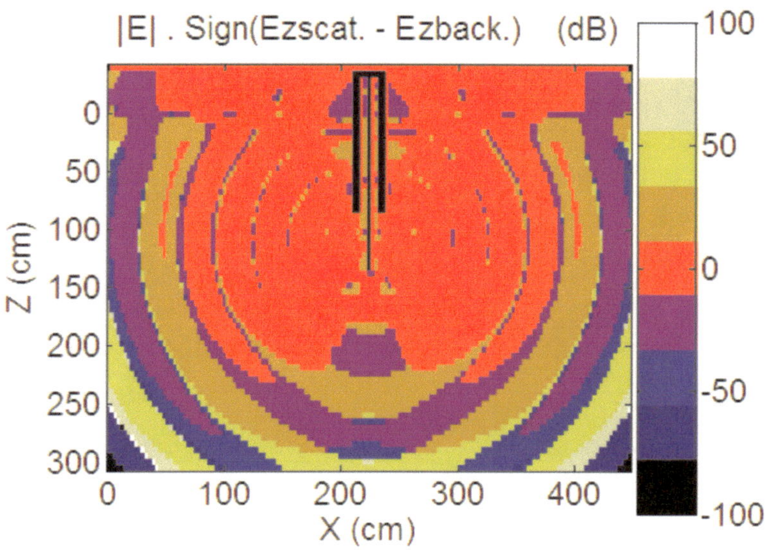

(a)

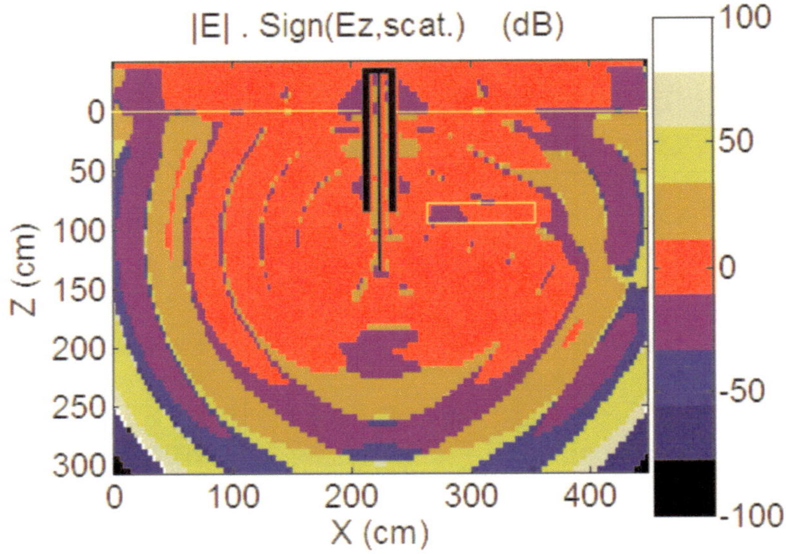

(b)

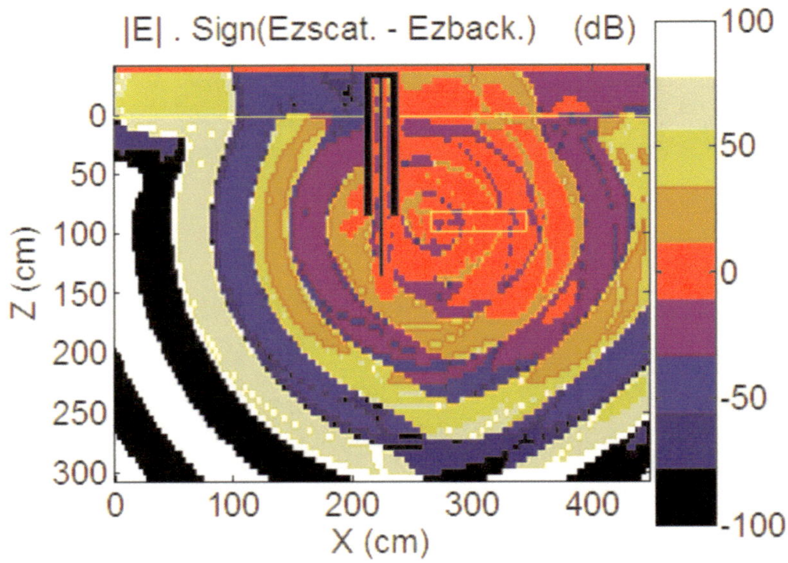

(c)

Fig. 9. Electric field, $[Sign(E_{z,Total} - E_{z,Incident})] \times 20\log_{10}(|E| \text{ or } |E_x \vec{i} + E_z \vec{k}|)$, on the depth slice (XZ-plane) at time t = 90 nsec (the extent of the DNAPL pool and soil-air interface are marked in yellow): a) Incident image, b) Total image, and c) Scattered image.

As before, for each slice and each electric field component, three figures are shown: (i) incident (background) field, (ii) total field, and (iii) scattered field. As seen in Figs. 8(a) and 8(d), background results are symmetric. The total fields of Figs. 8(b) and 8(e) and the scattered field shown in Figs. 8(c) and 8(f) are asymmetric. The interesting and encouraging point is the visibility of the DNAPL pool over the entire medium within the total and scattered fields, even on the far side of the pool from the antenna. This predicts that dipole antenna boreholes of the CWR method can be drilled far from DNAPL-contaminated zones to reduce the risk of further vertically downward penetration of DNAPLs associated with drilling through contaminated zones. This appears to be more valid at higher degrees of water-saturation. Again, this potential can also be demonstrated in a different form as shown in Fig. 9.

Previously, in the case of the monopole transmitter, the received total and incident signals were computed at a receiver located right above the DNAPL pool. Now, the two are computed for a receiver located 175 cm above the pool to examine the possibility of minimizing the destructive effect of placing the receiving antenna too close to the pool. This figure again demonstrates a strong magnitude and travel time difference between the total and incident fields received at the receiver located far above the DNAPL pool. The strong magnitude difference (around 40%) and time difference (around 2.5 nsec.) between the two signals, once again, embraces the potential of the use of the cross-borehole GPR method to detect DNAPL pools. As in the monopole case, the early arrival of the total field and its higher magnitude can be respectively justified by the higher velocity of EM waves through the DNAPL pool due to its lower dielectric permittivity (relative to the water-saturated soil) and the lower loss due to the pool's lower electrical conductivity. The dry soil case has been studied (it does not fit in the extent of this chapter) and proved to have weaker scattered signals, but still strong enough to have the potential to detect DNAPL pools using relatively widely spaced antennae.

As expected for the radiation pattern of dipoles, most of the energy is transmitted perpendicularly to the antenna through the mid-part of the antenna into the soil. The modeled dipole antenna is Z-polarized, and thus the Z-component of the electric field is the major component. The monopole antenna behaves similarly to the dipole antenna, the only difference being the strong signature of the DNAPL pool on the Z-component as well as X and Y components in the case of the dipole antennae. This wider spread perturbation due to the clutter promises more potential detection key-points for the dipole antenna installation. It is noteworthy that the perturbation due to the object on the X and Y components of the electric field of both monopole and dipole antennae appears more to the sides of the contaminated zone and perpendicular to the line connecting the center of the contaminated zone and the antennae. The perturbation on the Ez component in the monopole antenna case is distributed throughout the contaminated zone, while for the dipole one, this perturbation spreads to the far side across the contaminated zone as well.

6. Comparison with monopole experimentation

In this section, the numerically simulated monopole case is compared with experimentation for validation. The experimental setup (for more information refer to Farid et al., 2006) uses two PVC-cased ferrite-bead-jacketed monopole antennae connected to a vector network analyzer (Agilent 8714ES), and frequency-response measurements were collected for a homogeneous water-saturated sandy soil background. Fig. 11 shows a schematic of the experiment.

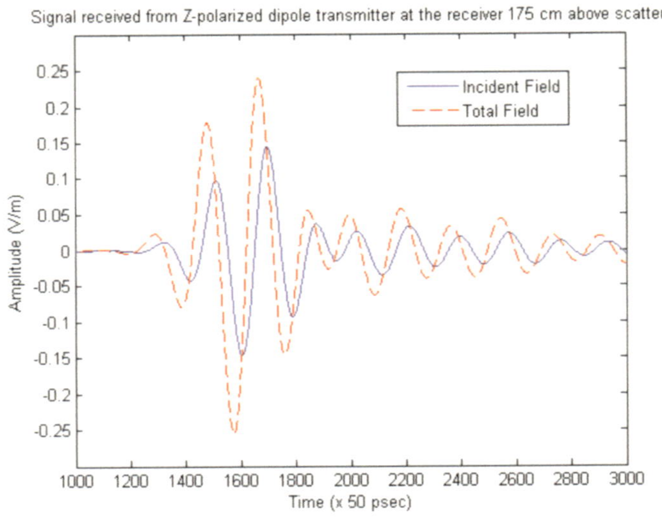

Fig. 10. Z-component of total and incident electric fields due to the dipole antenna, received at a receiver located 175 cm above the DNAPL pool.

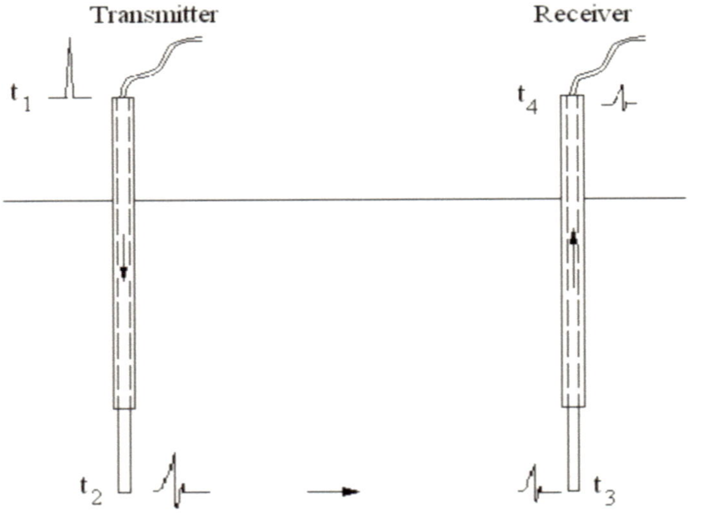

Fig. 11. Pulse traveling through transmitter, soil and then receiver

The frequency-response measurements are collected across the frequency range of 0.4 to 2.2 GHz. The frequency-response measurements are, then, transformed to the time domain using an inverse fast Fourier transform (IFFT) and an assumption of a narrow-width, wideband Gaussian pulse as the transmitted signal. Both the experiment and the FDTD model use the same Gaussian pulse source. Due to the frequency range used in the experimentation (0.4 GHz to 2.2 GHz), the width of the Gaussian signal should not exceed a maximum of: $W_{min.} = 1 / f_{max.} = 1 / 2.2$ GHz = 0.455 nsec. Using narrower signals seems to create a noise. Therefore, the upper limit (0.455 nsec.) is selected as the width of the signal. Fig. 12 shows the transmitted signal.

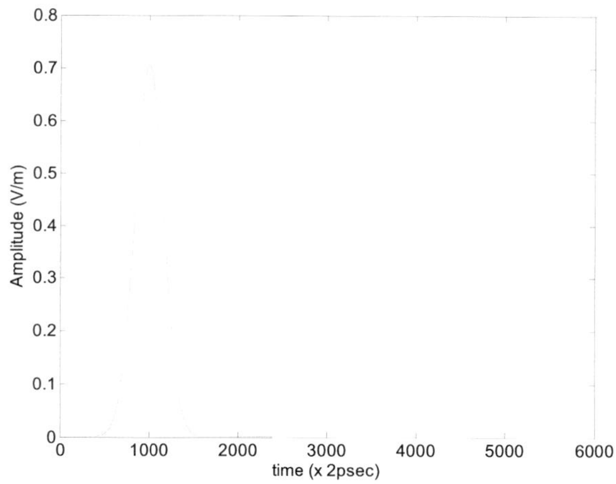

Fig. 12. X- and Y-components of the radial excitation at the end of the feed cable into the top of the transmitting antenna ($E_{x1} = E_{y1}$, = 0.7 V/m, $E_{z1} = 0$, and $|E| = 1$ V/m)

The FDTD-simulated received signal at the bottom of the receiving antenna is shown in Fig. 13(a). As seen, this received signal is distorted and does not resemble the Gaussian transmitted one. Therefore, its peak is not easily distinguishable, since the received signal is modulated and noisy. To resolve this issue, the received signal should first be demodulated and then low-pass-filtered. The demodulation frequency can be found by observing the received signal in the frequency domain (computed via a fast Fourier transform). This demodulation frequency is observed to be dependent on the separation between the transmitting and receiving antennae. A MATLAB code was prepared to automatically observe the received signals in the frequency domain, find the proper demodulation frequencies, and find the proper low-pass filter to filter the noise. The processed received signal is shown in Fig. 13(b).

The travel time of the FDTD model can be simply calculated as the difference between the peak times of the received (Fig. 13(b)) and transmitted (Fig. 12) signals.

To transform the experimental frequency-response to the travel-time, the transmitted signal is fast Fourier transformed to the frequency domain and multiplied by the frequency-response (Fig. 14) to find the received signal at the receiver in the frequency domain. Then, the result is transformed back to the time domain using an inverse fast Fourier transform. The result (received signal in the time domain) is shown in Fig. 15(a).

This signal does not resemble the transmitted Gaussian signal. Therefore, it needs to be processed (demodulated and low-pass-filtered). The processed received signal is shown in Fig. 15(b). The demodulation frequency and filter design vary with the distance between the transmitting and receiving antennae, which is automatically calculated using the above-mentioned MATLAB code.

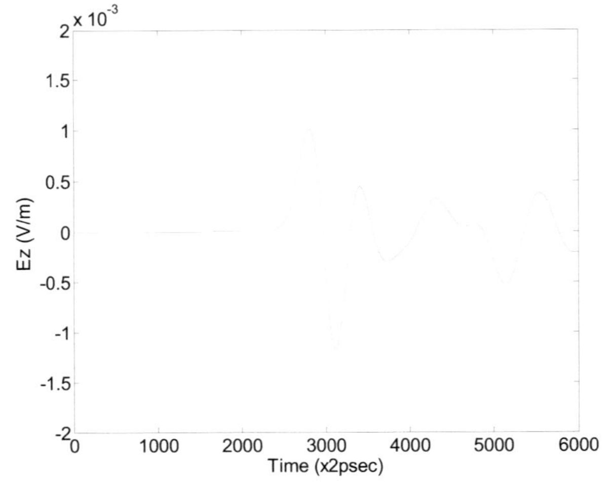

(a)

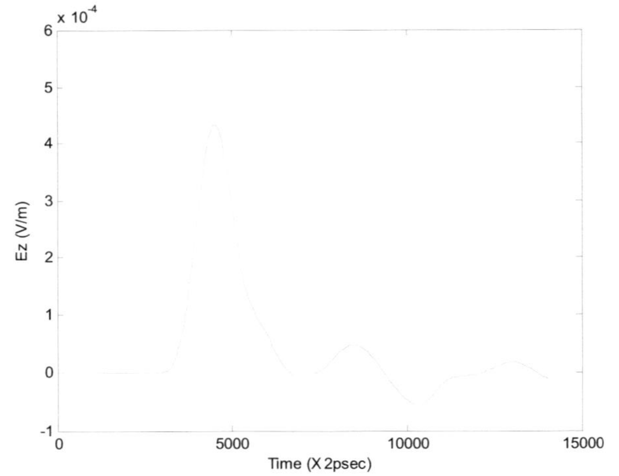

(b)

Fig. 13. Received signal (E_{z3}) at the bottom of the receiver in the saturated background from the FDTD simulation at time t_3: a) Unprocessed, and b) Processed.

The travel time of the experimental model can be simply calculated as the difference between the peak times of the received (Fig. 15(b)) and transmitted (Fig. 12) signals.
Since the vector network analyzer is calibrated at the end of the cables (the connection points to the monopole antennae), the experimental travel-time is measured between these two points. On the other hand, the FDTD travel-time (t_3 $-t_1$) is computed between the feed

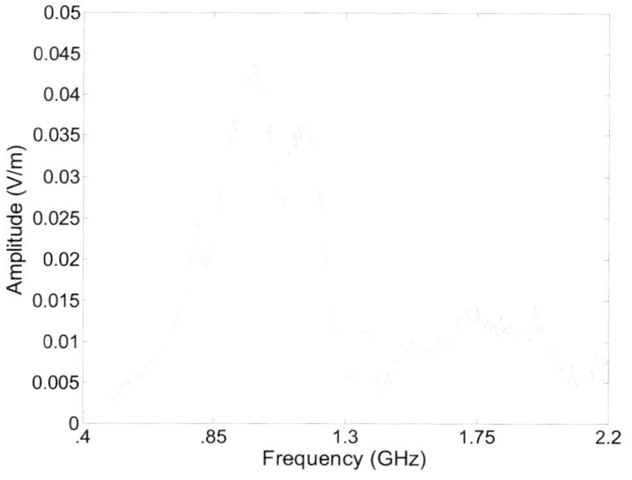

Fig. 14. Experimental frequency-response in water-saturated background soil

cable and the bottom of the receiving antenna. Therefore, it needs to be adjusted for this difference.

Up to this point, the FDTD travel-time (t_3 – t_1) from the feed cable to the tip of the receiving antenna is computed. The travel time through the receiving antenna (t_4 – t_3), which is by symmetry equal to (t_2 – t_1), should be added to (t_3 – t_1) to find the total travel time between the feed and receiver cables (t_4 – t_1) for the FDTD model. The resulting travel time from the FDTD simulation can be used for comparison with the experimental results.

The travel time computed from the forward model is *(4500 + 900 - 1000) × 2 psec = 8.8 nsec*, which closely agrees with the one indirectly computed from the experimentally collected frequency-response data: *(5700 - 1000) × 1.87 psec = 8.6 nsec*. The difference is due to the slight, potential discrepancy between the dielectric constant assigned to the forward model (used from the results of another work by the authors (Zhan et al., 2007)) and the real values of the experimentation.

The intensities of the unprocessed received signals from the FDTD simulation (Fig. 13(a)) and experimentation (Fig. 15(a)) agree relatively well, but not perfectly. The reason is the potential slight discrepancy between the electrical conductivity assigned to the FDTD model compared to the actual one of the experiment. However, due to the difference between the necessary processing methods (different filters), the intensity of the processed received signals for the FDTD simulation (Fig. 13(b)) and the one of the experiment (Fig. 15(b)) do not agree as closely.

This comparison consists of the incident field for the homogeneous background soil. The comparison for the total and scattered fields at the presence of any anomalies (e.g., dielectric objects) will be conducted in the future.

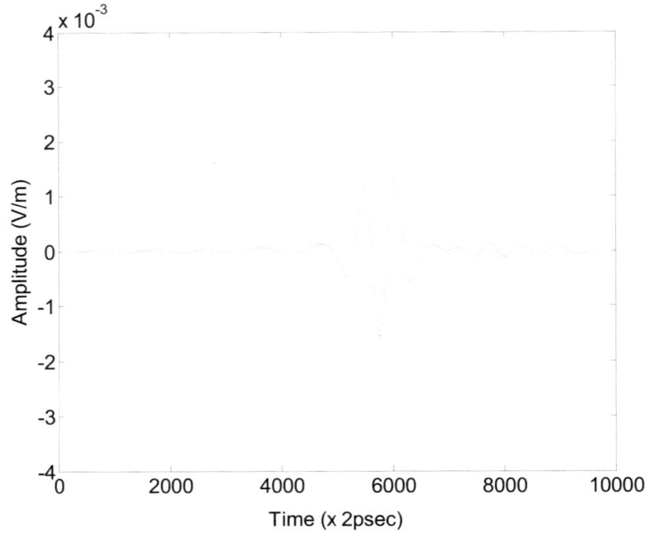

(a)

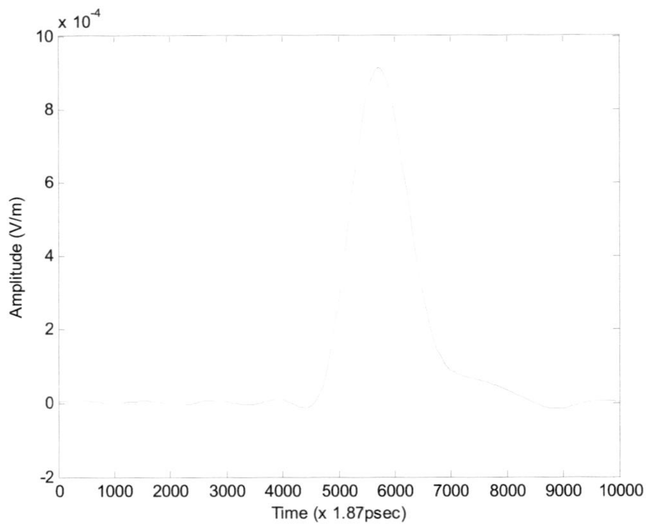

(b)

Fig. 15. Received signal (E_{z4}) at the top of the receiver in the saturated background, indirectly computed from the experimental frequency-response: a) Unprocessed, and b) Processed.

7. Conclusion

A finite difference time domain (FDTD) model was developed for monopole and dipole antennae. Then, the scattering due to dielectric materials (to simulate DNAPL pools) in soils was modeled and analyzed. Results of the two simulated cases using the FDTD model demonstrate strong perturbation by the DNAPL pool on the electric field in the fully water-saturated sandy soil. In the case of the monopole antenna, the DNAPL pool target is more visible on the X and Y components of the electric field compared to the major component Z. The perturbation on the intensity of the electric field ($|E|$) transmitted by the monopole antenna is not as strongly visible as in the dipole case. In the dipole case, X and Y components are those parallel to likely hydraulic-conductivity contrast planes (*e.g.*, usually horizontal clay lenses within a thick sand layer), which are potential locations to accumulate DNAPLs.

Different components of the electric field can selectively be collected using receiving antennae with different polarizations from the polarization of the transmitting antenna (*e.g.*, a horizontally-polarized receiving monopole antenna and a vertically-polarized transmitting monopole antenna). Therefore, designing the receiving antenna alignment and polarization to selectively collect electric field components parallel to a possible DNAPL pool may help to compensate for a stronger perturbation on the minor components (X and Y) of the electric field emitted from a Z-polarized monopole antenna. These minor components should be of a high enough signal to noise ratio.

In the case of the dipole antenna, all three components of the electric field in the fully water-saturated soil have almost equal detection potential. In both of the above cases, there is a strong dielectric contrast between the DNAPL pool and the water-saturated soil. However, different radiation patterns of the dipole antenna compared to the monopole antenna may make the dipole antenna more desirable for DNAPL detection.

Field problems can be scaled down in size along with scaling up the frequency in non-dispersive soils to achieve the proper geometry and frequency for simulation purposes. This linear scaling of frequency and size may not work as well for dispersive soils, since frequency-dependent dielectric properties of dispersive soils add nonlinearity to the scaling problem. Other conclusions follow.

- Images provided by such simulations show the field distribution that exists throughout the subsurface (i.e., similar to filling the entire volume with receiver antennae), but the field can only be observed practically by placing a reasonable number of receiving antennae at key underground positions with the appropriate polarization. This research can be used to find the radiation patterns of different antenna types and the interaction of the radiated field with soil heterogeneities, which leads to a better understanding of subsurface wave behavior at these key positions and aids the selection of optimum antenna patterns to cover these key positions.
- While the depth of contamination is a problem for surface-reflection methods (e.g., GPR), there are no theoretical depth limitations for CWR, except practical drilling limitations and cost. The separation limitations between transmitting and receiving antennae used for CWR still exist. However, CWR has the advantage of using a one-way traveling path (transmission), unlike the two-way traveling path of surface-reflection GPR. In addition, the strong reflecting air-soil interface in the

surface-reflection GPR technique is eliminated in the CWR technique and replaced with a better-controlled coupling between the borehole antennae and surrounding soil.

- The perturbation due to the DNAPL target is stronger for the greater dielectric permittivity contrast between DNAPL pools and highly moist soil, as opposed to DNAPL plumes with low DNAPL saturation and dryer soils.
- The signal to noise ratio of the scattered field by DNAPL pools should be high enough for measurements. As seen in the figures, the scattered field is comparable to the incident field. Therefore, if the signal to noise ratio of the incident field is high enough for measurement, the scattered field will probably have a large enough signal to noise ratio to be measurable as well.
- The results of this forward model with monopole and dipole antennae show that the field perturbation (scattered = total - incident) for relatively large DNAPL pools at high enough DNAPL saturation, is of the same order of magnitude as the incident signal. This proves DNAPL detection using CWR in water-saturated soils feasible. The simulation tool can also be used as a forward model to develop an inverse scheme for DNAPL imaging.
- Armed with the background data as well as the radiation patterns of different antennae (via simulations like those in this chapter), the existence of DNAPL pools can be confirmed with efficient inverse models and judicious placement of receiving antennae (i.e., pattern of antenna installation) where stronger perturbation and reception by receiving antennae are expected.

CWR may be a feasible and reasonable method to monitor DNAPL pools in a suitable environment. This most suitable environment is a medium consisting of a low-loss, low-heterogeneity porous material. In other media, it is more difficult to distinguish DNAPL accumulation from geologic variations, which are more complicated due to heterogeneity. Nevertheless, soil heterogeneity may not pose a crucial problem under water-saturated conditions since different soils behave similarly at relatively high degrees of water-saturation and high frequencies (the case is different for low frequencies). Monitoring DNAPL movement may well be possible or easier in an even less saturated heterogeneous environment because of the static nature of stratigraphic events and the dynamic nature of DNAPL flow. Several features of DNAPL pools may help to distinguish them from stratigraphic events, such as their irregular shapes with sharp lateral boundaries.

Finally, the FDTD model was compared for the incident field due to the monopole case in a homogeneous water-saturated sandy soil background with the experimental results. The reasonable agreement between both the travel time and intensity of the unprocessed, simulated and experimental results validates the FDTD model. The comparison and validation for the total and scattered fields at the presence of any anomalies (e.g., dielectric objects) need to be studied in the future.

8. Acknowledgement

This research was supported in part by the Gordon Center for Subsurface Sensing and Imaging Systems (CenSSIS), under the Engineering Research Centers Program of the National Science Foundation (NSF: Award Number EEC-9986821).

The authors would like to express gratitude for financial and scientific support provided by the Gordon CenSSIS and NSF.

9. References

Ajo-Franklin, J. B., Geller, J. T. & Harris, J. M. (2004). The dielectric properties of granular media saturated with DNAPL/water mixtures. *Geophysical Research Letters (GRL)*, Vol. 31, No. 17, L17501

Anderson, J. & Peltola, J. (1996). *Ground Penetrating Radar as a tool for detecting contaminated areas: Groundwater Pollution Premier*, CE 4594 Soil and Groundwater Pollution, Civil Engineering Department, Virginia Tech., Date of Access: Feb/2011, Available from: <http://www.cee.vt.edu/program_areas/environmental/teach/gwprimer/gprjp/gprjp.html#Intro>

Arulanandan, K. (1964). Dielectric method for prediction of porosity of saturated soils. *ASCE Journal of Geotechnical Engineering*, Vol. 117, No. 2, pp. 319-330

Arulanandan, K. & Smith, S. S. (1973). Electrical dispersion in relation to soil structure. *ASCE Journal of Soil Mechanics and Foundation Div.*, Vol. 99, No. 12, pp. 1113-1133

Balanis, C. A. (1989). *Advanced engineering electromagnetic,* John Wiley & Sons, New York, 1008p

Belli, K., Rappaport, C., Zhan, H. & Wadia-Fascetti, S. (2009). Effectiveness of 2D and 2.5D FDTD Ground Penetrating Radar Modelling for Bridge Deck Deterioration Evaluated by 3D FDTD. *IEEE Transactions on Geoscience and Remote Sensing*, Vol. 47, No. 11, pp. 3656 - 3663.

Belli, K., Rappaport, C. & Wadia-Fascetti, S. (2009a). Forward Time Domain Ground Penetrating Radar Modelling of Scattering from Anomalies in the Presence of Steel Reinforcements. *Research in Nondestructive Evaluation*, Vol. 20, No. 4, pp. 193 - 214.

Binley, A., Winship, P. & Middleton, R. (2001). High resolution characterization of vadose zone dynamics using Cross-Borehole Radar. *Water Resource Research*, Vol. 37, no. 11, pp. 2639-2652

Blackhawk Geoservices Inc. (2008). *Integrated geophysical detection of DNAPL source zones*, Final Report, Date of Access: Feb/2011, Available from: <http://handle.dtic.mil/100.2/ADA409159>

Bradford, J. H. & Wu, Y. (2007). Instantaneous spectral analysis; time-frequency mapping via wavelet matching with application to 3D GPR contaminated site characterization. *The Leading Edge*, Vol. 26, pp. 1018-1023

Brewster, M. L. & Annan, A. P. (1994). GPR monitoring of a controlled DNAPL release, 200 MHz radar. *Geophysics*, Vol. 59, No. 8, pp. 1211-1221

Daniels, J. J., Roberts, R. & Vendl, M. (1992). Site studies of Ground Penetrating Radar for monitoring petroleum product contaminants. *Proceedings of SAGEEP (Symposium of the Applications of Geophysics to Engineering and Environmental Problems)*, Society of Engineering Mine Exploration, pp. 597–609

Dobson, M. C., Ulaby, F. T., Hallikainen, M. T. & El-Rayes, M. A. (1985). Microwave dielectric behavior of wet soil, Part II: Dielectric mixing models. *IEEE Transaction on Geoscience and Remote Sensing*, GE- 23, No. 1, pp. 35–46

Farid, A. M., Alshawabkeh, A. N. & Rappaport, C. M. (2006). Calibration and Validation of a Laboratory Experimental Setup for CWR in Sand. *ASTM, Geotechnical Testing Journal*, Vol. 29, Issue 2, pp. 158-167

Firoozabadi, R., Miller, E., Rappaport, C. & Morgenthaler, A. (2007). A New Inverse Method for Subsurface Sensing of Object under Randomly Rough Ground Using Scattered Electromagnetic Field Data. *IEEE Transactions on Geoscience and Remote Sensing*, Vol. 45, No. 1, pp. 104-117.

Gandhi, O. (1993). A frequency dependent FDTD (Finite Difference Time Domain) formulation for general dispersive media. *IEEE Transactions on Microwave Theory and Techniques*, Vol. 41, pp. 658-665

Geller, J. T., Kowalsky, M. B., Seifert, P. K. & Nihei, K. T. (2000). Acoustic detection of immiscible liquids in sand. *Geophysical Research Letters*, Vol. 27, No. 3, pp. 417-420, 2000

Grant, I. S. & Philips, W. R. (1990). *Electromagnetism*, John Wiley & Sons, New York, 525 pp

Grimm, R. & Olhoeft, G. (2004). Cross-hole complex resistivity survey for PCE at the SRS A-014 outfall. *Proceedings of SAGEEP*, Colorado Springs, Colorado, 2004, pp. 455-464

Hallikainen, M. T., Ulaby, F. T., Dobson, M. C., El-Rayes, M. A. & Lin-Kun, W. (1985). Microwave dielectric behavior of wet soil: Part I- Empirical models and experimental observations. *IEEE Transaction on Geoscience and Remote Sensing*, GE-23, No. 1, pp. 25–34

Hipp, J. (1974). Soil electromagnetic parameters as functions of frequency, soil density, and soil moisture. *Proceedings of IEEE*, Vol. 62, No.1, pp. 98-103

Hoekstra, P. & Doyle, W. T. (1971). Dielectric relaxation of surface adsorbed water. *Journal of Colloid and Interface Science*, Vol. 36, No. 4, pp. 513-521

Hoekstra, P. & Delaney, A. (1974). Dielectric properties of soils at UHF and microwave frequencies. *Journal of Physics Research*, Vol. 79, No. 11, pp. 1699–1708

Interstate Technology and Regulatory Cooperation (ITRC) (2000). *Work Group DNAPLs/Chemical Oxidation Work Team, Dense Non-Aqueous Phase Liquids (DNAPLs)*, Review of Emerging Characterization and Remediation Technologies, Technology Overview, <http://www.itrcweb.org/DNAPL-1.pdf>

Kashiwa, T. & Fukai, I. (1990). A treatment by the FDTD method for the dispersive characteristics associated with electronic polarization. *Microwave and Guided Wave Letters*, Vol. 16, pp. 203-205

Kosmas, P. (2002). *Three-dimensional finite difference time domain modeling for Ground Penetrating Radar applications*, M.Sc. Thesis, Northeastern University, Boston, MA

Kunz, K. & Luebbers, R. (1993). *The FDTD (Finite Difference Time Domain) method for electromagnetic*, CRC Press, Boca Raton, Florida

Mur, G. (1981). Absorbing boundary conditions for the finite-difference approximation of the time-domain electromagnetic field equations. *IEEE Transactions on Electromagnetic Compatibility*, EMC- 23, pp. 377-382

Rappaport, C. M. & Winton, S. (1997). Modeling dispersive soil for FDTD computation by fitting conductivity parameters. *12th Annual Review of Progress in Applied Computational Electromagnetics Symposium Digest*, pp. 112-118

Rappaport, C. M., Wu, S. & Winton, S. C. (1999). FDTD wave propagation in dispersive soil using a single pole conductivity model. *IEEE Transactions on Magnetics*, Vol. 35, pp. 1542-1545

Rinaldi, V. A. & Francisca, F. M. (1999). Impedance analysis of soil dielectric dispersion (1 MHz - 1 GHz). *ASCE Journal of Geotechnical and Geoenvironmental Engineering*, Vol. 125, No. 2, pp. 111-121

Sachs, S. B. & Spiegler, K. S. (1964). Radio-frequency measurements of porous plugs, ion exchange resin-solution systems. *Journal of Physical Chemistry*, Vol. 68, pp. 1214-1222

Selig, E. T. & Mansukhani, S. (1975). Relationship of soil mixture to the dielectric property. *ASCE Journal of Geotechnical Division*, Vol. 101, No. 8, pp. 755–770

Sen, P. N., Scala, C. & Cohen, M. H. (1981). A self-similar model for sedimentary rocks with application to the dielectric constant of fused glass beads. *Geophysics*, Vol. 46, pp. 781-795

Sheriff, R. E. (1989). *Geophysical methods,* Prentice Hall, New Jersey, 605 p

Smith-Rose, R. L. (1933). The electrical properties of soils for altering current at radio frequencies. *Proceedings of Royal Society*, Vol. 140, No. 841A, pp. 359-377

Smith-Rose, R. L. (1935). The electrical properties of soils at frequencies up to 100 megacycles per second; with a note on resistivity of ground in the United Kingdom. *Proceedings of Physical Society*, Vol. 47, No. 262, pp. 923-931

Sneddon, K. W., Olhoeft, G. R. & Powers, M. H. (2000). Determining and mapping DNAPL saturation values from noninvasive GPR measurements. *Proceedings of SAGEEP*, Arlington, Virginia, pp. 293-302

Talbot, J. & Rappaport, C. M. (2000). An efficient Mur-type ABC for lossy scattering media. *Progress in Electromagnetics Research Symposium*, Vol. 194

Thevanayagham, S. (1995). Frequency-domain analysis of electrical dispersion of soils. *ASCE Journal of Geotechnical Engineering*, Vol. 121, No. 8, pp. 618-628

Von Hippel, A. R. (1953). *Dielectric materials and applications,* Technology Press of M.I.T. and John Wiley, New York

Weast, R. C. (1974). *CRC Handbook of Chemistry and Physics,* 55th edition, CRC Press, Cleveland, OH

Weedon, W. & Rappaport, C. M. (1997). A general method for FDTD modeling of wave propagation in arbitrary frequency-dispersive media. *IEEE Transactions on Antenna and Propagation*, pp. 401-410

Wikipedia, Date of Access: Feb/2011, Available from: <http://en.wikipedia.org/wiki/Dipole_antenna>

Yee, K. (1966). Numerical solution of initial boundary value problems involving Maxwell's equations in isotropic media. *IEEE Transaction on Antennae and Propagation*, Vol. 14, No. 3, pp. 302-307

Zhan, S. H., Farid, A., Alshawabkeh, A. N., Raemer, H. & Rappaport, C. M. (2007). Validated Half-Space Green's Function Formulation for Born Approximation for Cross-Well Radar Sensing of Contaminants. *IEEE, Transaction of Geoscience and Remote Sensing*, Vol. 45, No. 8, pp. 2423-2428, August

Part 2

Extended Einstein's Field Equations for Electromagnetism

6

General Relativity Extended

Gregory L. Light
Providence College, Providence Rhode Island
USA

1. Introduction

We extend Einstein's General Relativity in two ways:
(1) Einstein Field Equations ("*EFE*") explain gravity by energy distributions over space-time, but they can also explain electromagnetism by charge distributions in like manner. This is not to be confused with the well-known Einstein-Maxwell equations, in which electromagnetic fields' energy contents are added onto those as attributed to the presence of matter, to account for gravitational motions; in short, we are here substituting the term "electric charge" for energy, and electromagnetism for gravity, i.e., a geometrization of the electromagnetic force.
(2) *EFE* describe *one* space-time, but we propose *two*: one for "particles" and the other for "waves;" to wit, there are two gravitational constants and we have unified the gravitational motions in a "combined space-time 4-manifold."
In Section 2, we shall prove that electromagnetic fields as produced by charges, in analogy with gravitational fields as produced by energies, cause spacetime curvatures, not because of the energy contents of the fields but because of the Coulomb potential of the charges; as a result, we shall derive a special constant of proportionality between an electromagnetic energy-momentum tensor and Einstein tensor, to arrive at

$$R_{\mu\nu,em} - \frac{1}{2} R_{em} \cdot g_{\mu\nu,em}^{att;\ rep} = -\frac{16\pi G}{\left(1 - \gamma_{grav}^{-2} \cdot g_{11,grav}\right) c^5} T_{\mu\nu,em}^{att;\ rep}. \tag{1}$$

The geodesics of the resultant electromagnetic 4-manifold represent the same dynamics as that given by the classical Lagrangian resulting in the Lorentz force law of motion.
In Section 3, we define "combined manifold" $\mathcal{M}^{[3]}$ as the graph of a diffeomorphism from one manifold $\mathcal{M}^{[1]}$ to another $\mathcal{M}^{[2]}$, akin to the idea of a diagonal map. We derive the values for: (1) the energy distribution between a particle in $\mathcal{M}^{[1]}$ and its accompanied electromagnetic wave in $\mathcal{M}^{[2]}$, for the combined entity $[particle, wave]$, and (2) the gravitational constant $G2$ for $\mathcal{M}^{[2]}$, where there exist only electromagnetic waves and gravitational forces. Because of a large $G2$, an astronomical black hole B arose in $\mathcal{M}^{[2]}$, branching out $\mathcal{M}^{[1]}$ (the *Big Bang*), with a portion of a wave energy in $\mathcal{M}^{[2]}$ transferred to $\mathcal{M}^{[1]}$ as a photon, which collectively were responsible for the subsequent formation of matter. Being within the Schwarzschild radius, B in $\mathcal{M}^{[2]}$ is a complex (sub) manifold, which furnishes exactly the geometry for the observed quantum mechanics; moreover, B provides an energy interpretation to quantum probabilities in $\mathcal{M}^{[1]}$. In brief, our $\mathcal{M}^{[3]}$ casts quantum mechanics in the framework of General Relativity.
In Section 4, we draw a summary.

2. *EFE* for Electromagnetism

2.1 Background

In this Section 2 we derive Einstein Field Equations for electromagnetism and unite it with gravity in one common explicit form of *EFE*. Since Einstein's success in geometrizing gravity in General Relativity, a major drive has been the search for a unified geometric field theory (for some of the latest many attempts, see, e.g., [14, 24, 33]). A brief account here is in order. In about 1920 Kaluza and Klein proposed a 5-dimensional manifold combining Maxwell equations with *EFE*; the idea was soon put aside due to the emergence of quantum mechanics, which revealed two other fundamental forces of nature: the strong and the weak nuclear forces. Nevertheless, the construct of a "curled-up" dimension eventually resurfaced later in string theories.

In about the same time, Weyl introduced the idea of gauge invariance of conformal Riemannian geometry, which later led to Yang-Mills theory, supersymmetry, quantum field theories, and the unified M string theory by Witten (cf. [36]). A basic premise underlying these developments has been that in order to deal with the periodic nature as inherent in electrodynamics a complex structure is indispensable, thus opening up Clifford algebra, Finsler geometry, *Kähler* manifolds (see, e.g., [25]), and Calabi-Yau spaces, all involving dimensions higher than $\mathbb{R}^4$ - - the suitability of which in describing the physical universe has been increasingly questioned in recent literature (cf. e.g., [33]).

Amid the above intensive elaborate mathematical research, as is well known, gravity remains resistant to unification, where the electroweak theory has been established by Winberg, Glashow and Salem since the late 1960s and the electrostrong theory has been treated under the subject of quantum chromodynamics.

A distinct feature of gravity is the existence of the principle of equivalence between inertial masses and gravitational masses, so that the two cancel out and the size of the inertial mass does not need to be addressed explicitly. Here we shall solve the problem of the lack of the same principle for electromagnetism (cf. [5]) via the denominator of the constant of proportionality

$$\kappa_{em} = -\frac{16\pi G}{\left(1 - \gamma_{grav}^{-2} g_{11,grav}\right) c^5}. \tag{2}$$

In this connection, we also make a distinct identification of $T_{11,em}^{att;rep}$ with the norm of the Poynting vector (cf. [1] for a discussion of the Poynting vector), and as a result the derived geodesics correspond to the least action by Feynman. In that we have demonstrated a Poynting vector on the right-hand-side of *EFE* being in direct correspondence with a minimization of the integral of kinetic energy minus potential energy over all trajectories on the left, we see the reasons why any other identifications of $T_{\mu\nu,em}$ have resulted in difficulties in geometrizing electromagnetism or else have led to the above-mentioned other geometries. In this regard, our $T_{11,em}^{att;rep}$ has unit $joule/\left(second \cdot meter^2\right)$, representing energy flows in a specific direction across an area of square meter per second, and yet the common identification of $T_{11,em}$ with the energy densities has unit $joule/\left(meter^3\right)$ (see, e.g., [35], 45, equation (2.8.10)), representing stationary energies, but the energy-momentum tensor is defined for energy flows. Here we cite [7]: "An important problem is to determine the flow energy along a given direction for a given physical field. This description uses a 2-covariant symmetric tensor field T_{ij}, called the energy-momentum tensor. The energy flow in the X direction is given by the expression

$$T(X,X) = T_{ij}X^iX^j.\text{" } ([7], 75, \text{equation } 5.3.25) \tag{3}$$

As such, it comes as no surprise that our $T^{att;rep}_{11,em}$ directly leads to the least action, from which follows the Lorentz force law governing the general nonquantum electrodynamics (see [15], II-19-7).

Our approach here in this paper is to pay careful attention to the intricate details laying the foundation of Special Relativity, General Relativity, and electromagnetism and to underscore the essential logic that connects these three topics. Following Einstein, we make use of the differential geometric property of Einstein tensor

$$\mathcal{E}_{\mu\nu} := R_{\mu\nu} - \frac{1}{2}R \cdot g_{\mu\nu} \tag{4}$$

being proportional to energy-momentum tensor $T_{\mu\nu}$ (cf. [21], 858) and apply weak field approximations (see [12], 814-818) to establish the constant of proportionality κ_{em} as based on weakly attractive or repulsive electromagnetic fields (cf. [35], 151-157 for a derivation of EFE). As such, there will be numerous "approximately-equal" signs in our derivation of κ_{em}; nevertheless, the derived value of κ_{em} is *exact*.

The significance of our results is that the distribution of electric charges in space-time results in a 4-manifold $\mathcal{M}^4_{em}$ of curvatures and charges move along geodesics of $\mathcal{M}^4_{em}$, i.e., a geometrization of the electromagnetic force, which is a step toward a unified field theory (for related work integrating electromagnetism with EFE, cf. e.g., [29, 31]).

Our derivation below will first aim at deriving g_{em} (proving that the associated geodesics are exactly the classical electromagnetic Lagrangian), then $\mathcal{E}_{em}$, and finally

$$\frac{\mathcal{E}_{12,em}}{\mathcal{E}^{att;rep}_{11,em}} = \frac{-\|\bar{\mathbf{g}}\|\,\mathbf{V}_{Q,x}}{\pm\|\bar{\mathbf{S}}\|} \equiv \frac{T_{12,em}\,(\text{momentum})}{T^{att;rep}_{11,em}\,(\text{energy})}, \tag{5}$$

to obtain

$$\kappa_{em} = \frac{\mathcal{E}_{11}}{T_{11}}. \tag{6}$$

To go one step further, we will also unite electromagnetism with gravity in one set of *EFE* to arrive at

$$\mathcal{E}_{\mu\nu} := R_{\mu\nu} - \frac{1}{2}R \cdot g_{\mu\nu} = -\frac{8\pi G}{c^2}T_{\mu\nu,grav} \mp \frac{16\pi G}{\left(1-\gamma^{-2}g_{11,grav}\right)c^5}T^{att;rep^*}_{\mu\nu,em}. \tag{7}$$

2.2 Derivations

Definition 1. *The Minkowski space*

$$\mathbb{R}^{1+3} \; : = \{\, (t, \mathbf{x} \equiv (x,y,z)) \in \mathbb{R}^4 \mid \textit{the inner product} \tag{8}$$

$$\left\langle \mathbf{e}_i, \mathbf{e}_j \right\rangle \; : = \mathbf{e}_i^T \eta \mathbf{e}_j,\; i,j = 1,2,3,4, \tag{9}$$

$$\eta \; : = diag\left(1, -c^{-2}, -c^{-2}, -c^{-2}\right)_{\mathbf{E}}, \tag{10}$$

$$\mathbf{E} \equiv \left(\mathbf{e}_i \equiv (\textit{Kronecker } \delta_{i1}, \delta_{i2}, \delta_{i3}, \delta_{i4})\right)_{i=1}^4, \tag{11}$$

$$c \equiv \textit{the speed of light in the vacuum}\}. \tag{12}$$

The proper time τ_o of any reference frame O

$$\textit{is such that } \tau_o(O) \equiv (\tau_o, 0, 0, 0). \tag{13}$$

Remark 1. *If $\mathcal{M}^4 = \mathbb{R}^{1+3}$, then f = the Lorentz transformation $\mathcal{L}$; $\mathcal{L} : S \longrightarrow \tilde{S}$ has the following matrix representation if $(t, x, y, z) = (0,0,0,0) = (\tilde{t}, \tilde{x}, \tilde{y}, \tilde{z})$ and $\mathcal{L}(1, V, 0, 0) = (\tilde{t}_o, 0, 0, 0)$, $V \in \mathbb{R}$:*

$$L = \gamma \begin{pmatrix} 1 & -\frac{V}{c^2} \\ -V & 1 \end{pmatrix}_{(\mathbf{e}_1, \mathbf{e}_2)}, \tag{14}$$

where $(V, 0, 0)$ is the velocity of $\tilde{S}$ relative to S and

$$\gamma \equiv \left(1 - \left(\frac{V}{c}\right)^2\right)^{-\frac{1}{2}} \in [1, \infty) \tag{15}$$

is the Lorentz factor. Consider an emission of light at $t_o = 0 = \tilde{t}_o$ in the direction of $V \in \mathbb{R}$; then $\forall t_o, \tilde{t}_o > 0$ S observes $(t_o, t_o c)$ and $\tilde{S}$ observes $(\tilde{t}_o, \tilde{t}_o c)$; further,

$$L (t_o, t_o c)^T = \gamma \left(1 - \frac{V}{c}\right) \cdot (t_o, t_o c)^T = (\tilde{t}_o, \tilde{t}_o c)^T ; \tag{16}$$

thus,

$$\frac{\tilde{t}_o}{t_o} = \gamma \left(1 - \frac{V}{c}\right) = \lambda, \text{ an eigenvalue of } L. \tag{17}$$

Note that

$$\gamma \left(1 - \frac{V}{c}\right) \cdot \gamma \left(1 + \frac{V}{c}\right) = 1; \tag{18}$$

i.e., L has two eigenvalues

$$\lambda_{\max} = \gamma \left(1 + \frac{|V|}{c}\right) > 1, \text{ and} \tag{19}$$

$$\lambda_{\min} = \gamma \left(1 - \frac{|V|}{c}\right) < 1. \tag{20}$$

Remark 2. *At this point, we alert the reader to be aware of the existence of* three *identities: (1) the reader (or the analyst), who serves as the laboratory frame O and sets up a local parametrization*

$$f : U_{(0,\mathbf{0})} \subset \mathbb{R}^{1+3} \longrightarrow \text{ the space-time } 4 - manifold\ \mathcal{M}^4, \tag{21}$$

(2) S, and (3) $\tilde{S}$.

Remark 3. *In the above Equation (14), if $V = 0$, then $\mathcal{L} = I$. Consider now $V(t) \equiv 0\ \forall t \in (-\infty, 0]$; however, $\forall t \in (0, T]$, we have $V(t) \approx at$ for some $T > 0$ and some constant acceleration $a > 0$, due to the existence of some force. Then*

$$\lambda = \frac{\tilde{t}_o}{t_o} \approx \gamma(t) \left(1 - \frac{V(t)}{c}\right) \tag{22}$$

measures the curvatures of $\mathcal{M}^4$ over $(0, T]$. This treatment of λ will play a vital role in our subsequent derivations. Since $V(t) \approx at > 0$ on $(0, T]$, we have

$$\lambda \approx \sqrt{\frac{c - V(t)}{c + V(t)}} < 1. \tag{23}$$

By Einstein's General Relativity, a clock undergoing a gravitational free fall slows down (e.g., consider a clock approaching a black hole). As such, we conclude that $\lambda < 1$ for attractive forces; by a reversal of time in the preceding dynamics, we deduce that $\lambda > 1$ for repulsive forces. We will thus make the following distinction and notation:

$$\lambda_{att} := \gamma\left(1 - \frac{|V|}{c}\right) < 1, \text{ and} \tag{24}$$

$$\lambda_{rep} := \gamma\left(1 + \frac{|V|}{c}\right) > 1. \tag{25}$$

Further, note that $\forall \left(\frac{V}{c}\right) \approx 0$, one uses

$$\frac{m_o}{\lambda_{att}} \approx m_o\gamma \text{ and} \tag{26}$$

$$\frac{m_o}{\lambda_{rep}} \approx m_o\gamma^{-1} \tag{27}$$

for (Special) relativistic adjustment of a mass. Also, a metric g on $\mathcal{M}^4$ by definition is such that

$$g_{11} \approx \left(\frac{\tilde{t}_o}{t_o}\right)^2 \approx \left(\lambda_{att;\,rep}\right)^2 = \lambda_{att}^{\pm 2}. \tag{28}$$

Remark 4. *Let $p_1, p_2 \in \mathcal{M}^4$; then a maximization of*

$$\int_{f^{-1}(p_1)}^{f^{-1}(p_2)} \frac{d\tilde{t}_o}{dt_o} dt_o \tag{29}$$

over all trajectories $\{(t, x(t), y(t), z(t))\}$ derives the geodesic from p_1 to p_2 maximizing the proper time elapsed in $\tilde{S}$.

Proposition 1. *Let g be a local metric of $\mathcal{M}^4$ and express g as a matrix in the basis of $\mathbf{B} \equiv \left\{\frac{\partial f}{\partial t}, \frac{\partial f}{\partial x}, \frac{\partial f}{\partial y}, \frac{\partial f}{\partial z}\right\}$; if $f \approx \mathcal{L}$ (i.e., $\mathcal{M}^4$ is near flat), then*

$$\frac{d\tilde{t}_o}{dt_o} = (1,0,0,0)\ g_{\mathbf{B}}\ \left(\mp 1, \mathbf{V}_x, \mathbf{V}_y, \mathbf{V}_z\right)^T. \tag{30}$$

Proof. Without loss of generality, consider

$$L = \gamma\begin{pmatrix} 1 & \pm\frac{\mathbf{V}}{c^2} \\ \pm\mathbf{V} & 1 \end{pmatrix} \tag{31}$$

and calculate $(1,0)\ g_{\mathbf{B}}\ (\mp 1, \mathbf{V})$

$$= (1,0)\left(\left(L^{-1}\right)^T\right)^{-1}\left[\left(L^{-1}\right)^T g_{\mathbf{B}}\, L^{-1}\right] L\left(\mp 1, \mathbf{V}\right)^T \tag{32}$$

$$\approx (1,0)\left(\gamma\begin{pmatrix} 1 & \pm\mathbf{V} \\ \pm\frac{\mathbf{V}}{c^2} & 1\end{pmatrix}\right)\begin{pmatrix} 1 & 0 \\ 0 & -\frac{1}{c^2}\end{pmatrix}\begin{pmatrix}\Delta\tilde{t}_o \\ 0\end{pmatrix} \tag{33}$$

$$= \left(\gamma, \mp\frac{\gamma\mathbf{V}}{c^2}\right)\begin{pmatrix}\Delta\tilde{t}_o \\ 0\end{pmatrix} \text{ (observe that } L : (\mp 1, \mathbf{V})^T \longmapsto (\Delta\tilde{t}_o, 0)^T, \tag{34}$$

where $\Delta\tilde{t}_o$ is the proper time of $\tilde{S}$ by definition)

$$= \frac{\Delta\tilde{t}_o}{\sqrt{1-\left(\frac{\mathbf{V}}{c}\right)^2}} = \frac{\Delta\tilde{t}_o}{\left\|(\mp 1, -\mathbf{V})^T\right\|_\eta} = \frac{\Delta\tilde{t}_o}{\left\|L^{-1}(\mp 1, -\mathbf{V})^T\right\|_\eta} \tag{35}$$

$$= \frac{\Delta\tilde{t}_o}{\Delta t_o} \approx \frac{d\tilde{t}_o}{dt_o}, \text{(where } L^{-1} : (\mp 1, -\mathbf{V})^T \longmapsto (\Delta t_o, 0)^T, \tag{36}$$

analogous to the above Equation (34)).

■

The Setup - -

We consider the dynamics of a charge Q at $(0,0,0,0) \in U$ that attracts or repels a charge q at $(0,x,y,z) \in U$, where

$$r_\infty \equiv \sqrt{(x^2+y^2+z^2)} \text{ is such that } r_\infty^{-1} \approx 0. \tag{37}$$

Theorem 1. *(Feynman [15], II-28-2) The field momentum produced by Q is*

$$\mathbf{P}(t) = \frac{Q^2}{4\pi\epsilon_o r_o c^2}\mathbf{V}_Q(t), \tag{38}$$

where $\epsilon_o \equiv$ the permittivity constant $\approx \frac{1}{9\times 4\pi}\times 10^{-9}\times\frac{coulomb^2\cdot second^2}{kilogram\cdot meter^3}$, $r_o \equiv$ the "classical electron radius" $\approx 2.82\times 10^{-15}$ meter, and $\mathbf{V}_Q(t) << c$ is the velocity of Q at t.

Remark 5. *We note that the above Equation (38) was derived in [15] by an integration over the (continuous) field energy densities (cf. [15], II-28-2 and II-8-12). Thus, to apply Equation (38) to the above* Setup *of exactly two (discrete) point charges, we must have*

$$Q = q = \text{ the smallest charge } = \text{ an electron.} \tag{39}$$

Definition 2.

$$\text{The average } \textit{field momentum density } \bar{\mathbf{g}}(t) := \mathbf{P}(t) \,/\, \left(\frac{4\pi r_\infty^3}{3}\right). \tag{40}$$

Theorem 2. *(Feynman [15], II-27-9) The Poynting vector $\mathbf{S}$ is related to the momentum density $\mathbf{g}$ by*

$$\mathbf{g} = \frac{1}{c^2}\mathbf{S}. \tag{41}$$

Corollary 1.

$$\mathbf{P}(t) = \left(\frac{4\pi r_\infty^3}{3}\right)\bar{\mathbf{g}}(t) \tag{42}$$

$$= \left(\frac{4\pi r_\infty^3}{3}\right)\frac{\bar{\mathbf{S}}(t)}{c^2}. \tag{43}$$

$$\textit{where } \bar{\mathbf{S}}(t) \equiv \textit{the } \text{average } \textit{field energy flow in the direction} \tag{44}$$

$$\textit{of } \mathbf{V}_Q(t)\textit{, with unit equal to } \left(\frac{joule}{second \cdot meter^2}\right). \tag{45}$$

Theorem 3. *(Feynman [15], II-27-11: Conservation of the total momentum of particles and field)*

$$\mathbf{P}(t) \equiv m_{Q,o}\mathbf{V}_Q(t) = -m_{q,o}\mathbf{V}(t), \tag{46}$$

where $m_{Q,o}$ and $m_{q,o}$ are respectively the rest masses of Q and q.

Remark 6. *The Newton's law of motion as adjusted for the effect of Special Relativity is*

$$\mathbf{F}^{att;\, rep} = \left(\gamma^{\pm 1} m_o\right)\left(\gamma^{\pm 2}\mathbf{a}\right) \tag{47}$$

respectively for attractive and repulsive force $\mathbf{F}^{att;\, rep}$ if $\mathbf{a}$ is in the direction of $\mathbf{V}$ (cf. [23], Equation (13.31), 272-273; also, Equations (26),(27) above).

Proposition 2. *Let $v(t) := \|\mathbf{V}(t)\|$ and $v_Q(t) := \|\mathbf{V}_Q(t)\|$; then*

$$\gamma^{\pm 2}\left(\frac{v(t)}{c}\right) = \frac{\textit{the electric potential energy } PE_e \textit{ of } Q \textit{ and } q}{\textit{the rest energy } RE \textit{ of } q}. \tag{48}$$

Proof. By Theorems 1 and 3,

$$\left(\frac{v(t)}{c}\right) = \left(\frac{1}{m_{q,o}c^2}\right)\cdot q\left(\frac{Q}{q}\frac{v_Q(t)}{c}\frac{r_\infty}{r_o}\right)\cdot\frac{Q}{4\pi\epsilon_o r_\infty} \tag{49}$$

$$\equiv \frac{1}{RE}\cdot K\cdot\frac{qQ}{4\pi\epsilon_o r_\infty}, \tag{50}$$

where

$$K \equiv \frac{Q}{q}\frac{v_Q(t)}{c}\frac{r_\infty}{r_o} = \frac{v_Q(t)\cdot\left(\frac{r_\infty}{c}\right)}{r_o} \text{ (cf. Remark 5)} \tag{51}$$

is an electrodynamic adjustment factor of the electrostatic potential (cf. [15], II-15-14, 15);

$$K = 1 \text{ if } v_Q(t)\cdot\left(\frac{r_\infty}{c}\right) \equiv v_Q(t)\cdot t = r_o, \tag{52}$$

i.e., the point charge Q travels to the boundary of the "classical electron," or equivalently, Q is a stationary electron. Thus, taking into account the effect of Special Relativity, we have

$$\gamma^{\pm 2}\left(\frac{v(t)}{c}\right) = \frac{\gamma^{\pm 2}KQq/4\pi\epsilon_o r_\infty}{RE} = \frac{PE_e}{RE}. \tag{53}$$

■

Corollary 2.

$$-\gamma^{\pm 2}\left(\frac{v(t)}{c}\right)\left(\frac{v(t)\,v_Q(t)}{c^2}\right)=\frac{q\mathbf{V}(t)\cdot\mathbb{A}(t)}{RE}, \tag{54}$$

where $\mathbb{A}(t) :=$ *the vector potential, or curl* $\mathbb{A}(t) =$ *the magnetic field* $\mathbb{B}$.

Proof. Since

$$-v(t)\,v_Q(t) = \mathbf{V}(t)\cdot\mathbf{V}_Q(t) \text{ and} \tag{55}$$

$$\frac{\gamma^{\pm 2}KQ\mathbf{V}_Q(t)}{4\pi\epsilon_o r_\infty c^2} = \mathbb{A}(t)\ \ ([15]\,,\text{II-14-4}), \tag{56}$$

we have

$$-\gamma^{\pm 2}\left(\frac{v(t)}{c}\right)\left(\frac{v(t)\,v_Q(t)}{c^2}\right) \tag{57}$$

$$=\frac{\gamma^{\pm 2}KQ\,q\mathbf{V}(t)\cdot\mathbf{V}_Q(t)}{RE\cdot 4\pi\epsilon_o r_\infty c^2}=\frac{q\mathbf{V}(t)\cdot\mathbb{A}(t)}{RE}. \tag{58}$$

■

Definition 3. *We call an electromagnetic field attractive if the total potential energy is negative, and repulsive if the total potential energy is positive.*

Proposition 3. *For any weakly attractive or repulsive electromagnetic field, the metric* $g_{em}^{att;\,rep}$ *has the following matrix representation in the basis of* **B** *(refer to Proposition 1 above):*

$$g_{em}^{att;\,rep}=\begin{pmatrix} \lambda_{em}^{\pm 2} & -\frac{2\gamma^{\pm 2}v_Q\mathbf{V}_x}{c^3} & -\frac{2\gamma^{\pm 2}v_Q\mathbf{V}_y}{c^3} & -\frac{2\gamma^{\pm 2}v_Q\mathbf{V}_z}{c^3} \\ -\frac{2\gamma^{\pm 2}v_Q\mathbf{V}_x}{c^3} & o\left(\frac{v}{c}\right)-c^{-2} & o\left(\frac{v}{c}\right)^3 & o\left(\frac{v}{c}\right)^3 \\ -\frac{2\gamma^{\pm 2}v_Q\mathbf{V}_y}{c^3} & o\left(\frac{v}{c}\right)^3 & o\left(\frac{v}{c}\right)-c^{-2} & o\left(\frac{v}{c}\right)^3 \\ -\frac{2\gamma^{\pm 2}v_Q\mathbf{V}_z}{c^3} & o\left(\frac{v}{c}\right)^3 & o\left(\frac{v}{c}\right)^3 & o\left(\frac{v}{c}\right)-c^{-2} \end{pmatrix}. \tag{59}$$

Proof. First, we note that besides being symmetric, $g_{em}^{att;\,rep}\longrightarrow\eta$, as $\mathbf{V},\mathbf{V}_Q\longrightarrow\mathbf{0}$. Second,

$$g_{11,em}^{att;\,rep}=\lambda_{em}^{\pm 2}\approx\left(\frac{\tilde{t}_o}{t_o}\right)^2_{att;\,rep}\quad\text{(cf. Equation (28))}. \tag{60}$$

Third, by Proposition 1 we have

$$\frac{d\tilde{t}_o}{dt_o}=(1,0,0,0)\ g_{em}^{att;\,rep}\ \left(\mp 1,\mathbf{V}_x,\mathbf{V}_y,\mathbf{V}_z\right)^T \tag{61}$$

$$=\mp\lambda^{\pm 2}-\frac{2\gamma^{\pm 2}v_Q v^2}{c^3} \tag{62}$$

$$\approx\mp\gamma^{\pm 2}\left(1\mp\frac{2v}{c}\right)+\frac{2q\mathbf{V}\cdot\mathbb{A}}{RE}\text{ (by Corollary 2)} \tag{63}$$

$$=\mp\gamma^{\pm 2}+2\gamma^{\pm 2}\left(\frac{v}{c}\right)+\frac{2q\mathbf{V}\cdot\mathbb{A}}{RE} \tag{64}$$

$$\equiv\mp\left(\frac{1}{1-\left(\frac{v}{c}\right)^2}\right)^{\pm 1}+\frac{2\left(PE_e+q\mathbf{V}\cdot\mathbb{A}\right)}{RE}\text{(by Proposition 2);} \tag{65}$$

here we note that $q\mathbf{V}\cdot\mathbb{A}$ is not to be identified with the magnetic potential energy since the magnetic force being always orthogonal to the velocity of q does not do any work; nevertheless, we will henceforth set $PE_e + q\mathbf{V}\cdot\mathbb{A} \equiv PE_{em}$ for presentation brevity (cf. [30], 84, where PE_{em} is noted for the term "generalized potential" energy). To continue, we thus have

$$\frac{d\tilde{t}_o}{dt_o} \approx \mp\left(\frac{1}{1-\left(\frac{v}{c}\right)^2}\right)^{\pm 1} + \frac{2\left(PE_e + q\mathbf{V}\cdot\mathbb{A}\right)}{RE} \tag{66}$$

$$\approx \mp\left(1 \pm \left(\frac{v}{c}\right)^2\right) + \frac{2PE_{em}}{RE} \tag{67}$$

$$= \mp 1 - \frac{m_o v^2}{m_o c^2} + \frac{2PE_{em}}{RE} \tag{68}$$

$$= \mp 1 - \frac{2\left(\text{kinetic energy } KE - PE_{em}\right)}{RE}, \tag{69}$$

which is equivalent to Feynman's least action for the classical electrodynamics since a maximization of

$$\int_{f^{-1}(p_1)}^{f^{-1}(p_2)} \frac{d\tilde{t}_o}{dt_o} dt_o = \int (PE - KE)\, dt_o \tag{70}$$

is equivalent to a minimization of $\int (KE - PE)\, dt_o$ (cf. Equation (29), and [15], II-19-7). ∎

Remark 7. *Applying the same proof as above, we can also incidentally derive for any weak gravitational field the following results (which will be used later):*

$$g_{grav} \approx diag\left(\lambda^2_{grav}, -c^{-2}, -c^{-2}, -c^{-2}\right)_{\mathbf{B}}, \tag{71}$$

and

$$(1,0,0,0) \circ g_{grav,4\times 4,\mathbf{B}} \circ \left(-1, \mathbf{V}_x, \mathbf{V}_y, \mathbf{V}_z\right)^T$$

$$= -1 - 2\cdot\left(\frac{KE_{grav} - PE_{grav}}{RE}\right), \textit{with} \tag{72}$$

$$\frac{PE_{grav}}{RE} := \frac{m_o \cdot \left(\frac{\gamma^2 GM}{r^2}\right)\cdot r}{m_o c^2} = \gamma^2\left(\frac{a_{grav}}{c}\right) t \tag{73}$$

$$= \gamma^2\left(\frac{v}{c}\right). \tag{74}$$

We note that in the literature (e.g., [23], 288, 294) one finds that

$$\frac{d\tilde{t}_o}{dt} = \left(1, \mathbf{V}_x, \mathbf{V}_y, \mathbf{V}_z\right) \circ g_{4\times 4,\mathbf{E}} \circ \left(1, \mathbf{V}_x, \mathbf{V}_y, \mathbf{V}_z\right)^T, \tag{75}$$

where $g_{4\times 4,\mathbf{E}}$ measures the norm of the motion $\left(1, \mathbf{V}_x, \mathbf{V}_y, \mathbf{V}_z\right)$ on the parameter domain U and pass it onto $\|\cdot\|_{T_p\mathcal{M}^4} \equiv \left\|\left(\Delta\tilde{t}_o, 0, 0, 0\right)^T_{\tilde{\mathbf{E}}}\right\|_{T_p\mathcal{M}^4}$; we instead adhere to the standard treatment in differential geometry to express g as $g_{4\times 4,\mathbf{B}}$ on $T_p\mathcal{M}^4$, to project $\frac{\partial f}{\partial t}$ onto the proper time $\Delta\tilde{t}_o$ in the tangent space.

Corollary 3. *The Einstein tensor*

$$\mathcal{E}_{em}^{att;\,rep} \approx \begin{pmatrix} \mp\frac{6v}{r_k^2 c} & -\frac{6v_Q \mathbf{V}_x}{r_k^2 c^3} & -\frac{6v_Q \mathbf{V}_y}{r_k^2 c^3} & -\frac{6v_Q \mathbf{V}_z}{r_k^2 c^3} \\ -\frac{6v_Q \mathbf{V}_x}{r_k^2 c^3} & -O\left(r_k^{-2}\right) & O\left(r_k^{-2}c^{-4}\right) & O\left(r_k^{-2}c^{-4}\right) \\ -\frac{6v_Q \mathbf{V}_y}{r_k^2 c^3} & O\left(r_k^{-2}c^{-4}\right) & -O\left(r_k^{-2}\right) & O\left(r_k^{-2}c^{-4}\right) \\ -\frac{6v_Q \mathbf{V}_z}{r_k^2 c^3} & O\left(r_k^{-2}c^{-4}\right) & O\left(r_k^{-2}c^{-4}\right) & -O\left(r_k^{-2}\right) \end{pmatrix}_{\mathbf{B}}. \tag{76}$$

Proof. $\mathcal{E}_{\mu v} := R_{\mu v} - \frac{1}{2}R \cdot g_{\mu v}; \forall \mathcal{M}^4 \approx \mathbb{R}^{1+3}$ we have

$$\left(R_{\mu v}\right) \approx diag\left(-\frac{3}{r_K^2}, -\frac{1}{r_K^2}, -\frac{1}{r_K^2}, -\frac{1}{r_K^2}\right) \text{ and} \tag{77}$$

$$R \approx -\frac{6}{r_K^2}, \tag{78}$$

where $r_K \equiv$ the radius of sectional curvatures (cf. [21], 860; [35], 154). Thus, substituting Equation (59) into $\left(g_{\mu v}\right)$ in $\left(\mathcal{E}_{\mu v}\right)$, we arrive at the conclusion. ■

Lemma 4. *Denote the mass density of q by*

$$\bar{m}_{q,o} \equiv \frac{m_{q,o}}{\left(4\pi r_\infty^3/3\right)}; \tag{79}$$

then we have

$$\bar{m}_{q,o} r_\infty^2 \approx \left(1 - \gamma_{grav}^{-2} g_{11,grav}\right) \cdot \frac{3c^2}{8\pi G}, \tag{80}$$

where

$$g_{11,grav} \approx \lambda_{grav}^2 \approx \gamma_{grav}^2 \left(1 - \frac{2\mathbf{V}_\alpha}{c}\right), \tag{81}$$

with $\mathbf{V}_\alpha \equiv$ the radial velocity (> 0) of any arbitrary particle α gravitating toward q at a distance of r_∞, and $G \equiv$ the universal gravitational constant.

Proof.

$$g_{11,grav} \approx \lambda_{grav}^2 \approx \gamma_{grav}^2 \left(1 - \frac{2\mathbf{V}_\alpha}{c}\right) \text{ (refer to Eq. (24), (28))} \tag{82}$$

$$\approx \gamma_{grav}^2 \left(1 - \frac{2\mathbf{a}_\alpha t}{c}\right) \text{ (cf. Remark 2)} \tag{83}$$

$$= \gamma_{grav}^2 \left(1 - \frac{2G\bar{m}_{q,o}}{r_\infty^2 c} \cdot \frac{4\pi r_\infty^3}{3} \cdot \frac{r_\infty}{c}\right); \tag{84}$$

thus,

$$\bar{m}_{q,o} r_\infty^2 \approx \left(1 - \gamma_{grav}^{-2} g_{11,grav}\right) \cdot \frac{3c^2}{8\pi G}. \tag{85}$$

■

Remark 8. *The above lemma expresses the gravitating mass density of q in terms of its effect on $\mathcal{M}^4$ as measured by $g_{11,grav}$; by the principle of equivalence, $\bar{m}_{q,o}$ is also the inertial mass density, and in the next theorem $\bar{m}_{q,o}$ is to be treated as such. Also, note that as $r_\infty^{-1} \longrightarrow 0$, we have $\left|r_\infty^{-2} - r_K^{-2}\right| \longrightarrow 0$.*

Theorem 5.

$$\mathcal{E}_{\mu\nu,em}^{att;\,rep} := R_{\mu\nu,em} - \frac{1}{2}R_{em} \cdot g_{\mu\nu,em}^{att;\,rep} = -\frac{16\pi G}{\left(1-\gamma_{grav}^{-2} \cdot g_{11,grav}\right)c^5} T_{\mu\nu,em}^{att;\,rep}. \tag{86}$$

Proof.

$$\frac{\mathcal{E}_{12,em}}{\mathcal{E}_{11,em}^{att;\,rep}} = \pm\frac{1}{c^2}\left(\frac{v_Q}{v}\right)\mathbf{V}_x \text{ (by Equation (76))} \tag{87}$$

$$= \pm\frac{1}{c^2} \cdot \left(\frac{m_{q,o}}{m_{Q,o}}\right) \cdot \left(-\frac{m_{Q,o}}{m_{q,o}}\mathbf{V}_{Q,x}\right) \text{ (by Equation (46))} \tag{88}$$

$$= \frac{-\frac{\|\bar{\mathbf{S}}\|}{c^2}\mathbf{V}_{Q,x}}{\pm\|\bar{\mathbf{S}}\|} = \frac{-\|\bar{\mathbf{g}}\|\,\mathbf{V}_{Q,x}}{\pm\|\bar{\mathbf{S}}\|} \text{ (by Equation (43))} \tag{89}$$

$$\equiv \frac{T_{12,em}}{T_{11,em}^{att;\,rep}}, \tag{90}$$

where $T_{11,em}^{att;\,rep}$ and $T_{1j,em}$, $j = 2,3,4$, are respectively the energy-flow and the momentum densities. Thus,

$$\mathcal{E}_{em}^{att;\,rep} = \kappa_{em} T_{em}^{att;\,rep} \text{ has} \tag{91}$$

$$\kappa_{em} = \frac{\mathcal{E}_{11,em}^{att;\,rep}}{T_{11,em}^{att;\,rep}} = \mp\frac{6v}{r_k^2 c} / \pm\|\bar{\mathbf{S}}\| \text{ (by Equations (76), (90)),} \tag{92}$$

but

$$\|\bar{\mathbf{S}}\| = \frac{3c^2}{4\pi r_\infty^3} \cdot m_{q,o}v \text{ (by Equations (43), (46)),} \tag{93}$$

so

$$\kappa_{em} = -\frac{6}{r_K^2 c} \cdot \frac{4\pi r_\infty^3}{3c^2 m_{q,o}} \tag{94}$$

$$= -\frac{6}{r_\infty^2 c} \cdot \frac{1}{c^2 \bar{m}_{q,o}} \text{ (cf. Remark 8)} \tag{95}$$

$$= -\frac{6}{c^3} \cdot \frac{8\pi G}{\left(1-\gamma_{grav}^{-2} g_{11,grav}\right) \cdot 3c^2} \text{ (by the preceding Lemma 4)} \tag{96}$$

$$= -\frac{16\pi G}{\left(1-\gamma_{grav}^{-2} \cdot g_{11,grav}\right)c^5}. \tag{97}$$

■

Remark 9. $T_{11,em}^{att;\,rep} \equiv \pm\|\bar{\mathbf{S}}\|$ *has unit (recalling from Equation (45))*

$$\frac{joule}{second \cdot meter^2} \tag{98}$$

$$= \frac{kilogram \cdot meter^2}{second^2} \cdot \frac{1}{second \cdot meter^2} \tag{99}$$

$$= \frac{kilogram}{second^3}, \tag{100}$$

so that $\left(\kappa_{em} \cdot T_{11,em}^{att;rep}\right)$ *has unit*

$$= \frac{[G]}{[c^5]} \cdot \frac{kilogram}{second^3} \tag{101}$$

$$= \frac{meter^3}{kilogram \cdot second^2} \cdot \frac{second^5}{meter^5} \cdot \frac{kilogram}{second^3} \tag{102}$$

$$= \frac{1}{meter^2} = \left[\frac{1}{r_k^2}\right], \tag{103}$$

measuring the local curvatures of $\mathcal{M}_{em}^4$. *We emphasize that our* $T_{11,em}$ *represents energy flows in a specific direction across an area of square meter per second, which is different from the common identification of* $T_{11,em}$ *with stationary energy densities with unit:* $\left[joule/\left(meter^3\right)\right]$ *(see, e.g., [35], 45, equation (2.8.10)).*

Remark 10. *We can now obtain a geometric union of gravitation and electromagnetism to arrive at*

$$\mathcal{E}_{\mu\nu} := R_{\mu\nu} - \frac{1}{2}R \cdot g_{\mu\nu} = -\frac{8\pi G}{c^2}T_{\mu\nu,grav} \mp \frac{16\pi G}{\left(1-\gamma^{-2}g_{11,grav}\right)c^5}T_{\mu\nu,em}^{att;rep^*}, \tag{104}$$

where for expository neatness we set:

$$g_{\mu\nu,em}^{rep^*} \equiv g_{\mu\nu,em}^{rep}\ \forall\mu\nu \neq 1,\ g_{11,em}^{rep^*} \equiv -g_{11,em}^{rep} = -\lambda_{em}^{-2}; \tag{105}$$

$$T_{\mu\nu,em}^{rep^*} \equiv T_{\mu\nu,em}^{rep}\ \forall\mu\nu \neq 1,\ T_{11,em}^{rep^*} \equiv -T_{11,em}^{rep} = \left\|\bar{\mathrm{S}}(t)\right\|. \tag{106}$$

Theorem 6. *The set of Einstein Field Equations*

$$\mathcal{E}_{\mu\nu} := R_{\mu\nu} - \frac{1}{2}R \cdot g_{\mu\nu} = -\frac{8\pi G}{c^2}T_{\mu\nu,grav} \mp \frac{16\pi G}{\left(1-\gamma^{-2}g_{11,grav}\right)c^5}T_{\mu\nu,em}^{att;rep^*} \tag{107}$$

has solutions:

$$R_{\mu\nu} = R_{\mu\nu,grav} \pm R_{\mu\nu,em}, \tag{108}$$

$$R = R_{grav} + R_{em}, \tag{109}$$

$$\text{and } g_{\mu\nu} = w_{grav} \cdot g_{\mu\nu,grav} \pm w_{em} \cdot g_{\mu\nu,em}^{att;rep^*}, \tag{110}$$

$$\text{with } w_{grav} \equiv \frac{R_{grav}}{R} \text{ and } w_{em} \equiv \frac{R_{em}}{R} \equiv 1 - w_{grav}. \tag{111}$$

Proof. Consider the operation $\mathcal{E}_{\mu\nu,grav} \pm \mathcal{E}_{\mu\nu,em}^{att;rep}$ and denote

$$\frac{R_{grav} \cdot g_{\mu\nu,grav}}{R_{grav} + R_{em}} \pm \frac{R_{em} \cdot g_{\mu\nu,em}^{att;rep}}{R_{grav} + R_{em}} \tag{112}$$

by $g_{\mu\nu}\left(\equiv w_{grav} \cdot g_{\mu\nu,grav} \pm w_{em} \cdot g_{\mu\nu,em}^{att;rep}\right)$; we see that the operation of $\mathcal{E}_{\mu\nu,grav} \pm \mathcal{E}_{\mu\nu,em}^{att;rep}$ is valid if and only if $g_{\mu\nu}$ is form-invariant with respect to measuring geodesics, possessing the same energy interpretations as g_{grav} and $g_{em}^{att;rep}$. Here we have:

$$\begin{aligned}
&(1,0,0,0)\circ\left(w_{grav}\cdot g_{grav}+w_{em}\cdot g_{em}^{att}\right)\circ\left(-1,\mathbf{V}_x,\mathbf{V}_y,\mathbf{V}_z\right)^T && (113)\\
&\text{(cf. Equation (30) in Proposition 1)} && (114)\\
=\;& w_{grav}\cdot\left(-1-2\cdot\left(\frac{KE_{grav}}{RE}\right)+2\cdot\left(\frac{PE_{grav}}{RE}\right)\right)\\
&+w_{em}\cdot\left(-1-2\cdot\left(\frac{KE_{em}^{att}}{RE}\right)+2\cdot\left(\frac{PE_{em}^{att}}{RE}\right)\right) && (115)\\
&\text{(by equations (72) and (69))} && (116)\\
\equiv\;& -1-\frac{2KE_{gravem}^{att}}{RE}+\frac{2PE_{gravem}^{att}}{RE}, && (117)
\end{aligned}$$

where

$$KE_{gravem}^{att}\equiv w_{grav}\cdot KE_{grav}+w_{em}\cdot KE_{em}^{att},\text{ and} \quad (118)$$

$$PE_{gravem}^{att}\equiv w_{grav}\cdot PE_{grav}+w_{em}\cdot PE_{em}^{att}. \quad (119)$$

Now since

$$(-R_{11,em})-\frac{1}{2}R\cdot g_{11,em}^{rep^*}=\frac{-16\pi G}{\left(1-\gamma^{-2}g_{11,grav}\right)c^5}T_{11,em}^{rep^*} \quad (120)$$

and

$$\begin{aligned}
&(1,0,0,0)\circ g_{em}^{rep^*}\circ\left(-1,\mathbf{V}_x,\mathbf{V}_y,\mathbf{V}_z\right)^T\\
\equiv\;&(1,0,0,0)\circ g_{em}^{rep}\circ\left(1,\mathbf{V}_x,\mathbf{V}_y,\mathbf{V}_z\right)^T, && (121)
\end{aligned}$$

we have

$$\begin{aligned}
&(1,0,0,0)\circ\left(w_{grav}\cdot g_{grav}-w_{em}\cdot g_{em}^{rep^*}\right)\circ\left(-1,\mathbf{V}_x,\mathbf{V}_y,\mathbf{V}_z\right)^T && (122)\\
=\;& w_{grav}\cdot\left(-1-2\cdot\left(\frac{KE_{grav}}{RE}\right)+2\cdot\left(\frac{PE_{grav}}{RE}\right)\right)\\
&-w_{em}\cdot\left(1-2\cdot\left(\frac{KE_{em}^{rep}}{RE}\right)+2\cdot\left(\frac{PE_{em}^{rep}}{RE}\right)\right)\;\text{(equation (69))} && (123)\\
\equiv\;& -1-\frac{2KE_{gravem}^{rep}}{RE}+\frac{2PE_{gravem}^{rep}}{RE}, && (124)
\end{aligned}$$

where

$$KE_{gravem}^{rep}\equiv w_{grav}\cdot KE_{grav}-w_{em}\cdot KE_{em}^{rep},\text{ and} \quad (125)$$

$$PE_{gravem}^{rep}\equiv w_{grav}\cdot PE_{grav}-w_{em}\cdot PE_{em}^{rep}. \quad (126)$$

Consequently, $g_{\mu\nu}=w_{grav}\cdot g_{\mu\nu,grav}\pm w_{em}\cdot g_{\mu\nu,em}^{att;rep^*}$ is form-invariant in measuring geodesics, with identical interpretations of energies to that of $g_{\mu\nu,grav}$ and $g_{\mu\nu,em}^{att;rep}$. I.e.,

$$\mathcal{E}:=\mathcal{E}_{grav}\pm\mathcal{E}_{em}^{att;rep}=-\frac{8\pi G}{c^2}T_{grav}\mp\frac{16\pi G}{\left(1-\gamma^{\mp 2}g_{11,grav}\right)c^5}T_{em}^{att;rep^*} \quad (127)$$

results in a metric $g_{\mu\nu}$ that renders

$$g_{1\cdot} \circ (-1, \mathbf{V})^T = -1 - \frac{2KE_{gravem}}{RE} + \frac{2PE_{gravem}}{RE}. \tag{128}$$

■

Corollary 4.

$$\frac{\tilde{t}_o}{t_o} \approx 1 + \frac{KE_{gravem}}{RE} - \frac{PE_{gravem}}{RE}, \tag{129}$$

where

$$KE_{gravem} \equiv w_{grav} \cdot KE_{grav} \pm w_{em} \cdot KE_{em}^{att;rep} \tag{130}$$

$$\text{and } PE_{gravem} \equiv w_{grav} \cdot PE_{grav} \pm w_{em} \cdot PE_{em}^{att;rep}. \tag{131}$$

Proof. By Equation (28), $\lambda^2_{att;\,rep} \approx \left(\frac{\tilde{t}_o}{t_o}\right)^2$, but

$$g_{11,grav} \approx \lambda^2_{grav} \approx 1 + 2 \cdot \frac{KE_{grav}}{RE} - 2 \cdot \frac{PE_{grav}}{RE} \tag{132}$$

(cf. equation (72))

and

$$g_{11,em}^{att;rep^*} \approx \pm\lambda_{em}^{\pm 2} \text{ (cf. equation (59) and notation (105))} \tag{133}$$

$$= \pm\left(1 \pm 2 \cdot \frac{KE_{em}^{att;rep}}{RE} \mp 2 \cdot \frac{PE_{em}^{att;rep}}{RE}\right) \tag{134}$$

(cf. equation (69));

thus,

$$\left(\frac{\tilde{t}_o}{t_o}\right)^2 \approx g_{11} = w_{grav} \cdot g_{11,grav} \pm w_{em} \cdot g_{11,em}^{att;rep^*} \tag{135}$$

$$= w_{grav} \cdot \lambda^2_{grav} \pm w_{em} \cdot \left(\pm\lambda_{em}^{\pm 2}\right) \tag{136}$$

$$= w_{grav} \cdot \left(1 + 2 \cdot \frac{KE_{grav}}{RE} - 2 \cdot \frac{PE_{grav}}{RE}\right) + w_{em} \cdot \left(1 \pm 2 \cdot \frac{KE_{em}^{att;rep}}{RE} \mp 2 \cdot \frac{PE_{em}^{att;rep}}{RE}\right) \tag{137}$$

$$= 1 + 2 \cdot \frac{w_{grav} \cdot KE_{grav} \pm w_{em} \cdot KE_{em}^{att;rep}}{RE} - 2 \cdot \frac{w_{grav} \cdot PE_{grav} \pm w_{em} \cdot PE_{em}^{att;rep}}{RE}, \tag{138}$$

so that

$$\frac{\tilde{t}_o}{t_o} \approx 1 + \frac{KE_{gravem}}{RE} - \frac{PE_{gravem}}{RE}. \tag{139}$$

■

Remark 11. *In General Relativity the spacetime proportionality* $\left(\frac{\tilde{t}_o}{t_o}\right)$ *is a major point of interest, and we have derived the above analogous equation that integrates gravity with electromagnetism.*

3. *EFE* for the Quantum Geometry

3.1 Description

In this section we construct a "combined space-time 4-manifold $\mathcal{M}^{[3]}$" as the graph of a diffeomorphism from one manifold $\mathcal{M}^{[1]}$ to another $\mathcal{M}^{[2]}$, akin to the idea of a diagonal map. $\mathcal{M}^{[2]}$ consists solely of electromagnetic waves as described by Maxwell Equations for a free space (from matter), which with all its (continuous) field energy can exist independently; $\mathcal{M}^{[2]}$ predates $\mathcal{M}^{[1]}$. Due to a large gravitational constant $G^{[2]}$ in $\mathcal{M}^{[2]}$, an astronomical black hole $\mathbf{B} \subset \mathcal{M}^{[2]}$ came into being (cf. e.g., [10, 34], for formation of space-time singularities in Einstein manifolds), and resulted in $\mathcal{M}^{[1]} \times \mathbf{B}$ (i.e., the *Big Bang* - - when $\mathcal{M}^{[2]}$ branched out $\mathcal{M}^{[1]}$; cf. e.g., [16], for how a black hole may give rise to a macroscopic universe): photons then emerged in $\mathcal{M}^{[1]}$ with their accompanied electromagnetic waves existing in **B**. Any energy entity j in $\mathcal{M}^{[1]}$ is a particle resulting from a superposition of electromagnetic waves in **B** and

$$\text{the combined entity} \equiv [particle, wave] \tag{140}$$

$$\text{has energy } E_j^{[3]} = E_j^{[1]} + E_j^{[2]} \tag{141}$$

(where the term "particle wave" was exactly used in Feynman [15], "ghost wave - - Gespensterfelder" by Einstein [28, p. 287-288], and "pilot wave" by de Broglie). Particles in $\mathcal{M}^{[1]}$ engage in electromagnetic, (nuclear) weak, or strong interactions via exchanging virtual particles. Both particles and waves engage in gravitational forces separately and respectively in $\mathcal{M}^{[1]}$ and $\mathcal{M}^{[2]}$. Being within the Schwarzschild radius, **B** in $\mathcal{M}^{[2]}$ is a complex (sub) manifold, which furnishes exactly the geometry for the observed quantum mechanics in $\mathcal{M}^{[3]}$; moreover, **B** provides an energy interpretation to quantum probabilities in $\mathcal{M}^{[1]}$. In summary, $\mathcal{M}^{[3]}$ casts quantum mechanics in the framework of General Relativity and honors the most venerable tenet in physics - - the conservation of energy - - from the Big Bang to mini black holes.

3.2 Derivations

Definition 4. *Let* $j \in \mathbb{N}$*; a combined energy entity is* $E_j^{[3]} := E_j^{[1]} + E_j^{[2]}$*, where* $\forall i \in \{1,2\}$ $E_j^{[i]}$ *exerts and receives gravitational forces on and from* $\left\{E_k^{[i]} \mid k \in \mathbb{N} - \{j\}\right\}$.

Lemma 7. $\forall i \in \{1,2\}$ $\left\{E_j^{[i]} \mid j \in \mathbb{N}\right\}$ *form a space-time 4-manifold* $\mathcal{M}^{[i]}$ *that observes EFE:*

$$R_{\mu\nu}^{[i]} - \frac{1}{2}R^{[i]}g_{\mu\nu}^{[i]} = -\frac{8\pi G^{[i]}}{c^2}T_{\mu\nu}^{[i]}. \tag{142}$$

Proof. (By General Relativity.) ∎

Remark 12. *The long existing idea of dual mass is fundamentally different from that of our* $[particle, wave]$*; dual mass (see* [22, 27]*) is a solution of the above EFE for* $i = 1$ only.

Definition 5. *A combined space-time 4-manifold is*

$$\mathcal{M}^{[3]} := \left\{ \left(p^{[1]}, p^{[2]}\right) \in \mathcal{M}^{[1]} \times \mathcal{M}^{[2]} \mid h\left(p^{[1]}\right) = p^{[2]}, h = \textit{any diffeomorphism} \right\}. \quad (143)$$

Proposition 4. $\left\{E_j^{[3]} \mid j \in \mathbb{N}\right\}$ *form* $\mathcal{M}^{[3]}$.

Proof. $\forall j \in \mathbb{N}$ $E_j^{[3]}$ can be assigned with a coordinate point $\mathbf{u}_j \in U \subset \mathbb{R}^{1+3} \equiv$ the Minkowski space. Since $\forall i \in \{1,2\}$ $\mathcal{M}^{[i]}$ is a manifold, there exists a diffeomorphism $f^{[i]} : U \longrightarrow f^{[i]}(U) \subset \mathcal{M}^{[i]}$; i.e., $f^{[i]}\left(\mathbf{u}_j\right) = \mathbf{p}_j^{[i]} \in \mathcal{M}^{[i]}$, so that $\mathbf{p}_j^{[2]} = f^{[2]}\left(\mathbf{u}_j\right) = f^{[2]}\left(f^{[1]^{-1}}\left(\mathbf{p}_j^{[1]}\right)\right) = h\left(\mathbf{p}_j^{[1]}\right)$, with $h \equiv f^{[2]} \circ f^{[1]^{-1}}$ being a diffeomorphism. ∎

Theorem 8. *Any metric* $g_{\mu\nu}^{[3]}$ *for* $\mathcal{M}^{[3]}$ *is such that*

$$g_{\mu\nu}^{[3]} = \frac{G^{[2]}}{G^{[1]} + G^{[2]}} \cdot g_{\mu\nu}^{[1]} + \frac{G^{[1]}}{G^{[1]} + G^{[2]}} \cdot g_{\mu\nu}^{[2]}. \quad (144)$$

Proof. Since $g_{\mu\nu}^{[3]}$ is the inner product of the direct sum of the tangent spaces: $T_{p^{[1]}}\mathcal{M}^{[1]} \oplus T_{p^{[2]}}\mathcal{M}^{[2]}$, we have $g_{\mu\nu}^{[3]} = a \cdot g_{\mu\nu}^{[1]} + b \cdot g_{\mu\nu}^{[2]}$ for some $a, b \in \mathbb{R}$. Since $\forall i \in \{1,2,3\}$ $g_{11}^{[i]}$ is the *time* $\times$ *time* component of $g^{[i]}$, we have the well-known relation

$$g_{11}^{[i]} = 1 - \frac{2G^{[i]}M^{[i]}}{rc^2}, \quad (145)$$

implying at once that $a = w_1 \in (0,1)$ and $b = 1 - w_1$. Thus,

$$g_{11}^{[3]} = 1 - \frac{2G^{[3]}M^{[3]}}{rc^2} \quad (146)$$

$$= w_1\left(1 - \frac{2G^{[1]}M^{[1]}}{rc^2}\right) + (1 - w_1)\left(1 - \frac{2G^{[2]}M^{[2]}}{rc^2}\right) \quad (147)$$

$$= 1 - \frac{2w_1 G^{[1]}M^{[1]} + 2(1 - w_1)G^{[2]}M^{[2]}}{rc^2}, \quad (148)$$

implying that

$$G^{[3]}M^{[3]} \equiv G^{[3]}M^{[1]} + G^{[3]}M^{[2]} \quad (149)$$

$$= w_1 G^{[1]}M^{[1]} + (1 - w_1)G^{[2]}M^{[2]}. \quad (150)$$

Since $M^{[1]}$ and $M^{[2]}$ are arbitrary, we have

$$w_1 G^{[1]} = G^{[3]} = (1 - w_1)G^{[2]}, \quad (151)$$

$$\text{i.e., } w_1\left(G^{[1]} + G^{[2]}\right) = G^{[2]}, \quad (152)$$

$$\text{or } w_1 = \frac{G^{[2]}}{G^{[1]} + G^{[2]}} \text{ and } 1 - w_1 = \frac{G^{[1]}}{G^{[1]} + G^{[2]}}. \quad (153)$$

∎

Corollary 5.

$$G^{[3]} = \left(\frac{G^{[1]}G^{[2]}}{G^{[1]} + G^{[2]}} \right). \tag{154}$$

Corollary 6. *If* $\frac{G^{[1]}}{G^{[2]}} \approx 0$*, then:*

(1) $G^{[3]} \approx G^{[1]}$ and $w_1 \approx 1$;

(2) if $\left\{ E_j^{[2]} \mid j \in \mathbb{N} \right\}$ are contained within a radius R such that

$$g_{11}^{[2]} = 1 - \frac{2G^{[2]} \sum_j E_j^{[2]}}{Rc^4} < 0, \tag{155}$$

then the proper time ratio

$$\frac{\Delta t_0^{[2]}}{\Delta t_0^{[1]}} = \sqrt{g_{11}^{[2]}} \in \mathbb{C}, \tag{156}$$

i.e., $t_0^{[2]}$ carries the unit of $\sqrt{-1}$ second (by analytic continuation).

Remark 13. *If in addition to* $\left\{ E_j^{[3]} = E_j^{[1]} + E_j^{[2]} \mid j \in \mathbb{N} \right\}$ *there exist dark energies as defined by*

$$\left\{ \left(0, E_l^{[2]}\right) \mid l \in \mathbb{N} \right\} \tag{157}$$

in $\mathcal{M}^{[2]}$*, then the above Schwarzschild radius R is even larger.*

Remark 14. *Without our setup of* $\mathcal{M}^{[2]}$*, the subject of black holes necessarily has been about gravitational collapses within* $\mathcal{M}^{[1]}$ *due to high concentrations of matter. By contrast, our geometry is about a large* $G^{[2]}$ *that causes* $g_{11}^{[2]} < 0$ *over* $\mathbf{B} \subset \mathcal{M}^{[2]}$*; in* [16] *the authors showed the possibility that the interior of a black hole could "give rise to a new macroscopic universe;" that macroscopic universe is just our* $\mathcal{M}^{[1]}$*, and the black hole is* $\mathbf{B} \subset \mathcal{M}^{[2]}$*. As such, studies of the black hole interior are of great relevance to our construct of* $\mathcal{M}^{[1]} \times \left(\mathbf{B} \subset \mathcal{M}^{[2]} \right)$ *provided however that the analytic framework is free from the familiar premise of material crushing, or particles entering/escaping a black hole (as in Hawking radiation, see, e.g.,* [23]*; for a review of some of the research in the black hole interior, see, e.g.,* [2, 6, 8, 17]*).*

Corollary 7. $\forall \left\{ M^{[3]}, m^{[3]} \right\}$ *one has the following Newtonian limit:*

$$\begin{aligned} m^{[3]}\mathbf{a}^{[3]} = -[&\left(\frac{G^{[2]}}{G^{[1]} + G^{[2]}} \right) \left(\frac{G^{[1]}M^{[1]}m^{[1]}}{\|\mathbf{r}\|^2} \right) \\ &+ \left(\frac{G^{[1]}}{G^{[1]} + G^{[2]}} \right) \left(\frac{G^{[2]}M^{[2]}m^{[2]}}{\|\mathbf{r}\|^2} \right)] \cdot \frac{\mathbf{r}}{\|\mathbf{r}\|}, \end{aligned} \tag{158}$$

or

$$\mathbf{a}^{[3]} = -\frac{G^{[3]}M^{[3]}}{\|\mathbf{r}\|^2} \left(\frac{M^{[1]}}{M^{[3]}} \cdot \frac{m^{[1]}}{m^{[3]}} + \frac{M^{[2]}}{M^{[3]}} \cdot \frac{m^{[2]}}{m^{[3]}} \right) \frac{\mathbf{r}}{\|\mathbf{r}\|}. \tag{159}$$

Corollary 8. *If* $\frac{M^{[1]}}{M^{[3]}} = \frac{m^{[1]}}{m^{[3]}} \equiv \mu_1 \in (0,1)$, *then the laboratory-measured mass as denoted by* $\hat{M}$ *is such that*

$$\hat{M} = M^{[3]} \left({\mu_1}^2 + (1-\mu_1)^2\right). \tag{160}$$

Proof.

$$\mathbf{a}^{[3]} = -\frac{G^{[3]}\hat{M}}{\|\mathbf{r}\|^2}\frac{\mathbf{r}}{\|\mathbf{r}\|} \tag{161}$$

$$= -\frac{G^{[3]}M^{[3]}\left({\mu_1}^2 + (1-\mu_1)^2\right)}{\|\mathbf{r}\|^2}\frac{\mathbf{r}}{\|\mathbf{r}\|} \text{ (by Equation (159)).} \tag{162}$$

■

Corollary 9.

$$M^{[3]} = \frac{\hat{M}}{{\mu_1}^2 + (1-\mu_1)^2}, \tag{163}$$

$$M^{[1]} = \frac{\hat{M}\mu_1}{{\mu_1}^2 + (1-\mu_1)^2} \equiv \hat{M}\phi^{[1]}, \text{ and} \tag{164}$$

$$M^{[2]} = \frac{\hat{M}(1-\mu_1)}{{\mu_1}^2 + (1-\mu_1)^2} \equiv \hat{M}\phi^{[2]}. \tag{165}$$

Notation 1. *The above notation of an overhead caret, e.g.,* $\hat{E} = E^{[3]}({\mu_1}^2 + (1-\mu_1)^2)$ *for a laboratory-measured energy, will be used throughout the remainder of our Chapter; note in particular that a quantity multiplied by* $\phi^{[2]} \equiv \frac{(1-\mu_1)}{{\mu_1}^2+(1-\mu_1)^2}$, *e.g.,* $\hat{E}\phi^{[2]}$, *indicates a conversion from a laboratory established quantity into that part of the quantity as contained in* $\mathbf{B} \subset \mathcal{M}^{[2]}$.

Hypotheses (We will assume the following in our subsequent derivations:)

(1) $\frac{G^{[1]}}{G^{[2]}} \approx 0$ is such that

$$(a)\ g^{[3]}_{\mu\nu} = \frac{G^{[2]}}{G^{[1]}+G^{[2]}} \cdot g^{[1]}_{\mu\nu} + \frac{G^{[1]}}{G^{[1]}+G^{[2]}} \cdot g^{[2]}_{\mu\nu} \approx g^{[1]}_{\mu\nu}, \text{ and} \tag{166}$$

$$(b)\ g^{[2]}_{11} = \left(\frac{\Delta t^{[2]}_0}{\Delta t^{[1]}_0}\right)^2 < 0 \text{ throughout } \mathbf{B} \subset \mathcal{M}^{[2]}, \tag{167}$$

implying that $\Delta t^{[2]}_0$ has unit $\sqrt{-1}$ second (by analytic continuation; cf. e.g., [4] for the inherent necessity of the unit of i in standard quantum theory, and [20] for analytic continuation of Lorentzian metrics).

(2) $\forall j \in \mathbb{N}$ $E^{[2]}_j$ is either a single electromagnetic wave of length λ_j or a superposition of electromagnetic waves, and $E^{[2]}_j$ engages in gravitational forces with $\left\{E^{[2]}_k \mid k \in \mathbb{N} - \{j\}\right\}$ only.

(3) $\forall j \in \mathbb{N}$ $E^{[1]}_j$ is a particle (a photon if $E^{[2]}_j$ is a single electromagnetic wave) and engages in gravitational forces with $\left\{E^{[1]}_k \mid k \in \mathbb{N} - \{j\}\right\}$; in addition, $E^{[1]}_j$ may engage in electromagnetic, weak, or strong interactions with $E^{[1]}_{k\neq j}$ via exchanging virtual particles in $\mathcal{M}^{[1]}$.

Notation 2. *$\ddot{h} \equiv \frac{h}{second^2}$, $h \equiv$ Planck constant; $NLT \equiv$ nonlinear terms.*

Theorem 9.

$$G^{[2]} = \frac{c^5}{4\ddot{h}\phi^{[2]}}.$$

Proof. In order to apply General Relativity in our derivation, we set the Planck length as the lower limit of electromagnetic wave lengths under consideration, i.e., $\lambda \geq \lambda_P :\approx 10^{-35}$meter, or equivalently, $\nu \equiv \frac{c}{\lambda} \in \left(0\,\text{Hz}, 10^{43}\text{Hz} \equiv \nu_P\right)$ (which covers a spectrum from infrared to ultraviolet, to well beyond gamma rays, $\nu_{gamma} \approx 10^{21}$Hz). Thus, let $E^{[1]}_j$ be a photon with frequency $\nu^{[1]}_j \in (0\,\text{Hz}, \nu_P)$ as observed from a laboratory frame $S^{[1]}$ (in $\mathcal{M}^{[1]}$). Consider $E^{[2]}_j$ ($\equiv \hat{E}_j\phi^{[2]}$) within its wave length λ_j, i.e., $E^{[2]}_j$ as contained in a ball B of radius $\frac{\lambda_j}{2}$, and consider a reference frame $S^{[2]}$ on the boundary of B. Since the gravitational effect of $E^{[2]}_j$ on $S^{[2]}$ is as if the ball B of energy $E^{[2]}_j$ were concentrated at the ball center, we have

$$g^{[2]}_{11} = 1 - \frac{2G^{[2]}E^{[2]}_j}{\frac{\lambda_j}{2}\cdot c^4} \equiv 1 - \frac{4G^{[2]}E^{[2]}_j\nu^{[1]}_j}{c^5}. \tag{168}$$

Since the frequency $\nu^{[2]}_j$ of $E^{[2]}_j$ relative to frame $S^{[2]}$ is exactly 1 cycle and by Hypothesis (1)(b) the unit of $t^{[2]}_0$ is $\sqrt{-1}$ second, we have

$$\nu^{[2]}_j = \frac{1\text{ (cycle)}}{i\cdot\text{second}}, \tag{169}$$

so that

$$g^{[2]}_{11} := \left(\frac{\partial t^{[2]}_0}{\partial t^{[1]}_0}\right)^2 := \lim_{\Delta t^{[1]}_0\to 0}\left(\frac{\Delta t^{[2]}_0}{\Delta t^{[1]}_0}\right)^2 \tag{170}$$

$$= \left(\frac{\Delta t^{[2]}_0}{\Delta t^{[1]}_0 = 1\text{ second}}\right)^2 - NLT \text{ (where the nonlinear terms} \tag{171}$$

$NLT > 0$ due to the gravitational attraction of $S^{[2]}$ toward $E^{[2]}_j$)

$$\equiv \left(\frac{\nu^{[1]}_j}{\nu^{[2]}_j}\right)^2 - NLT \equiv \left(\frac{\nu^{[1]}_j}{1/(i\cdot\text{second})}\right)^2 - NLT \tag{172}$$

$$= -\nu^{[1]2}_j\text{second}^2 - NLT \tag{173}$$

$$= 1 - \frac{4G^{[2]}E^{[2]}_j\nu^{[1]}_j}{c^5} \text{ (from Equation (168));} \tag{174}$$

by the preceding Equations, (173) and (174), we have

$$-\nu_j^{[1]}\text{second}^2 - \frac{NLT+1}{\nu_j^{[1]}} = -\frac{4G^{[2]}E_j^{[2]}}{c^5}, \tag{175}$$

or

$$\frac{c^5}{4G^{[2]}}\left(\nu_j^{[1]}\text{second}^2 + \frac{NLT+1}{\nu_j^{[1]}}\right) = E_j^{[2]} \equiv \hat{E}_j\phi^{[2]}, \tag{176}$$

or

$$\hat{E}_j = \left(\frac{c^5\text{second}^2}{4G^{[2]}\phi^{[2]}}\right)\cdot\nu_j^{[1]} + \left(\frac{c^5}{4G^{[2]}\phi^{[2]}}\right)\cdot\frac{NLT+1}{\nu_j^{[1]}} \tag{177}$$

$$\equiv h\nu_j^{[1]} + \ddot{h}\cdot\frac{NLT+1}{\nu_j^{[1]}} \text{ (refer to Notation 2)}, \tag{178}$$

where

$$\ddot{h}\cdot\frac{NLT+1}{\nu_j^{[1]}} \equiv \ddot{h}\cdot\frac{NLT+1}{c}\cdot\lambda_j \equiv \Delta\hat{E}_j \tag{179}$$

is the uncertainty energy. (180)

Thus, comparing Equations (177) with (178), we have

$$G^{[2]} = \frac{c^5}{4\ddot{h}\phi^{[2]}}. \tag{181}$$

■

Remark 15. *The above factor* $\left(1/\phi^{[2]}\right) \equiv \left[\mu_1{}^2 + (1-\mu_1)^2\right]/(1-\mu_1)$ *from Corollary 9 and Equation (165) has a U-shaped graph as a function of* $\mu_1 \equiv m^{[1]}/m^{[3]}$*: as* μ_1 *increase from 0 to 0.29* $\left(\approx 1-\frac{\sqrt{2}}{2}\right)$*, 0.5 and 1,* $\left(\frac{1}{\phi^{[2]}}\right)$ *decreases from 1 to the minimum 0.83* $\left(\approx 2\left(\sqrt{2}-1\right)\right)$*, then rises to 1 and approaches* ∞*. Incidentally, we have also provided a derivation of* $\hat{E} = h\nu$ *from the above Equation (178); we note that* $g_{11}^{[2]} = 1 - \frac{4G^{[2]}E_j^{[2]}\nu_j^{[1]}}{c^5}$*, being a derivative, contains quantum uncertainties as* $\Delta t_0^{[1]} \to 0$.

We now cast quantum mechanics in General Relativity.

Claim Let $U \subset \mathbb{R}^{1+3}$ be a parameter domain of a laboratory frame; let $\rho : U \longrightarrow [0,\infty)$ be the probability density function of a particle $E_j^{[1]}$, and let $\mathbb{E} : U \longrightarrow \mathbb{C}^3$ be the electric field that contains $E_j^{[2]}$ in $\mathbf{B} \subset \mathcal{M}^{[2]}$ (which is complex by Hypothesis (1)(b)). Assume that ρ is of a positive constant proportionality β (of unit $\left(\frac{1}{\text{joule}}\right)$) to the electromagnetic field energy density of $\mathbb{E}$ (over U). Then the wave function $\psi : U \longrightarrow \mathbb{C}$ of $E_j^{[1]}$ is such that

$$\psi(t, \mathbf{x}) = z_0 \cdot \|\mathbb{E}(t, \mathbf{x})\|_{\mathbb{C}^3}, \tag{182}$$

where $z_0 \in \mathbb{C}$ is a constant and the complex norm (cf. e.g., [18], p. 221)

$$\|(z_1, z_2, z_3)\|_{\mathbb{C}^3}^2 := z_1^2 + z_2^2 + z_3^2 \in \mathbb{C}. \tag{183}$$

We back up the above *Claim* as follows: By the assumption in the *Claim*,

$$|\psi(t, \mathbf{x})|^2 = \rho(t, \mathbf{x}) = \beta \cdot \left| \|\mathbb{E}(t, \mathbf{x})\|_{\mathbb{C}^3} \right|^2 \cdot \epsilon_o \phi^{[2]}, \tag{184}$$

where $\epsilon_o \equiv$ the permittivity constant. Thus,

$$\psi(t, \mathbf{x}) = \sqrt{\beta \epsilon_o \phi^{[2]}} \cdot e^{i\theta} \|\mathbb{E}(t, \mathbf{x})\|_{\mathbb{C}^3} \tag{185}$$

$$= z_0 \cdot \|\mathbb{E}(t, \mathbf{x})\|_{\mathbb{C}^3}. \tag{186}$$

Remark 16. *From Hypotheses (1), (2), and (3), the $\mathbb{E}(t, \mathbf{x})$ in $\mathbf{B} \subset \mathcal{M}^{[2]}$ of the above Claim is only the effect or the consequence of the dynamics in $\mathcal{M}^{[1]}$; i.e., $\mathbb{E}(t, \mathbf{x})$ is formed by the forces in $\mathcal{M}^{[1]}$.*

Remark 17. *Also from Hypotheses (1), (2), and (3), any particle $p_i \in \mathcal{M}^{[1]}$ is formed by a superposition of electromagnetic fields in $\mathbf{B} \subset \mathcal{M}^{[2]}$; i.e., p_i has its distinct identity $\mathbb{E}_{p_i}(t, \mathbf{x})$, with*

$$\mathbb{E}_{p_i}(t, \mathbf{x}) = \sum_j \mathbb{E}_{i,j}(t, \mathbf{x_j}) = \begin{pmatrix} z_1(t, \mathbf{x}) \\ z_2(t, \mathbf{x}) \\ z_3(t, \mathbf{x}) \end{pmatrix}_i \in \mathbb{C}^3(t, \mathbf{x}), \tag{187}$$

i.e., composed of electromagnetic propagations through $(t, \mathbf{x})$ of multiple directions $\{\mathbf{x_j}\} \subset \mathbb{R}^3$, multiple frequencies $\{\omega_j\}$, and multiple phases $\{\theta_j\}$. This assertion is supported by the following three considerations:

(1) As is well known, traveling waves can sum to standing waves, and the sum of standing waves can approximate arbitrary functions by Fourier series.
(2) Physically, the pair creation process of antiparticles by photons such as

$$\gamma + \gamma \longrightarrow electron\ e^- + positron\ e^+, \tag{188}$$

has been well established (cf. [19], 164).
(3) We also note the possibility of engendering a new particle $\tilde{p}_i$ from an existing particle p_i via a field transformation

$$\Phi : \mathbb{E}_{p_i}(t, \mathbf{x}) \in \mathbb{C}^3(t, \mathbf{x}) \longmapsto \mathbb{E}_{\tilde{p}_i}(t, \mathbf{x}) \in \mathbb{C}^3(t, \mathbf{x}), \tag{189}$$

especially by the general principle of symmetry as associated with electric charge, spatial parity, and time direction.

Remark 18. *Historically Schrödinger had initially interpreted his* $|\Psi_{p_i}(t,\mathbf{x})|^2$ *as the electric charge density (cf. e.g., [15], III-21-6). Now the above Equation (184) shows that his interpretation was not too different from ours. In fact, the vector potential* $\mathbb{A}$ *in classical electrodynamics is the same as the wave function* Ψ *in quantum mechanics, so that the solutions of Maxwell Equations are identical to those of Schrödinger's Equation (cf. [15], II-15-8 and 20-3, also III-21-6). In short, Maxwell Equations, as applied to free spaces, already gave a description of the (quantum) fields* $\subset \mathbf{B} \subset \mathcal{M}^{[2]}$*, even though the way by which Maxwell derived his equations in 1861 was based on the electrodynamics of charges in* $\mathcal{M}^{[1]}$ *(see, e.g., [23], 40-47); i.e., his electromagnetic fields* $(\mathbb{E},\mathbb{B})$ *have always been in the complex* $\mathbf{B} \subset \mathcal{M}^{[2]}$*. That the complex quantum electrodynamics can assume a real classical form is simply due to the isomorphism*

$$\mathbb{R}/\langle 2\pi\rangle \approx \textit{the group of rotations; i.e.,} \tag{190}$$

$$\mathbb{E}_{oj}\cdot\cos\left(\omega_j t-\mathbf{k}_j\cdot\mathbf{x_j}+\theta_j\right)\approx\mathbb{E}_{oj}\cdot e^{-i\left(\omega_j t-\mathbf{k}_j\cdot\mathbf{x_j}+\theta_j\right)}=\mathbb{E}_j\left(t,\mathbf{x_j}\right). \tag{191}$$

Remark 19. *By the same assumption of* $\rho(t,\mathbf{x}) = \beta\cdot\left|\|\mathbb{E}(t,\mathbf{x})\|_{\mathbb{C}^3}\right|^2\cdot\epsilon_o\phi^{[2]}$ *(Equation (184)) as in the above Claim, i.e., quantum probability density in* $\mathcal{M}^{[1]} \equiv$ *electromagnetic field energy density in* $\mathbf{B}\subset\mathcal{M}^{[2]}$ (mod *joule of energy), we have analogously, probability current density in* $\mathcal{M}^{[1]} \equiv$ *the Poynting vector in* $\mathbf{B}\subset\mathcal{M}^{[2]}$ (mod *joule of energy), i.e.,*

$$\mathbf{j}(t,\mathbf{x})=\beta\cdot\mathbb{S}^{[2]}(t,\mathbf{x}). \tag{192}$$

We formalize this assertion by the following proposition.

Proposition 5. *The probability current density of a particle*

$$\mathbf{j}(t,\mathbf{x}) := \left(\frac{\hbar}{2\hat{m}i}\right)\left(\bar{\psi}(t,\mathbf{x})\cdot\nabla\psi(t,\mathbf{x})-\psi(t,\mathbf{x})\cdot\nabla\bar{\psi}(t,\mathbf{x})\right) \tag{193}$$

$$=\beta\cdot\mathbf{S}^{[2]}(t,\mathbf{x}), \tag{194}$$

where $\hbar \equiv \frac{h}{2\pi}$, $\hat{m} \equiv m^{[3]}_{measured} \equiv$ *the measured mass of the [particle, wave], and* $\mathbf{S}^{[2]}(t,\mathbf{x})$ *is the Poynting vector apportioned to* $\mathbf{B}\subset\mathcal{M}^{[2]}$.

Proof. Without loss of generality as based on (linear) superpositions of fields, consider a free photon that travels in the direction of $(x>0,0,0)$ with

$$\psi(t,\mathbf{x}) = z_0\cdot\|\mathbb{E}(t,\mathbf{x})\|_{\mathbb{C}^3} \tag{195}$$

$$= z_0\cdot\left\|\left(0,e^{-i(\omega t-kx)},0\right)^T\right\|_{\mathbb{C}^3} \tag{196}$$

$$= z_0 e^{-i(\omega t-kx)}. \tag{197}$$

Then

$$\nabla\psi = \left(z_0 e^{-i(\omega t-kx)}\cdot ki,0,0\right)^T \text{ and} \tag{198}$$

$$\nabla\bar{\psi} = \left(z_0 e^{i(\omega t-kx)}\cdot(-ki),0,0\right)^T, \tag{199}$$

so that $\mathbf{j} := \left(\frac{\hbar}{2\hat{m}i}\right)(\bar{\psi}\cdot\nabla\psi - \psi\cdot\nabla\bar{\psi}) = \frac{1}{2\hat{m}}\left(\bar{\psi}\cdot\frac{\hbar}{i}\nabla\psi - \psi\cdot\frac{\hbar}{i}\nabla\bar{\psi}\right)$

$$= \frac{1}{2\hat{m}}\left(\bar{\psi}\psi\cdot(\hbar k,0,0)^T + \psi\bar{\psi}\cdot(\hbar k,0,0)^T\right) \equiv \frac{1}{\hat{m}}\cdot|\psi|^2\cdot\hat{\mathbf{p}} \tag{200}$$

(where $\hat{\mathbf{p}}$ denotes the measured momentum vector of unit $[\frac{\text{kilogram}\cdot\text{meter}}{\text{second}}]$)

$$= \frac{1}{\hat{m}}\cdot\left(\beta\cdot\left(\hat{u}\phi^{[2]}\right)\right)\cdot\frac{\hat{\mathbf{S}}\cdot\text{meter}^3}{c^2} \tag{201}$$

(where $|\psi|^2$ equal to $\beta\cdot\left(\hat{u}\phi^{[2]}\right)$ is from the above Equation (184),

and $\hat{\mathbf{S}}$ denotes the measured Poynting vector, cf. [15], II-27-9, so that $\frac{\hat{\mathbf{S}}}{c^2}$ equals the momentum density of unit $\left[\frac{\text{kilogram}}{\text{second}\cdot\text{meter}^2}\right]$)

$$= \left(\frac{\hat{u}}{\hat{m}c^2/\text{meter}^3}\right)\cdot\beta\cdot\left(\hat{\mathbf{S}}\phi^{[2]}\right) = 1\cdot\beta\cdot\mathbf{S}^{[2]} \tag{202}$$

(due to the uniform probability density for a free photon).

■

Remark 20. *Our geometry of* $\mathcal{M}^{[1]}\times\mathbf{B}$ *serves to explain the following.*

(1) Quantum tunneling: A particle in $\mathcal{M}^{[1]}$ enters a mini black hole A, turns completely into a wave in $\mathbf{B}\subset\mathcal{M}^{[2]}$, and continue to travel in $\mathbf{B}\subset\mathcal{M}^{[2]}$ until mini black hole B, where it re-emerges in $\mathcal{M}^{[1]}$ (cf. [32]; for a recent study on wormholes, see [9]). Of course the above event is subject to the WKB probability approximation

$$\exp\left\{\left[-\frac{1}{\hbar}\int_{x_1}^{x_2}\sqrt{2m\,(V(x)-E)}dx\right][1+O(\hbar)]\right\}, \tag{203}$$

so that tunneling does not occur outside the quantum domain; however, as one of "the Top Ten Physics Newsmakers of the Decade" (*APS News*, February 2010), quantum teleportation of information between two atoms separated by more than one meter was achieved in February 2009.

(2) Vacuum polarization: Here we provide a different geometric structure for this phenomenon from that of the "infinite sea of invisible negative energy particles" by Dirac. We consider the negative spectrum $\left(-\infty,-mc^2\right]$ of the Dirac operator $D := \left(-ic\hbar\boldsymbol{\alpha}\cdot\boldsymbol{\nabla}+mc^2\beta\right)$ a pure mathematical artifact, and we claim that instead of being negative energies traveling backward in time, antiparticles differ from their (counterpart) ordinary particles in the order of the cross product $\mathbb{B}\times\mathbb{E}$ of the electromagnetic fields in $\mathbf{B}\subset\mathcal{M}^{[2]}$. Accordingly, we identify a vacuum (in $\mathcal{M}^{[1]}$) with a pre-existing electromagnetic field in $\mathbf{B}\subset\mathcal{M}^{[2]}$ (for a recent study on vacuum energy, see [11]).

(3) The existence of dark matter and energy $\left[0,E^{[2]}\right]$: The above (1) suggests that if a particle can engage in long-distance tunneling from point A to B, then between A and B the particle becomes dark matter/energy with total energy

$$E^{[3]} = \sqrt{\left(m_0^{[3]}\right)^2 c^4 + p^2c^2}, \tag{204}$$

where $m_0^{[3]} = m_0^{[2]}$ = the rest mass of the dark matter, and pc = the dark energy. Here we remark that whether $m_0^{[3]} > 0$ or $m_0^{[3]} = 0$ depends on the superposition of the electromagnetic waves in $\mathbf{B} \subset \mathcal{M}^{[2]}$ forming a standing wave or not.

Remark 21. *In [16] Frolov et al. showed the possibility that a black hole can give rise to a macroscopic universe. Our model of $\mathcal{M}^{[3]}$ precisely claims that our recognized universe of matter $\mathcal{M}^{[1]}$ is a black hole $\mathbf{B}$ in $\mathcal{M}^{[2]}$; this geometry renders the possibility that any black hole in $\mathcal{M}^{[1]}$ leads back to $\mathbf{B} \subset \mathcal{M}^{[2]}$ (as in the Kruskal-Szekeres scheme); i.e., geometric singularities in $\mathcal{M}^{[1]}$ serve to transfer energies between $\mathcal{M}^{[1]}$ and $\mathbf{B} \subset \mathcal{M}^{[2]}$ (see [32], also cf. [3] about the subject of how quantum gravity takes over a "naked singularity"), so that a point particle does not have an infinite mass density. As such, we claim that electrons are point particles in $\mathcal{M}^{[1]}$ that carry their electromagnetic waves in $\mathbf{B} \subset \mathcal{M}^{[2]}$ and hence they do not have self-interactions as implied in, e.g., the Maxwell-Dirac system (cf. [13])*

$$\left\{ \begin{array}{c} \left(i\gamma^{\mu}\partial_{\mu} - \gamma^{\mu}A_{\mu} - 1\right)\Psi = 0, \\ \partial\mu A^{\mu} = 0,\ 4\pi\partial_{\mu}\partial^{\mu}A_{\nu} = \left(\bar{\Psi}, \gamma^{\nu}\Psi\right) \end{array} \right\} \tag{205}$$

or the Klein-Gordon-Dirac system

$$\left\{ \begin{array}{c} \left(i\gamma^{\mu}\partial_{\mu} - \chi - 1\right)\Psi = 0, \\ \partial^{\mu}\partial_{\mu}\chi + M^2\chi = \frac{1}{4\pi}\left(\bar{\Psi}, \Psi\right) \end{array} \right\}. \tag{206}$$

Thus, our $\mathcal{M}^{[1]} \times \mathbf{B} \subset \mathcal{M}^{[1]} \times \mathcal{M}^{[2]}$ resolves the pervasive problem of singularities at $r = 0$ in both the classical and the quantum domains by considering a neighborhood N of $r = 0$ that transfers uncertainty energies between $\mathcal{M}^{[1]}$ and $\mathbf{B} \subset \mathcal{M}^{[2]}$, so that in calculating the electromagnetic energy of e^-, one stops at Bdry N.

Remark 22. *We also note that an electromagnetic field (being periodic in $\mathbf{B}$) renders itself a quotient space, displaying the phenomenon of "instantaneous communication," a feature serving as potential reference for quantum computing. To elaborate, the complex electric field related to a photon γ_j, $\mathbb{E}_j\left(t, \mathbf{x_j}\right) = \mathbb{E}_{oj} \cdot e^{-i\left(\omega_j t - \mathbf{k}_j \cdot \mathbf{x_j} + \theta_j\right)}$, results in a quotient space, i.e., $\forall\left(t, \mathbf{x}\right) \in U - \{0, \mathbf{0}\}$ we have*

$$t \equiv t_0 \left(\operatorname{mod} \frac{2\pi}{\omega_j} \equiv \frac{1}{\nu_j}\right) \tag{207}$$

for some $t_0 \in \left[0, \frac{1}{\nu_j}\right]$, and

$$\mathbf{x} \equiv \mathbf{x_0} \left(\operatorname{mod} \left(\frac{2\pi}{k_j}\right)\left(\frac{\mathbf{x}}{\|\mathbf{x}\|}\right) \equiv \lambda_j \left(\frac{\mathbf{x}}{\|\mathbf{x}\|}\right)\right) \tag{208}$$

for some $\mathbf{x_0}$ with $\|\mathbf{x_0}\| \in \left[0, \lambda_j\right]$; as such, $\forall \lambda_j \gtrapprox 0$ we have

$$t \doteqdot 0 \text{ and } \mathbf{x} \doteqdot \mathbf{0}, \tag{209}$$

resulting in "instantaneous communication" across U, which, among other things, accounts for the double-slit phenomenon: That is, to propagate γ_j along the direction of $(0, \mathbf{0})$ to, say,

$$\left(\frac{\sqrt{1+d^2}}{c}, 1 \text{ meter}, |d| \text{ meter (the "upper slit")}, 0\right) \tag{210}$$

is nearly the same as via $(0, \mathbf{0})$ *to* $(0, 0, y > 0, 0)$, *with*

$$\left\| \left\| \mathbb{E}_{oj} \cdot e^{-i\left(\omega_j t - k_j y\right)} \right\| \right\|^2 = \left\| \mathbb{E}_{oj} \right\|^2 > 0, \tag{211}$$

i.e., a nonzero (constant) probability density ρ *along the y-axis for* γ_j *to be observed; similarly, a switch to the lower slit*

$$\left(\frac{\sqrt{1+d^2}}{c}, 1 \; meter, -\left|d\right| \; meter, 0 \right) \tag{212}$$

is to result in the same conclusion. However, if both slits are open, then there exists a superposition of fields

$$\cos\left(\omega_j t - k_j y\right) + \cos\left(\omega_j t + k_j y\right) = 2\cos\omega_j t \; \cos k_j y \tag{213}$$

and the probability density of γ_j *equals zero* $\forall y$ *such that* $\cos k_j y = 0$.

Also, $\mathbb{F}(t, \mathbf{x}) \equiv \sum_i \mathbb{E}_{p_i}(t, \mathbf{x})$ = *the aggregate quantum field* in $\mathcal{M}^{[2]}$ (over all particles $\{p_i\}$ in $\mathcal{M}^{[1]}$) presents itself as *one* quantum field; as such, $\left\{\mathbb{E}_{p_i}(t, \mathbf{x})\right\}$ are correlated or "entangled," displaying global behavior such as the celebrated Einstein-Podolski-Rosen ("EPR") phenomenon.

Remark 23. *Our geometry of* $\mathcal{M}^{[1]} \times \mathbf{B}$ *thus has contributed physical logic to quantum mechanics, in particular, providing an energy interpretation to probabilities; as yet another demonstration, consider the fine structure constant,*

$$\alpha := \frac{\frac{e^2}{4\pi\epsilon_o}}{\hbar c} = \frac{\frac{e^2}{4\pi\epsilon_o}}{\frac{h}{2\pi} \cdot \nu\lambda} = \frac{\frac{e^2}{4\pi\epsilon_o\lambda}}{E^{[3]}_{measured}/2\pi} \tag{214}$$

= *(the electrostatic potential energy between two electrons separated by a distance of* λ*) / (the energy* $E^{[3]}_{measured}$ *of the virtual photon needed to mediate the two electrons divided by* 2π*) = the constant* α, *or,* $E^{[3]}_{measured} \cdot \lambda$ = *constant, i.e., a uniform probability for any two electrons to interact across all space.*

Although in the above we derived an expression for $G^{[2]}$, it contained an undetermined parameter

$$\phi^{[2]} \equiv \frac{1-\mu_1}{\mu_1{}^2 + (1-\mu_1)^2}. \tag{215}$$

Concerning $\mu_1 \equiv \frac{M^{[1]}}{M^{[3]}}$, we consider the discrepancy in the electromagnetic mass of an electron as measured in a stationary state versus in a moving state with a constant velocity of $\|\mathbf{V}\| << c$. In Feynman ([15], II-28-4), one finds (cf. e.g., [26], for this well-known problem)

$$m_{\mathbf{V}=\mathbf{0}} = \frac{3}{4} m_{\mathbf{V}\neq\mathbf{0}}; \tag{216}$$

by our Hypotheses (2) and (3), electromagnetic forces take place only in $\mathcal{M}^{[1]}$, but motions necessarily take place in $\mathcal{M}^{[3]}$; as such, it appears reasonable to attribute $m_{\mathbf{V}=\mathbf{0}}$ to $\mathcal{M}^{[1]}$ and $m_{\mathbf{V}\neq\mathbf{0}}$ to $\mathcal{M}^{[3]}$, i.e.,

$$\mu_1 = \frac{3}{4}. \tag{217}$$

If so, then

$$\phi^{[1]} = 1.2, \tag{218}$$

$$\phi^{[2]} = 0.4, \text{ and thus by Equation (181)} \tag{219}$$

$$G^{[2]} = \frac{c^5}{1.6\ddot{h}} \approx 2.3 \times 10^{75} \times \frac{\text{meter}^3}{\text{kilogram} \cdot \text{second}^2} \tag{220}$$

$$\approx 10^{85} G^{[1]} \approx 10^{85} G^{[3]}. \tag{221}$$

Here we note that the generally recognized Schwarzschild radius for $\mathcal{M}^{[1]}$ (readily found in textbooks) is:

$$g_{11}^{[1]} = 0 = 1 - \frac{2G^{[1]}M^{[1]}}{R^{[1]}c^2} \tag{222}$$

$$\approx 1 - \frac{2 \times 6.7 \times 10^{-11} \times 10^{51}}{R^{[1]} \times (3 \times 10^8)^2}, \tag{223}$$

i.e.,

$$R^{[1]} \approx 10^{24} \text{ meter} \tag{224}$$

$$< 10^{26} \text{ meter (the actual radius of } \mathcal{M}^{[1]}\text{);} \tag{225}$$

thus, with $\mu_1 = \frac{3}{4}$ and $G^{[2]} \approx 10^{85} G^{[1]}$, we have

$$g_{11}^{[2]} = 0 = 1 - \frac{2 \times 6.7 \times 10^{-11} \times 10^{85} \times 10^{51} \times (1/3)}{R^{[2]} \times 9 \times 10^{16}}, \tag{226}$$

i.e.,

$$R^{[2]} \approx 10^{108} \text{ meter} \tag{227}$$

$$> > 10^{26} \text{ meter,} \tag{228}$$

so that $\mathcal{M}^{[2]}$ could give rise to $\mathcal{M}^{[1]}$.

4. Summary

In Section 2 above, we have shown that the classical electromagnetic least action is a geodesic of our $\mathcal{M}^4_{em}$, but as Feynman indicated ([15], II-19-8,9), the least action in quantum electrodynamics is the same as that of the classical; thus, we have contributed a geometric underpinning of both the classical and the quantum electrodynamics.

Then in Section 3, we have shown that our construct of the combined space-time 4-manifold $\mathcal{M}^{[3]}$ provides quantum mechanics with a more complete geometric framework, which can resolve many outstanding conceptual and analytical problems. In this regard, we envision a further development of our theory, to furnish more detailed analyses such as when a dark matter or energy $\left(0, E^{[2]}\right)$ becomes a combined particle of $\left(\frac{3}{4}E^{[2]}, \frac{1}{4}E^{[2]}\right) \in \mathcal{M}^{[3]}$ and how one may put our $\mathcal{M}^{[3]}$ to laboratory tests.

Thus, we have extended Einstein's General Relativity. We close our chapter with the following comment. In our view, the most crucial period in the development of modern physics was the decade from 1920 to 1930, when the new Einstein's General Relativity met the new quantum mechanics. What happened then was that those ten years was too short for General Relativity to be thoroughly digested and explored. For example, as mentioned in Section 2 the attempt of unifying electromagnetism with gravity in one set of *EFE* failed simply due to a hasty error in the identification of the electromagnetic energy-momentum tensor T. More fundamentally though, as mentioned in Section 3 the particle-wave duality as observed in the material space-time simply could not be explained satisfactorily, due to the self-imposed geometric constraint of a single set of *EFE* - - for the visible $\mathcal{M}^{[1]}$. As a result, waves became probabilities and concepts like "probability current" became necessities. Our $\mathcal{M}^{[3]}$ here interprets waves as energies and provides quantum mechanics with a more satisfying geometry.

5. References

[1] P.K. Anastasovski, T.E. Bearden, et al., Inconsistencies of the U(1) theory of electrodynamics: stress energy momentum tensor, Found. Phys. Lett., 12 No. 2 (1999), 187-192.

[2] R. Balbinot, L. Mazzacurati and A. Fabbri, Dilaton gravity black holes with a regular interior, Phys. Rev. D 57 (1998), 6185 (:1-5).

[3] E. Barausse, V. Cardoso and G. Khanna, Test bodies and naked singularities: is the self-force the cosmic censor? Phys. Rev. Lett., 105 (2010), 261102 (:1-4).

[4] J.B. Barbour, Time and complex numbers in canonical quantum gravity, Phys. Rev. D 47 (1993), 5422 (:1-27).

[5] H.-H. Borzeszkowski, H.-J. Treder, On metric and matter in unconnected, connected, and metrically connected manifolds, Found. Phys., 34 No. 10 (2004), 1541-1572.

[6] P.R. Brady, S. Droz and S.M. Morsink, Late-time singularity inside nonspherical black holes, Phys. Rev. D 58 (1998), 084034 (:1-15).

[7] O. Calin, D.-C. Chang, Geometric Mechanics on Riemannian Manifolds - - Applications to Partial Differential Equations, Birkhaüser, Boston, 2005.

[8] P. Candelas and B.P. Jensen, Feynman Green function inside a Schwarzschild black hole, Phys. Rev. D 33 (1986), 1596 (:1-7).

[9] M.B. Cantcheff, N.E. Grandi and M. Sturla, Wormhole solutions to Hořava gravity, Phys. Rev. D 82 (2010), 124034 (:1-9).

[10] J. Cheeger and G. Tian, Curvature and injectivity radius estimates for Einstein 4-manifolds, J. Amer. Math. Soc., 19 (2006), 487-525.

[11] G. Duplančić, D. Glavan and H. Štefančić, Probability distribution of the vacuum energy density, Phys. Rev. D 82 (2010), 125008 (:1-5).

[12] A. Einstein, Die Grundlage der allgemeinen Relativitätstheorie, Annalen der Physik, 49 (1916), 769-822.

[13] M.J. Esteban, M. Lewin and E. Séré, Variational methods in relativistic quantum mechanics, Bull. Amer. Math. Soc., 45 (4) (2008), 535-593.

[14] M.W. Evans, The Evans lemma of differential geometry, Found. Phys. Lett., 17 No. 5 (2004), 433 (:1-22).

[15] R.P. Feynman, R.B. Leighton, M. Sands, The Feynman Lectures on Physics, Addison-Wesley, Reading, 1963.

[16] V.P. Frolov, M.A. Markov and V.F. Mukhanov, Black holes as possible sources of closed and semiclosed worlds, Phys. Rev. D 41 (1990), 383 (:1-11).

[17] J. Hansen, A. Khokhlov and I. Novikov, Physics of the interior of a spherical, charged black hole with a scalar field, Phys. Rev. D 71 (2005), 064013 (:1-25).

[18] V.G. Ivancevic and T.T. Ivancevic, Complex Dynamics - - Advanced System Dynamics in Complex Variables, Springer, Dordrecht, 2007.

[19] M. Kaku, Quantum Field Theory - - A Modern Introduction, Oxford University Press, Oxford, 1993.

[20] P. Kraus, H. Ooguri and S. Shenker, Inside the horizon with AdS/CFT, Phys. Rev. D 67 (2003), 124022 (:1-15).

[21] G.L. Light, A clear logic of Einstein field equations, Int. J. Appl. Math, 20 No. 6 (2007), 843-866.

[22] G.L. Light, Energy inquiry, Can. J. Phys., 68 (1990), 242-243.

[23] M.S. Longair, Theoretical Concepts in Physics, Cambridge University Press, Cambridge, 1986.

[24] D.R. Lunsford, Gravitation and electrodynamics over SO(3,3), Int. J. Theo. Phys., 43 No. 1 (2004), 161-177.

[25] G. Maschler, Central Kähler metrics, Trans. Amer. Math. Soc., 355 (2003), 2161-2182.

[26] P. Moylan, An elementary account of the factor of 4/3 in the electromagnetic mass, Am. J. Phys. 63 (1995), 818-820.

[27] E. Newman, L. Tamburino and T. Unti, Empty-space generalization of the Schwarzschild metric, J. Math. Phys. 4 (1963), 915-923.

[28] A. Pais, Niels Bohr's Times in Physics, Philosophy, and Polity, Oxford University Press, Oxford, 1982.

[29] R.K. Sachs and H. Wu, General relativity and cosmology, Bull. Amer. Math. Soc. 83 (1977), 1101-1164.

[30] R. Shankar, Principles of Quantum Mechanics, 2nd ed., Springer, New York, 1994.

[31] J. Smoller, A. Wasserman, S.T. Yau and B. McLeod, Smooth static solutions of the Einstein-Yang/Mills equation, Bull. Amer. Math. Soc. 27 (1992), 239-242.

[32] K. Srinivasan and T. Padmanabhan, Particle production and complex path analysis, Phys. Rev. D 60 (1999), 024007 (:1-20).

[33] I. Suhendro, A unified field theory of gravity, electromagnetism, and the Yang-Mills gauge field, Prog. Phys., 1 (2008), 31-37.

[34] D. Töben, Parallel focal structure and singular Riemannian foliations, Trans. Amer. Math. Soc., 358 (2006), 1677-1704.

[35] S. Weinberg, Gravitation and Cosmology, Principles and Applications of the General Theory of Relativity, Wiley, New York, 1972.

[36] E. Witten, From superconductors and four-manifolds to weak interactions, Bull. Amer. Math. Soc., 44 No. 3 (2007), 361-391.

Part 3

High Frequency Techniques

7

Field Estimation through Ray-Tracing for Microwave Links

Ada Vittoria Bosisio
National Research Council of Italy, CNR/IEIIT c/o Politecnico di Milano
Italy

1. Introduction

Engineers using microwave radio links have to address the effects of multipath propagation that arise when several rays arrive at the receiver after travelling along different paths from the transmitter. Rays, along their propagation, undergo reflections at the earth's surface or at variations in the refractive index or its gradient.

Since the '60s, standard ray theories for radio wave applications use effective earth radius concepts, ray bending based on Snell's law in a layered spherical atmosphere, or analytical method, limited to simple refractive index profiles (Du Castel, 1966; Livingston, 1970). At that time, scientists were interested in statistics of fading, distinguishing between *fast* or *slow fading* and the typical results were either nomograms of attenuation as a function of distance for a set of common frequency and height values or cumulative distribution functions representing the fraction of time that the signal was expected to be received at or below a given level (CCIR, 1978). Several techniques were employed to calculate the field strength, such as Cornu's spiral derived by the Huygens' principle together with the introduction of the specular and the diffuse reflection contribution (Hall, 1979). In that approach, the field amplitude was determined assuming the principle of conservation of energy, i.e. the flux of the energy along a ray is the same at every cross section of a narrow tube of rays (Keller, 1957).

Nowadays, radio communications live a renaissance due to the development of mobile and wireless communications and the increasing demand of wideband services. The need of greater rate of transmission is accomplished using higher frequencies for which effects like the tropospheric scatter can affect dramatically the quality of the received signal. Thus, aspects such as analysis of delay power spectra or of potential intersystem interferences require a more realistic modelling. As a matter of fact, local fluctuations in the refractive index n can cause scatter, while its abrupt changes with the height can cause reflection and originate ducting layers, in which the rays are reflected and refracted back in such a way that the field is trapped inside these layers (Bean & Dutton, 1968).

Although the goal of radio engineering is to transmit reliably from the source to the receiver, this is of no use if the received signal is unintelligible due to interference or if the transmission causes interference with other systems. So an analysis of the channel behaviour is mandatory to meet both link design and system requirements. Besides, the availability of geographic databases, digital elevation models and the increasing computing power let us face more realistic description to model the interaction of the propagating wave with the atmosphere and the surrounding scenario (Driessen, 2000; Kurner & Cichon, 1993; Lebherz et al., 1992).

In more recent developments, ray modelling of wave propagation addresses the dispersive effects of perturbed atmosphere on the performance of high-capacity digital radio channel (Akbarpour & Webster, 2005; Sevgi & Felsen, 1998).
Besides telecommunications applications, a proper modelling of atmospheric propagation is of concern in atmospheric sciences with particular focus on radio occultation data analysis for assimilation with numerical weather prediction models (Pany et al., 2001). Also, the decrease of propagation speed related to the atmospheric refractivity causes tropospheric delay, which influences applications of the Global Navigation Satellite System (Eresmaa et al., 2008).
The author presents a widely diffuse technique in the domain of seismic studies that accommodates lateral variations in the medium properties (Farra, 1993). As an asymptotic technique, based on high frequency approximation, it permits fast computation but provides a local solution of propagation problem (Červený et al., 1977). The technique was adapted for application with electromagnetic waves and specifically tailored for signals travelling in the atmosphere. The hypothesis of horizontal uniformity can be removed and no stratification is needed to calculate the ray trajectory. The terrain profile coordinates are mapped over a Cartesian reference through analytical transformation. Refractive index values are given in the same Cartesian reference. Hence, propagation is modelled in a two dimensional range-height scenario over irregular terrain through non-homogeneous atmosphere.
The field amplitude is evaluated by means of a perturbation technique using paraxial rays to observe the wave front structure along the path from the transmitter towards the receiver.
This approach allows to analyze system performance and the channel impulse response in presence of any kind of atmosphere, characterized by local values of refractive index n in the bi-dimensional panel including the antennae and the terrain profile. The variations of the refractive index n along the third dimension are taken equal to zero; nonetheless, the field amplitude is calculated correctly because of the paraxial approximation which takes place in a 3D domain. Median power strength of the numerical results was compared with predictions given by Friis' Formula (Balanis, 1996).

2. Modelling

The proposed ray tracing technique is widely used in seismic for subsoil investigation and it follows an approach based upon the integration of the Eikonal equation with Hamiltonian-Jacobi technique (Kružkov, 1975). According to this formulation, rays are defined by the vector $y(\tau)=(x(\tau), p(\tau))$, where $x(\tau)$ and $p(\tau)$ are the position vector and the slowness (inverse of phase velocity) vector along the ray, both function of the integration variable τ and of the initial conditions (launching point and direction). The slowness vector $\boldsymbol{p}$ is defined as $\mathbf{k}/\omega$ and $e^{j\mathbf{k}\cdot\mathbf{x}}$ is the phase function. The vector $\boldsymbol{y}(\tau)$ satisfies the Hamilton differential equations, whose solutions describe the wave propagation in the medium, under the asymptotic assumption (Abramovitz & Stegun, 1970):

$$\begin{cases} \dfrac{d\boldsymbol{x}}{d\tau} = \nabla_{p}\mathcal{H} \\ \dfrac{d\boldsymbol{p}}{d\tau} = -\nabla_{x}\mathcal{H} \end{cases} \tag{1}$$

where ∇_p and ∇_x represent the gradient computed versus the slowness vector and the position vector, respectively. $\mathcal{H}$ is the Hamiltonian function describing the wave propagation in the considered medium, i.e. the atmosphere, chosen as follows:

$$\mathcal{H}(\boldsymbol{x},\boldsymbol{p})=\frac{1}{2}\left(v^2(\boldsymbol{x})\boldsymbol{p}^2-1\right) \tag{2}$$

being $v(\boldsymbol{x})$ the propagation speed of the medium at the position vector $\boldsymbol{x}$.
The advantage of such an approach is twofold: it takes into account the medium inhomogeneities that originate multipath propagation and wavefronts folding (caustics), and it permits to consider both vertical and long range variations of the atmospheric model, without any approximation like flat-earth model (Hall, 1979). Also, it permits to keep separated the different field contributions due to the different interactions between the rays and the surrounding scenario, which impacts on the phase calculation of the overall received field.

2.1 Ray tracing

A suitable discussion for ray tracing can be found in (Farra, 1993); nevertheless, in this section some basic concepts and analytical details are reported.
Substituting (2) in (1), one obtains the equations for the vector $\boldsymbol{y}(\tau)$:

$$\begin{cases}\dfrac{d\boldsymbol{x}}{d\tau}=\boldsymbol{p}v^2(\boldsymbol{x})\\[2mm]\dfrac{d\boldsymbol{p}}{d\tau}=-\dfrac{\nabla_{\mathbf{x}}v(\boldsymbol{x})}{v(\boldsymbol{x})}\end{cases} \tag{3}$$

x and p define the 6D phase-space, where the solution of (3), i.e. the rays, are the characteristic lines of the Eikonal equation, thus enabling the interpolation without ambiguity on a single fold (Vinje et al., 1993). For this reason rays are also called bi-characteristics lines of the wave equation. By choosing as integration variable the ray propagation delay or travel time T, system (3) becomes:

$$\begin{cases}\dfrac{dx}{dT}=c^2(\boldsymbol{x})\boldsymbol{p}_x\\[2mm]\dfrac{dy}{dT}=c^2(\boldsymbol{x})\boldsymbol{p}_y\\[2mm]\dfrac{dz}{dT}=c^2(\boldsymbol{x})\boldsymbol{p}_z\\[2mm]\dfrac{dp_x}{dT}=\dfrac{1}{c^2(\boldsymbol{x})}\dfrac{\partial c(\boldsymbol{x})}{\partial x}\\[2mm]\dfrac{dp_y}{dT}=\dfrac{1}{c^2(\boldsymbol{x})}\dfrac{\partial c(\boldsymbol{x})}{\partial y}\\[2mm]\dfrac{dp_z}{dT}=\dfrac{1}{c^2(\boldsymbol{x})}\dfrac{\partial c(\boldsymbol{x})}{\partial z}\end{cases} \tag{4}$$

The solution of system (4) describes, under the asymptotic assumption, the propagation in the atmosphere in terms of ray trajectories and time delay.

2.2 Amplitude calculation

Anomalous conditions such as medium inhomogeneity or velocity model variations can affect dramatically the wavefront propagation. Under these circumstances the computation based upon the spherical divergence assumption results in erroneous evaluations.

One possible choice is to compute the amplitude through paraxial rays (first order perturbation theory), used to determine the flow tube, whose deformations, together with the theorem of the conservation of the flux along the ray path, allow the calculation of the amplitude A at the time τ.

Let us consider a reference ray with characteristic vector $\boldsymbol{y}_0(\tau)=(\boldsymbol{x}_0(\tau), \boldsymbol{p}_0(\tau))$. A paraxial ray is obtained from the reference one by applying the first order perturbation theory, so paraxial rays coordinates are defined by:

$$\begin{aligned} \boldsymbol{x}(\tau) &= \boldsymbol{x}_0(\tau) + \delta\boldsymbol{x}(\tau) \\ \boldsymbol{p}(\tau) &= \boldsymbol{p}_0(\tau) + \delta\boldsymbol{p}(\tau) \end{aligned} \tag{5}$$

Tracing paraxial rays consists in finding in the phase-space the canonical perturbation vector $\delta\boldsymbol{y}(\tau)=(\delta\boldsymbol{x}(\tau),\delta\boldsymbol{p}(\tau))$. These perturbations in the trajectory are due to small changes in the initial conditions of the ray.

The linear system for the calculation of paraxial rays is obtained by inserting the perturbation vector given by (5) in system (1) and developing to the first order:

$$\frac{d\delta\boldsymbol{y}}{d\tau} = \mathbf{A}(\tau)\cdot\delta\boldsymbol{y} \tag{6}$$

where

$$\mathbf{A} = \begin{bmatrix} \nabla_{\mathbf{x}}\nabla_{\mathbf{p}}H & \nabla_{\mathbf{p}}\nabla_{\mathbf{p}}H \\ -\nabla_{\mathbf{x}}\nabla_{\mathbf{x}}H & -\nabla_{\mathbf{p}}\nabla_{\mathbf{x}}H \end{bmatrix} \tag{7}$$

is a 6×6 matrix whose elements are the derivatives of the Hamiltonian function calculated on the reference ray.

Paraxial rays allow to define the *flow tube*, schematically represented in figure 1.

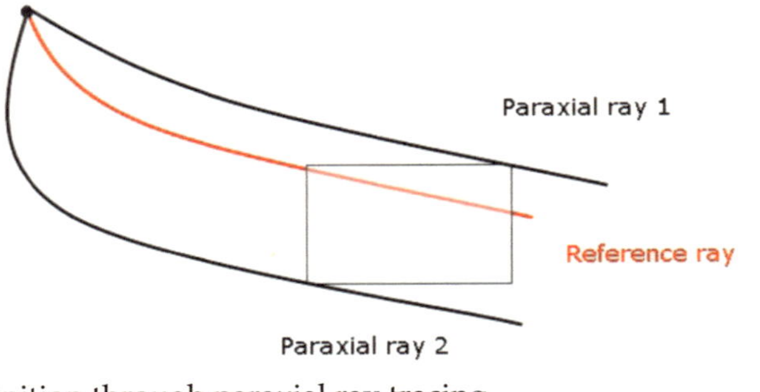

Fig. 1. Flow tube definition through paraxial ray tracing

The deformation of the flow tube along the ray path ℓ is given by the jacobian ***J***:

$$J = \frac{\partial(x,y,z)}{\partial(\tau,\vartheta,\phi)} = \det\begin{bmatrix} \frac{\partial x}{\partial \tau} & \frac{\partial x}{\partial \vartheta} & \frac{\partial x}{\partial \phi} \\ \frac{\partial y}{\partial \tau} & \frac{\partial y}{\partial \vartheta} & \frac{\partial y}{\partial \phi} \\ \frac{\partial z}{\partial \tau} & \frac{\partial z}{\partial \vartheta} & \frac{\partial z}{\partial \phi} \end{bmatrix} \quad (8)$$

where τ is the sampling parameter, θ the elevation angle and ϕ the azimuth angle. The amplitude of the plane wave associated to the ray is computed considering both the deformation of the flow tube and the conservation of power density flow law along the ray path. The volume element of the flow tube can be expressed as $dV = J\ d\tau\ d\vartheta\ d\phi$ or $dV = dS\ d\ell$. Thus, the elementary surface dS has the following expression:

$$dS = \frac{dV}{d\ell} = \frac{J\ d\tau\ d\vartheta\ d\phi}{d\ell} = \frac{J}{v} d\vartheta\ d\phi \quad (9)$$

From the conservation of the power density flow along the ray (Keller, 1957) and known the amplitude A_i at τ_i, it follows that the amplitude A_{i+1} at τ_{i+1} is:

$$A_{i+1} = A_i \frac{v_{i+1}}{v_i} \sqrt{\frac{J_i}{J_{i+1}}} \quad (10)$$

Equation (10) shows that the ray amplitude depends on the velocity model. Hence, paraxial rays take into account anomalous propagation conditions.

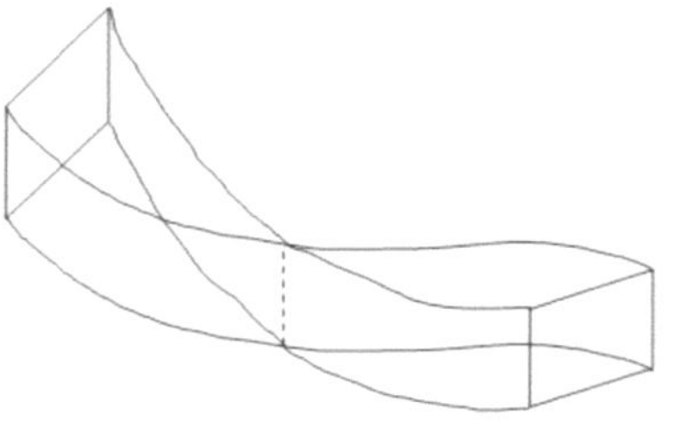

Fig. 2. Sketch of a caustics of the first order: the flow tube looses one dimension

The singularity of the Jacobian J is a singularity, known as *caustic*, of the plane wave associated to the ray (Kravtsov & Orlov, 1990). Caustics arise when the ray field folds. Caustics are of first order when the flow tube looses one dimension and of the second order when two are the dimensions lost. The wave front folding affects the phase of the field carried by the wave with a phase shift of $\pi / 2$ radians for each caustic order. These events are carefully accounted so that the proper phase shifts can be applied to the field.

2.3 Parameterization

The modelling is performed in a 2D panel, as far as the terrain profile and the atmosphere characteristics are concerned, while the ray trajectories are computed in a 3D space.

This assumption is based on the hypothesis that lateral variations in the propagation medium and in the terrain profile are negligible in the third dimension, which does not strictly holds.
Therefore, the propagation occurs in the so called *Earth* system *(s,h)*, where *s* is the range measured along the idealized spherical earth and *h* is the altitude taken along radial direction passing through the Earth centre, respectively. Instead, for sake of simplicity, all calculation are developed in a Cartesian reference system *(x,z)* related to the Earth by the following transformation relationships:

$$\begin{aligned} x &= (R_0 + h)\sin\left(\frac{s}{R_0}\right) \\ z &= (R_0 + h)\cos\left(\frac{s}{R_0}\right) - R_0 \end{aligned} \tag{11}$$

where R_0 is the Earth radius. The transformation from cartographic coordinates *(s,h)* to absolute coordinates *(x,z)* is given by the inverse transformation:

$$\begin{aligned} s &= R_0 \arctan\left(\frac{x}{z + R_0}\right) \\ h &= \sqrt{x^2 + (z + R_0)^2} - R_0 \end{aligned} \tag{12}$$

being $\theta = s / R_0$ the angle at the centre of the Earth, as shown in figure 3.
This way to proceed allows to take into account the actual geometry of the problem without introducing approximations like equivalent Earth radius. Also, it keeps separated the domain in which the radio link characteristics are defined and the computational one, where rays are traced.

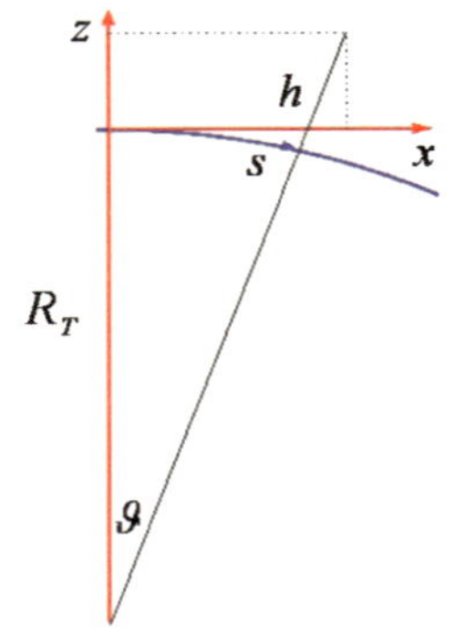

Fig. 3. The two reference systems: cartographic coordinates *(s,h)* and absolute ones *(x,z)*.

The characteristics of the atmosphere are taken into account in terms of its propagation velocity using several quantities that are generally function of the position vector $\boldsymbol{x}$:

$$v(x) = \frac{c}{n(x)} = \frac{1}{p(x)} \tag{13}$$

where c is the speed of the light and $n(x)$ is the refractive index of the atmosphere, which usually is expressed in terms of the refractivity N:

$$N = (n-1) \cdot 10^6 \tag{14}$$

N is a dimensionless quantity but in literature it is often measured in N-units or in parts per million (ppm). Its value depends upon the altitude and the range locally according to the atmospheric pressure P, the vapour pressure P_V and the temperature T as in the following (Bean and Dutton, 1968) :

$$N = 77.6\frac{P}{T} + 3.73 \cdot 10^5 \frac{P_V}{T^2} \tag{15}$$

Besides the variations in the refractive index, or its gradient, rays along their propagation undergo reflections and diffraction at the earth's surface. At the present stage of the modelling, neither the diffraction mechanisms nor any scattering features are included except reflection.
The trajectory of the reflected ray is given by the Snell's law under the hypothesis that locally the ground acts as a perfect smooth surface. The initial conditions with which the reflected ray is traced depend both on the geometric characteristics of the terrain profile and on the ray direction of incidence. The amplitude and the phase of the rays reflected from the ground are computed according to the chosen ground parameterization, i.e. the electric soil properties. These determine the value of the ground impedance η, which is a function of both soil permittivity ε_r and conductivity σ (Ulaby, 1999):

$$\eta = \sqrt{\frac{\mu}{\varepsilon_0 \varepsilon_r}\left(1 - j\frac{\sigma}{\omega \varepsilon_0 \varepsilon_r}\right)} \tag{16}$$

being ε_0 and ω the free space permittivity and the angular frequency respectively.
Once that the rays trajectories are computed, together with their complex amplitudes - weighted by the antenna pattern - and the travel times, they are classified into different wave fronts and then interpolated on the locations where the field is desired. The total field is obtained by addition of the single contribution of each wave front.

3. Received field estimation

The field reconstruction is described by means of a pilot example of a 4.5 GHz point-to-point radio link 80 km long affected by multipath. The aim of the technique is to evaluate the vertical electric field (or the received power) intensity at the receiver range. According to the considered geometry, this leads to the prediction of the vertical profile of the electric field $E(h)$ (or $P(h)$), where h is the receiver height ranging between $(h_0 - \Delta h/2)$ and $(h_0 + \Delta h/2)$.
In the following we refer to this interval of width Δh as *target zone* or simply *target*.
The modelled atmospheric condition is indicated in the panel of figure 4, while figure 5 shows the vertical profile of N at few chosen distances. The N value is comprised between

280 and 360 N-units and it shows variations in both dimensions. This atmospheric model - taken from (Bean and Dutton, 1968), p.325 - let us focus on several kind of interactions and their effects on the propagating signal that could possible occur in actual conditions.

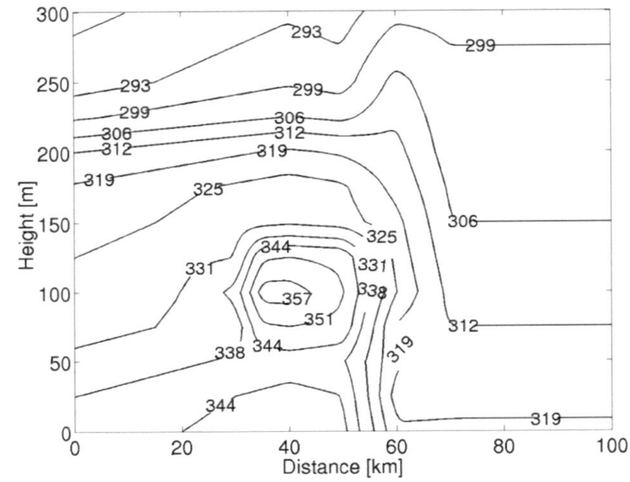

Fig. 4. 2D panel of the atmosphere charachteristics. Contour lines report the value of the refractivity index N

To appreciate the role of the variations in the refractivity, and without loss of generality, the ground profile was chosen as a plane surface characterized by the values of permittivity ε_r = 3 and conductivity σ = 0.001 [S/m] in the frequency band of interest.

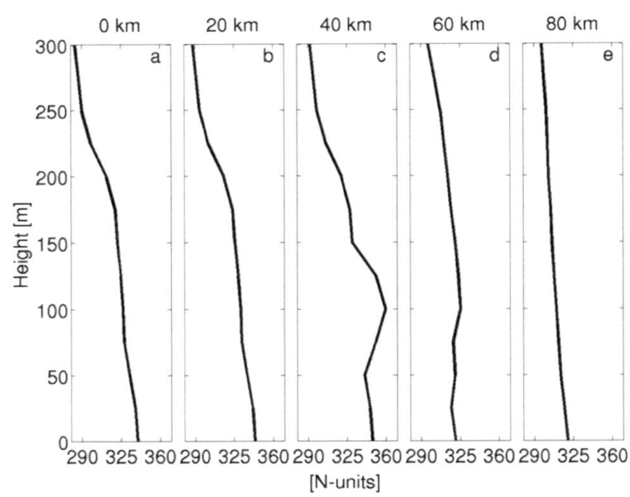

Fig. 5. Vertical profile at 5 chosen range distances.

The top image in figure 6 shows the trajectories of the rays belonging to a chosen elevation angular sector (-0.5° ÷ 0.5° with respect to the local horizon) traced in the cartographic system. The transmitter is in the origin of abscissas axis at 100 m above the ground. Besides the wavefront that carries the direct arrival of the signal, one may observe that several wavefronts are involved, resulting from the inhomogeneities of the atmosphere and ground reflection.

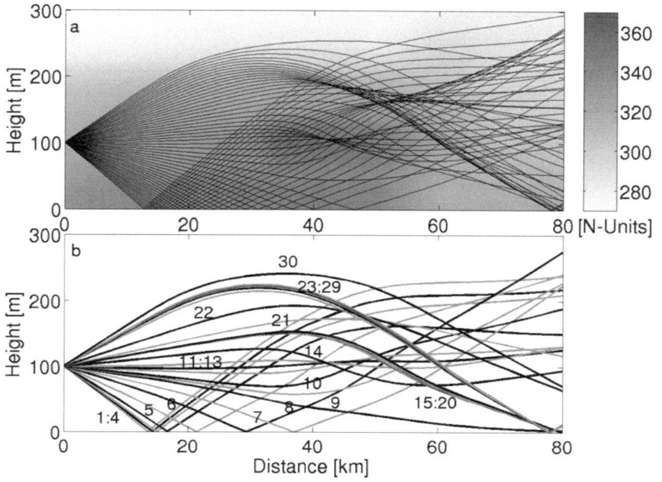

Fig. 6. Ray pattern (top) and wavefronts (bottom)

The bottom image of figure 6 shows the rays pertaining to the 30 wavefronts that results from the ray tracing process. The first step consists in grouping rays in different wave fronts, each characterized by the same *history* such as same number and location of caustics or reflections. Wavefronts are numbered and rays belonging to the same wavefront can be used to interpolate the field. As a matter of fact, rays arrive at the receiver range at discrete values so that the vertical profiling of the field intensity involves three steps: wavefronts separation; interpolation of the ray comb belonging to the same wavefront; coherent addition of the wavefronts themselves.

3.1 Wave front separation

The first criterion of wavefront separation that could come in mind is that of the travel time arrivals. Figure 7a shows the multi-valued behaviour of the wavefront delay of arrivals at different heights, while figure 7b shows how the parameterization of the wavefronts in the ray launching angle resolves the ambiguity: working in the angle domain seems the natural way for unfolding multi-valued ray fields (Operto et al., 2000).

The wavefronts separation is based on a broad and on a narrow selection process applied consecutively. The broad selection process involves the ray history: as for example, two rays belong to two different wavefronts if they are reflected a different number of times. The narrow selection process uses as criterion the *angular distance*: if two rays are reflected the same number of times they belong to the different wavefronts when their elevation starting angles difference is greater than $\delta\theta$. This angular distance has the role of guarantee the proper accuracy in the interpolation of the ray field and it depends on the velocity model of the medium (Sun, 1992).

The results of the wavefront separation process are organized in a database whose structure is schematically represented in figure 8. Wavefronts and rays are organized in a record frame that gives for the kth wavefront the propagation delay τ_k, the amplitude A_k and the phase ϕ_k obtained by interpolation of the travel time T_{ik}, a_{ik} and φ_{ik} of the rays at their arrival height, where i pertains to the individual rays in the wavefront k.

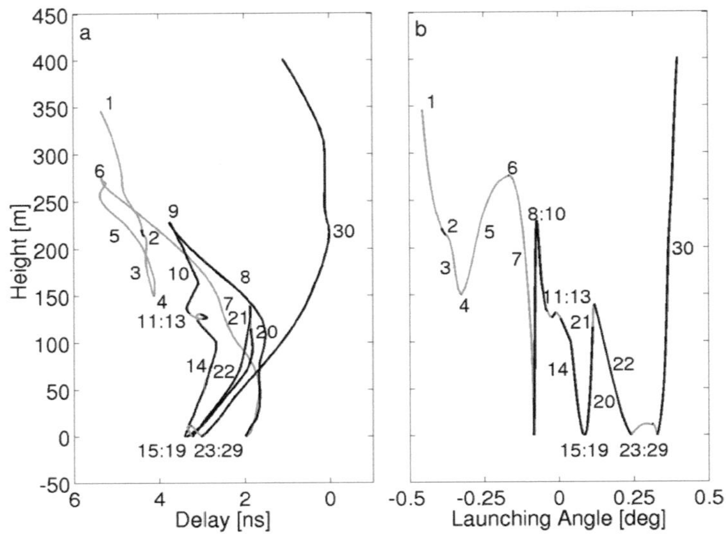

Fig. 7. Wavefronts time delay dispersion (a) and launching angle distribution (b)

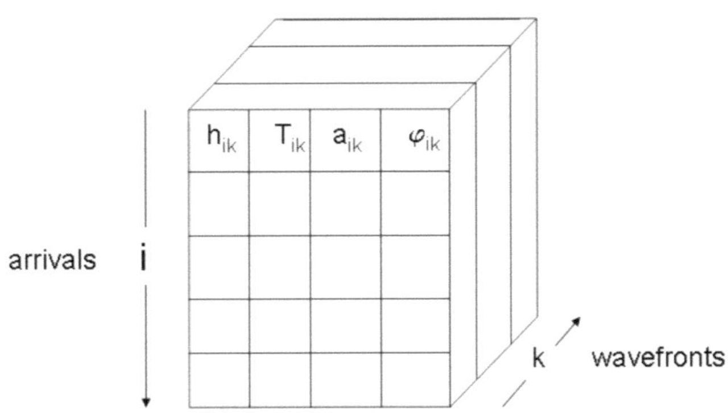

Fig. 8. Ray field database: the arrivals are organized according to their wavefronts

Each ray record refers is composed of the following parameters: the height of arrival, the travel time, which depend on the ray *history*; the amplitude and the phase, which depend on the ray *history*, on the reflection coefficients and on the system parameters.
The various quantities calculated by the ray tracing are referred to as Green's function attributes. Figure 9 shows among the attributes those that directly contribute to the field calculation.

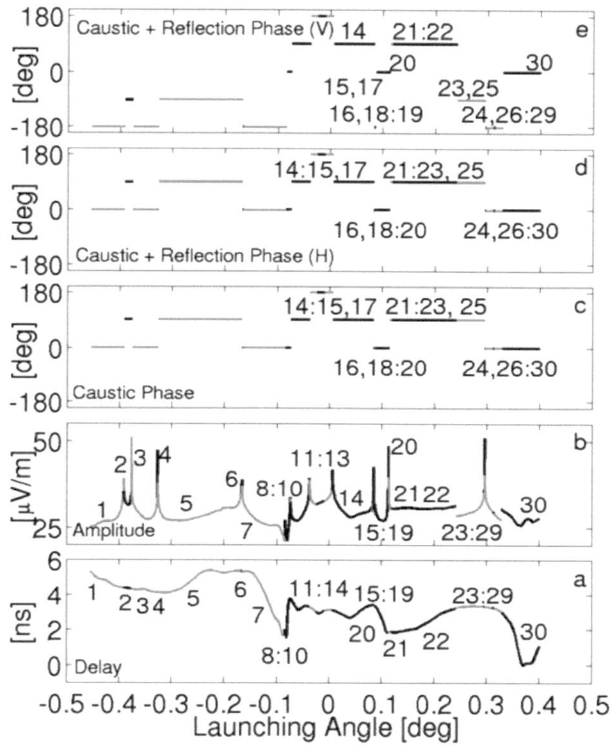

Fig. 9. Green's function attributes for the field calculation: from bottom to top: delay(a), amplitude (b), caustic phase (c), caustic +reflection phase (H polarization), caustic + reflection phase (V polarization)

3.2 Wave front interpolation

The field calculation along the vertical profile at the receiver range is performed through the interpolation of the different wavefronts contribution. Let K be the number of the wavefronts, each including N_k arrivals, the electric field associated to the ith arrival of the kth wavefront is:

$$\mathbf{E}(h_{ik}) = a_{ik} e^{j(2\pi f_0 T_{ik} + \phi_{ik})} \equiv \mathbf{E}_{ik} \tag{17}$$

where f is the frequency of operation of the link. The contribution of the kth wavefront to the total field is obtained through interpolation of the travel times, of the amplitudes and of the phases in the following way:

$$\mathbf{E}_k(h) = g_k(\mathbf{E}_{1k}, \mathbf{E}_{2k}, \ldots, \mathbf{E}_{N_k k}) \tag{18}$$

where g_k is a generic interpolating function and h is the height. As result, the interpolation of local values returns the field intensity in the target zone. Finally, the overall field is computed adding coherently all the wavefronts contributions:

$$\mathbf{E}(h) = \sum_{k=1}^{K} \mathbf{E}_k(h) \tag{19}$$

The vertical profile of the received power is obtained by applying locally the Poynting's theorem to a spherical surface with radius $\boldsymbol{R}$:

$$P(h) = \frac{4\pi R^2}{\eta} |\mathbf{E}(h)|^2 \tag{20}$$

being η the medium impedance and $\boldsymbol{R}$ the receiver range.

The solid line in the leftmost panel of figure 10 shows the power profile due to the direct arrival, while the dotted line represents the theoretical values predicted by Friis' formula (Balanis, 1996). Here, the vertical power profile is shown along the entire computational domain, from 0 up to 300 m, in order to show the effects of the tropospheric multipath on the received signal. The panel in the middle and the rightmost one of figure 10 show the power profile computed adding all the wavefront contribution at vertical (V) and horizontal (H) polarization.

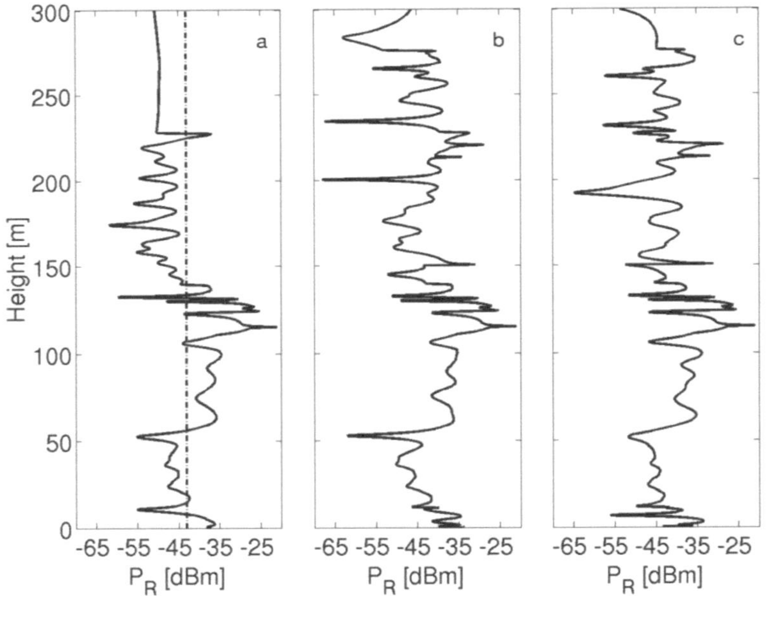

Fig. 10. Received Power vertical profile, from left to right: direct arrival contribution (solid line) and theoretical arrive (Friis); direct + reflected contribution (vertical polarization); direct + reflected contribution (horizontal polarization)

4. Radio channel analysis

4.1 Time domain approach

The aim of time domain approach is to estimate the radio channel response to a transmitted pulse with a given waveform $s(t)$. The use of the analytical extension concept let treat correctly the two distinct phase contributions of each arrival, i.e. the first one due to the propagation delay and the second one due to caustics and reflections, if any. The analytical extension of a base band signal $s(t)$ is given by:

$$\tilde{s}_{BB}(t) = s(t) + j\mathrm{H}\{\mathrm{s(t)}\} \tag{21}$$

where H represents the Hilbert transform operator. The corresponding pass band signal, which is transmitted on radio channel, is obtained with a frequency shift:

$$\tilde{s}_{RF} = \tilde{s}_{BB}(t)e^{j2\pi f_0 t} \tag{22}$$

The channel pulse response can be written as a sequence of ideal shifted pulses of the transmitted waveform with their own amplitude and phase:

$$w(t) = \sum_{k=1}^{K} A_K e^{j\phi_k}\delta(t-\tau_k) \tag{23}$$

where the amplitudes, the phases and the propagation delays have been calculated, respectively, through interpolation of a_{ik}, φ_{ik} and T_{ik} at height h. These procedure has low computational cost as the ray trajectories and propagation delay τ_k do not depend on the frequency and they can be computed once for all. In equation (23) some quantities such as amplitudes A_k and phases ϕ_k depends on the frequency of operation but using a narrow band approximation one can assume that amplitude and phase variations are negligible within the signal bandwidth. Under these hypothesis, the pass band received signal $\tilde{u}_{RF}(t)$ is given by the convolution between the pass band transmitted signal $\tilde{s}_{RF}(t)$ and the channel response (23):

$$\tilde{u}_{RF}(t) = \sum_{k=1}^{K} A_K e^{j\phi_k}\delta(t-\tau_k)\tilde{s}_{BB}(t-\tau_k)e^{j2\pi f_0(t-\tau_k)} \tag{24}$$

The received base band signal $\tilde{u}_{BB}(t)$ is obtained multiplying the pass band signal $\tilde{u}_{RF}(t)$ by the exponential $e^{-j2\pi f_0 t}$:

$$\tilde{u}_{BB}(t) = \sum_{k=1}^{K} A_k e^{j\phi_k}\tilde{s}_{BB}(t-\tau_k)e^{-j2\pi f_0 \tau_k} \tag{25}$$

whose real part gives the received signal $u(t)$:

$$u(t) = \mathrm{Re}\{\tilde{u}_{BB}(t)\} \tag{26}$$

The panel in figure 11 reports the estimated field and the time delay as a function of height ranging between $(h_0 - \Delta h/2)$ and $(h_0 + \Delta h/2)$, being h_0= 125 m and Δh = 25 m observed at the receiver range. This represents the impulse response of the system computed at vertical polarization. The input signal is a Nyquist wavelet and it is drawn in the left image of figure 11 as a dotted line. Also, figure 11 shows the received signal (solid line) at four given height of the panel. Two in-phase but slightly delayed arrivals are evident in panel a). In panel b) there are 2 arrivals, with one of them with a $\pi/4$ phase shift, while in panel d) the arrivals are of opposite phase.

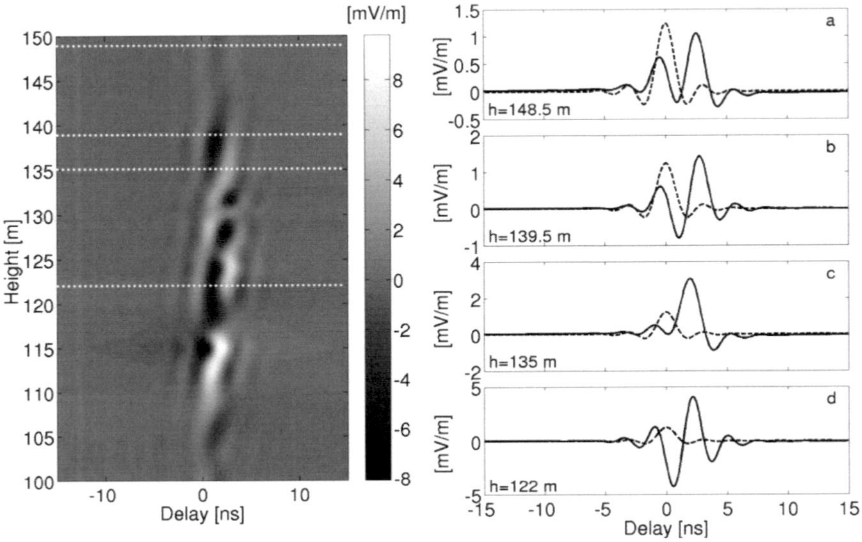

Fig. 11. Impulse response panel (left). The image on the right shows the channel response to the transmitted Nyquist wavelet at four height values, sketched as a dotted white line in the panel on the left.

4.2 Frequency domain approach

The computation of the electric field in the target zone for a given frequency f_0 as reported in §3.2 strictly holds under the hypothesis of the transmission of a sinusoidal signal, which has an infinite duration and an infinitesimal band. As radio systems transmit signals with finite duration and band, there is the need to evaluate the response of the radio channel to the effects of frequency variation inside the system bandwidth of operation.

Let us assume that the radio link is transmitting a signal with bandwidth B around the carrier frequency f_0. Therefore, given the frequency f and using the equation (17) principle, the received electric field expression is:

$$E(f) = \sum_{k=1}^{K} A_k e^{j(2\pi f \tau_k + \phi_k)} \tag{27}$$

Applying equation (27) for each frequency in the interval spanning from $(f_0 - B/2)$ to $(f_0 + B/2)$ one obtains an estimate of the pass band radio channel transfer function. The evaluation of equation (27) at different frequencies requires to update the amplitude terms

A_k and the phase terms ϕ_k, which both depends (directly or indirectly) on the frequency of operation *f*.
Frequency domain approach can lead also to a characterization of radio channel in the time domain applying Fourier transform to the radio channel transfer function provided by equation (27). However, when the bandwidth of the transmitted signal is narrow, that is when $f_0 >> B$, this procedure leads to a waste of computational resources. A more efficient way to manage narrow band channels is to face the problem in the time domain.
The transfer function of the radio channel is shown in figure 12 together with the frequency behaviour computed at the same four height levels highlighted in the time domain analysis. Again the target zone is comprised between 100 and 150 m and the signal has vertical polarization. One can notices how the transfer function may change from a smooth behaviour, panel c) to a sharper one with a notch deepness of about 30 dB, panel a).

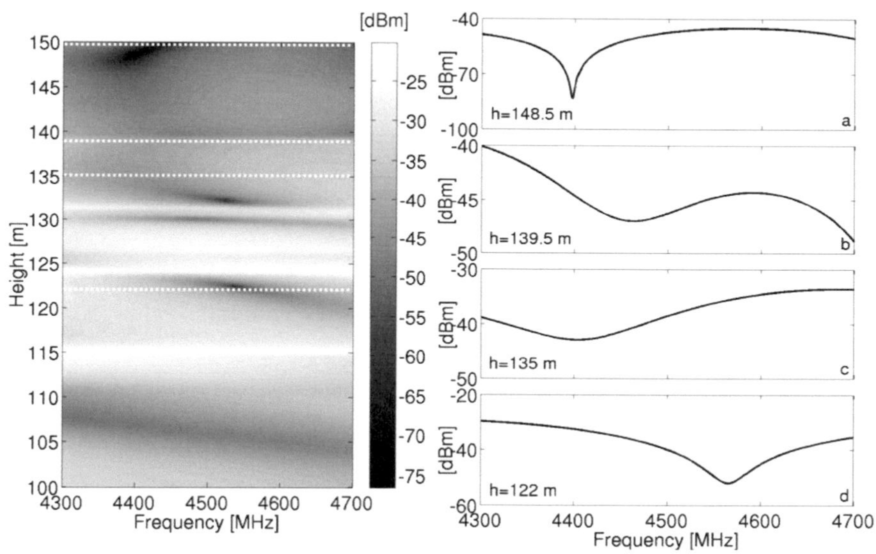

Fig. 12. Transfer function panel (left). The image on the right shows the channel response to the transmitted Nyquist wavelet at four height values, sketched as a dotted white line in the panel on the left.

5. Numerical validation

The ray tracing technique was developed to set up a prediction tool that is operative both as a MATLAB-based package and a C++ executable code. Unfortunately, beside an industrial application whose results are not available for publication at this stage, the computed field and power vertical profiles were not validated against measurements. Nevertheless the quality of the prediction performance can be observed when compared against canonical scenarios or other authors reference solutions.
Usually, when developing solver codes, one starts with the comparison of the predicted results with analytical or assessed solutions to check the correctness of the implementation. If this is the proper choice at that stage, it could lead to a flat and formal exposition of the methodology under study when describing the single steps towards the overall result. That

is why the particular example chosen to illustrate the technique throughout §3 and §4 is far complicated, as it appears from the number and variety of wavefronts and from the aspects of the radio link characteristics both in time and in frequency domain. As the complexity could conceal errors or misevaluations, the author chose as reference solution an example in which a standard atmosphere model describes the velocity of the propagation medium over a flat perfect electric conductor ground. Even in such a simple example, the importance of the wave front discrimination could induce in errors when reconstructing the field.

5.1 Two ray example

The transmitter, in the origin of the abscissa axis, is at 150 m above the ground and the receiver range is 50 km apart. The atmospheric model is a standard exponential one with a gradient of 40 ppm in the first 1000m. Figure 13 shows the trajectories together with the two wavefront, one associated with the direct arrival and another with the reflected one. The ground is a perfect conductor with reflection coefficient equal to -1.
The two arrivals, direct and reflected, are shown in figure 14: on the leftmost panel the wavefront time delay dispersion results in 2 distinct contribution at the same instant. As in §3.1, the parameterization in the angular domain nicely unfolds the integration path (rightmost panel) .
The predicted vertical received power profile is reported in figure 15, together with the direct arrival contributions given by Friis' formula. One can notice the typical pattern of zero and maxima of reception due to the rotating in-phase and counter-phase summation of the two arrivals.
Finally, figure 16 shows the impulse response panel calculated for the two ray model and vertical polarization. Again, the received signal shows different behaviour due to the atmospheric multipath as a function of the height.

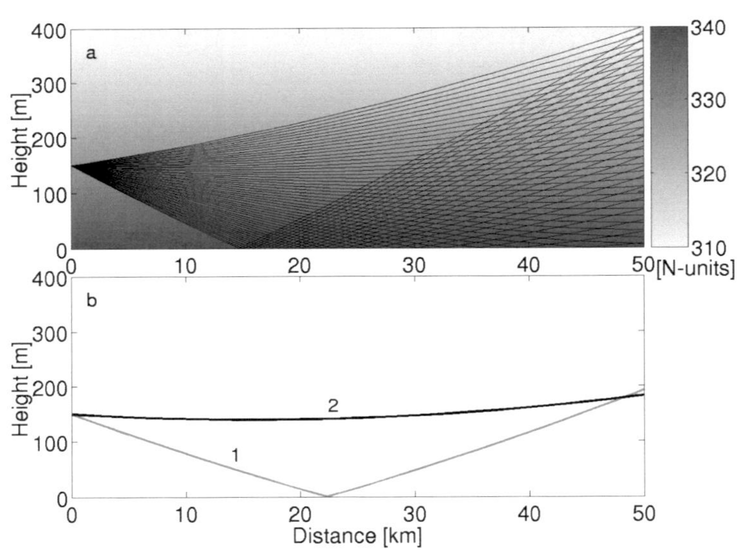

Fig. 13. Ray pattern (top) and wavefronts (bottom)

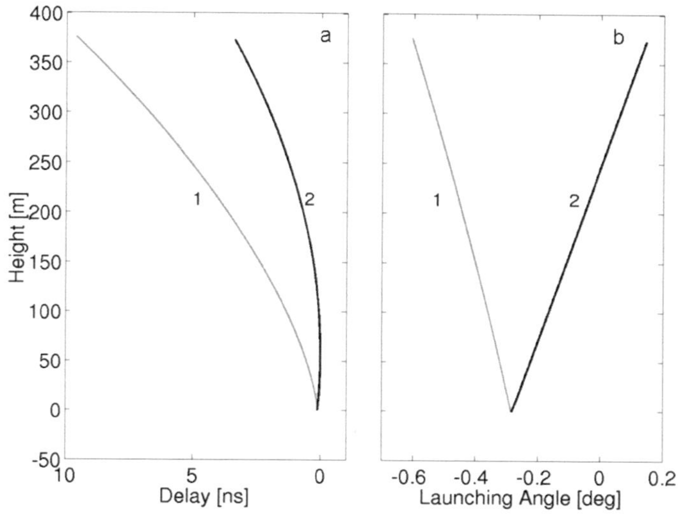

Fig. 14. Wavefronts time delay dispersion (a) and launching angle distribution (b)

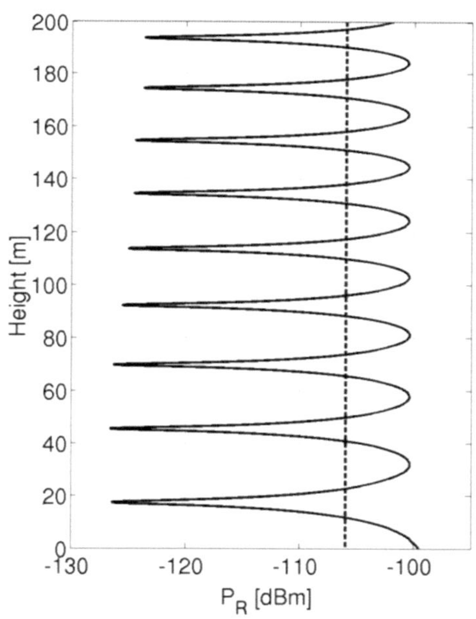

Fig. 15. Received Power vertical profile (solid line) and direct arrival (dotted line)

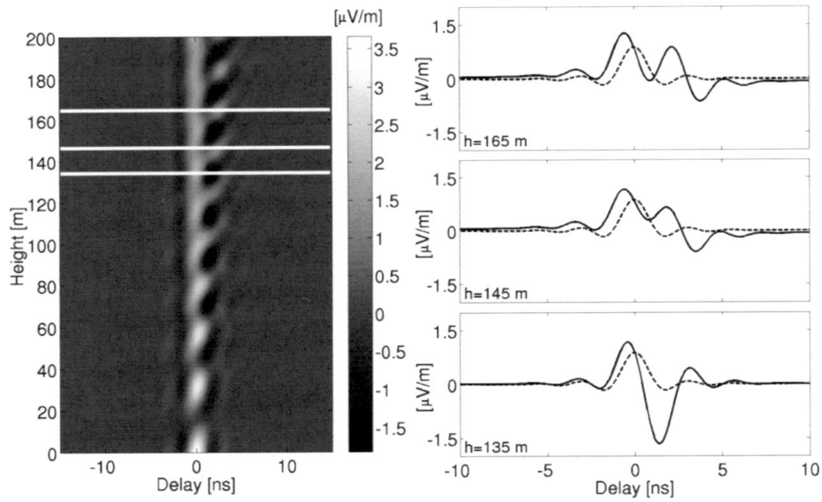

Fig. 16. Impulse response panel (left). The image on the right shows the channel response to the transmitted Nyquist wavelet at three height values, sketched as a white line in the panel on the left.

6. Conclusion

The ray tracing technique based on the resolution of the Eikonal equation with Hamilton-Jacobi technique is well suited to model the dispersive effects of perturbed atmosphere on the performance of high-capacity digital radio channel.
The advantage of such an approach is the possibility of describing the medium with its variation in the lateral domain in presence of atmospheric multipath and of non uniform terrain profile.
The field profiling at the receiving site is obtained through 2,5D propagation as, notwithstanding all variations in the third dimension are taken equal to zero, the field amplitude is calculated correctly because of the paraxial approximation which takes place in a 3D domain.
At the time of writing the author is refining the code towards a full 3D implementation, which requires a reformulation in spherical coordinates, and the modelling of the mechanism of ground scattering and diffraction.

7. Acknowledgment

The author would like to thank Professor Giuseppe Drufuca (Politecnico di Milano) for his useful comments and suggestions and Simone Re (former research associate, Politecnico di Milano) for his contribution to the code development and calculations.

8. References

Abramovitz M., Stegun I.A. (1970), *Handbook of mathematical functions*, Dover Publications, New York (USA).

Akbarpour R., Webster A.R., Ray-Tracing and parabolic Equation Methods in the Modeling of a Tropospheric Microwave Link, *IEEE Transactions on Antennas and Propagation*, Vol.53, No.11, (November 2005), pp. 2785-3791, ISSN 0018-926X.

Balanis C. A (1996), *Antenna Theory: Analysis and Design*, John Wiley & Sons, ISBN 0471592684, New York (USA).

Bean B. R., Dutton E.~J.(1968), *Radio Meteorology*, Dover Publications, New York (USA)

CCIR, Report 338, Propagation Data and Prediction Methods Required for Line-of-Sight Radio Relay Systems, *Study Group 5 of the International Consultative Radio Committee (CCIR)*, Geneva (CH), 1978.

Červenŷ V. I.A. Molotkov, I. Psencik (1977), *Ray method in Seismology*, Univerzita Karlova Praha.

Driessen P.F., Prediction of multipath delay profiles in mountainous terrain, *IEEE Journal on Selected Areas in Communications*, Vol.18, No.3, (March 2000), pp.336-346, ISSN 0733-8716.

Du Castel (1966), *Tropospheric Radiowave propagation beyond the Horizon*, Pergamon Press, Oxford (UK).

Eresmaa R., Healy S., Järvinen H., Salonen K., Implementation of a ray-tracing operator for ground-based GPS Slant Delay observation modelling, *Journal of Geophysical Research*, Vol.113, D11114, doi:10.1029/2007JD009256, (June 2008), 9 pages , ISSN 0148-0227.

Farra V., Ray Tracing in Complex Media, *Journal of Applied Geophysics*, Vol.30, No.1-2, (April 1993), pp. 55-73, ISSN 0926-9851.

Hall M. P. M.(1979), *Effects of the troposphere on radio communication*, IEEE Electromagnetic Waves Series, ISBN 0-906048-25-7, Exeter, UK.

Keller J.B., Diffraction by an Aperture, *Journal of Applied Physics*, Vol. 28, No. 4, (April 1957), pp 426-444.

Kravtsov Y.A., Orlov Y.I (1990), *Geometrical Optics of Inhomogeneous Media*, Springer-Verlag, ISBN 3-540-51944-0, New York (USA).

Kružkov S. N., Generalized solutions of the Hamilton-Jacobi equations of Eikonal type I. Formulation of the problems; existence, uniqueness and stability theorems; some properties of the solutions, *Mathematics of the USSR-Sbornik*, Vol. 27, No. 3, (1975), pp. 406-446.

Kurner T., Cichon D. J., Concepts and results for 3D digital terrain-based wave propagation models: an overview, *IEEE Journal on Selected Areas in Communication*, Vol.11, No.7, (September 1993), pp.1002-1012, ISSN 0733-8716.

Lebherz M., Wiesbeck W., Krank W., A Versatile wave propagation model for the VHF/UHF range considering three-dimensional terrain, *IEEE Transactions on Antennas and Propagation*, Vol.40, No.10 (October 1992), pp 1121-1131, ISSN 0018-926X.

Livingston D.C. (1970). *The physics of microwave propagation*, Prentice Hall, New Jersey (USA).

Operto M.S., Xu S., Lambaré G., Can we quantitatively image complex structures with rays?, *Geophysics*, Vol. 65, No. 4, (July August 2000), pp. 1223-1238, ISSN 0016-8033.

Pany T., Pesec P., G. Stangl, Atmospheric GPS Slant Path Delays and Ray Tracing through Numerical Wather Models, a Comparison, Physics and Chemistry of the Earth, Part A: Solid Earth and Geodesy, Vol.26, No.3, (2001), pp.183-188, ISSN 1474-7065.

Sevgi L., Felsen L.B., A New Algorithm For Ground Wave Propagation Based On A Hybrid Ray-Mode Approach, *International Journal of Numerical Modelling:*, Vol.11, No.2, (March-April 1998), pp. 87-103, Online ISSN 1099-1204.

Sun Y., Computation of 2D multiple arrival travel time fields by an interpolative shooting method, SEG Annual Meeting, Extended Abstracts (1992), pp 156-159, Tulsa (OK), USA.

Ulaby F.T. (1999), Fundamentals of applied electromagnetics, Prentice Hall, ISBN 0-13-011554-1, New Jersey (USA).

Vinje V., Iversen E., Gjoystdal H., Traveltime and amplitude estimation using wavefront construction, *Geophysics*, Vol. 58, No. 6, (August 1993), pp. 1157-1166, ISSN 0016-8033.

8

High Frequency Techniques: the Physical Optics Approximation and the Modified Equivalent Current Approximation (MECA)

Javier Gutiérrez-Meana, José Á. Martínez-Lorenzo and Fernando Las-Heras
University of Oviedo
Spain

1. Introduction

In most of the electromagnetic problems, the number of unknowns to evaluate the scattered fields grows whenever the size of the antenna, device or scenario increases or the working frequency becomes higher. In this context, the rigorous full-wave methods –e.g. Method of Moments (MoM), Fast Multipole Method (FMM) (Engheta et al., 1992), Finite-Difference Time-Domain (FDTD) (Taflove & Umashankar, 1987) or Finite-Difference Frequency-Domain (FDFD) (Rappaport & McCartin, 1991), Finite Element Method (FEM) (Kempel et al., 1998) – can not tackle the analysis of such problems beyond an upper limit determined by the computational requirements in terms of time and memory. High frequency techniques consist in the asymptotic evaluation of the Maxwell's equations. As a consequence, they provide good accuracy when dealing with electrically large geometries meanwhile the computational needs diminish with respect to the aforementioned methods.

Within the high frequency techniques, the Geometrical Optics (GO) and the Physical Optics (PO) approximation are the most extended methods due to the successful results obtained in various fields such as Radar Cross Section (RCS), design of reflector antennas or radioelectric coverage calculation. Since the Physical Optics approximation is detailed in the following section, the Geometrical Optics is briefly summarised.

The main interest in the GO lies in the fact that incident, reflected and transmitted electromagnetic waves are studied based on the conservation of the energy flux along a ray tube between a source and an observation point. Therefore, the Geometrical Optics is usually referred to as Ray Optics. The GO comprises two different methodologies (Rossi & Gabillet, 2002): Ray Tracing (Glassner, 1989) – the starting point is the receiver or observation point and a path to the source is sought analysing the reflections on walls, buildings, mountains – and Ray Launching – multiple rays are launched from the source, so they are independently followed until an observation point or the receiver is reached. One of the common applications of the GO is the evaluation of radio electric coverage or the channel characterization in urban scenarios.

Both the GO and the PO techniques require of an additional method to compute the contribution due to the diffraction phenomenon. The GO can be complemented by means of the Geometrical Theory of Diffraction (GTD) (Keller, 1962) or the Uniform Theory of

Diffraction (UTD) (Pathak & Kouyoumjian, 1974). On the other hand, the Physical Theory of Diffraction (PTD) (Ufimtsev, 1962) is applied in joint with the PO formulation.
This chapter focuses on the PO approximation and especially on its extension to dielectric and lossy materials, namely the Modified Equivalent Current Approximation (MECA) method. In section 2, the PO approximation is presented and the formulation is obtained from the equivalence principle. Then the MECA method is introduced, so the equivalent current densities are obtained at the end of section 3.1.3. In order to complete the expressions, the reflection coefficients are then calculated and the determination of the electromagnetic field levels at the observation points based on an analytical solution of the radiation integral is accomplished in section 3.2. A couple of validation examples are studied before continuing to describe the algorithms for solving the visibility problem and, specifically, the Pyramid method in section 4. Afterwards, some of the applications of MECA are listed in "*Application examples*" where an example of radio electric coverage evaluation is also shown. "*Conclusion*" and "*References*" closes the chapter.

2. The Physical Optics (PO) approximation

The Physical Optics (PO) approximation (Harrington, 2001; Balanis, 1989) is a well-known high frequency technique based on the determination of the equivalent current densities induced on the surface of an illuminated perfect electric conductor (PEC) plane. The final expressions can be achieved through the equivalence principle as it will be shown in section 2.1, so once these equivalent or PO current densities are obtained, both electric and magnetic field levels can be calculated from the corresponding radiation integrals.
In order to be in compliance with the constraints introduced by the PO approximation, some aspects have to be evaluated prior to the selection of this technique to tackle any electromagnetic problem:

- The geometry must be made of electrically large obstacles with a smooth variation of their surfaces.
- Given a radiating source, a distinction between lit and non-lit regions must be possible to perform.

The latter implies the need of a complementary algorithm to identify line of sight (LOS) directions from a specific point of view. Some methods will be commented in section 4.

Henceforth, for the sake of simplicity, a time-harmonic variation, $e^{j\omega t}$, of the electromagnetic fields and current densities is assumed. Likewise, the spatial dependence of those variables is not explicitly written.

2.1 PO formulation

The expressions for the electric and magnetic equivalent current densities, J^{PO} and M^{PO} respectively, can be derived from the physical equivalent (Balanis, 1989) in Fig. 1(left), where a scheme consisting of two different media is presented: a PEC on the left and a medium with permittivity ε_1, permeability μ_1 and conductivity σ_1 on the right. Consequently, the boundary S between the first and the second media is a PEC plane with the outward normal unit vector $\hat{n}$ pointing to the right.

The incident electric and magnetic fields due to external sources and in absence of any obstacle are E_1^{inc} and H_1^{inc}. On the other hand, the total fields inside the PEC are null ($E_{PEC}^{tot} = H_{PEC}^{tot} = \mathbf{0}$), while in the second medium E_1^{tot} and H_1^{tot} are calculated by adding those incident fields to the reflected ones denoted by E_1^{ref} and H_1^{ref}. Therefore the electric and magnetic induced current densities, J and M, at the boundary S can be obtained from the tangential components of the total fields as:

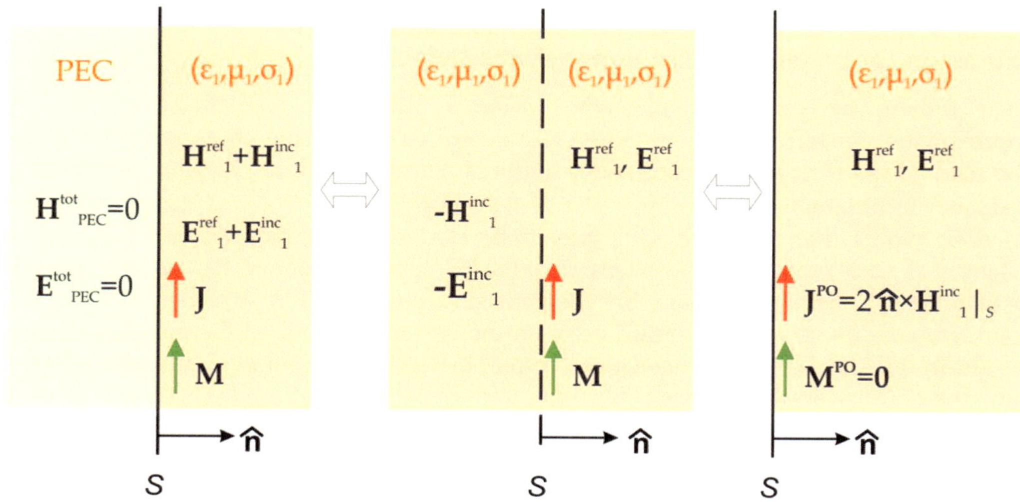

Fig. 1. Equivalence principle and Physical Optics approximation.

$$\begin{aligned} J &= \hat{n}\times\left(H_1^{tot}-H_{PEC}^{tot}\right)\Big|_S = \hat{n}\times H_1^{tot}\Big|_S = \hat{n}\times\left(H_1^{inc}+H_1^{ref}\right)\Big|_S \\ M &= -\hat{n}\times\left(E_1^{tot}-E_{PEC}^{tot}\right)\Big|_S = -\hat{n}\times E_1^{tot}\Big|_S = -\hat{n}\times\left(E_1^{inc}+E_1^{ref}\right)\Big|_S = \mathbf{0} \end{aligned} \tag{1}$$

Fig. 1 (centre) shows the equivalent problem for the non-PEC medium. In order to keep the same boundary conditions, this medium has been extended replacing the PEC part and the total fields are now $-E_1^{inc}$ and $-H_1^{inc}$ at the left side of S and E_1^{ref} and H_1^{ref} at its right side.

Considering the characterization of the boundary S as a PEC plane in Fig. 1(right) and taking into account the expressions in Eq. (1), the Physical Optics approximation states that H_1^{inc} and H_1^{ref} at the boundary S are in phase and also have the same amplitude. Thus, J^{PO} and M^{PO} can be expressed as:

$$\begin{aligned} J^{PO} &= \hat{n}\times\left(H_1^{inc}+H_1^{ref}\right)\Big|_S \approx 2\hat{n}\times H_1^{inc}\Big|_S \\ M^{PO} &= -\hat{n}\times\left(E_1^{inc}+E_1^{ref}\right)\Big|_S = \mathbf{0} \end{aligned} \tag{2}$$

At this point, an additional consideration to complete the PO formulation is included: when dealing with finite geometries, the PO current density is null in the regions not illuminated by the source:

$$J^{PO} = \begin{cases} 2\hat{n}\times H_1^{inc}\Big|_S & \text{lit region} \\ \mathbf{0} & \text{non-lit region} \end{cases} \tag{3}$$

This is the reason why the distinction between shadowed and illuminated parts of the scenario is one of the aforementioned constraints to correctly apply the PO approximation.

3. The Modified Equivalent Current Approximation (MECA) method: a PO extension for penetrable and non-metallic objects

Even though several electromagnetic problems fulfil the restrictions by the PO approximation in terms of electric size or radius of curvature, the characterization of the obstacles as perfect electric conductors reduces the scope of application of this high frequency technique.
Previous works (Rengarajan & Gillespie, 1988; Hodges & Rahmat-Samii, 1993, Sáez de Adana et al., 2004) agree that the extension of the PO approximation for penetrable and non-metallic objects have to account for the reflection coefficients R in order to apply the boundary conditions at the interface between the two media -one of the media uses to be the air. In this context, it is convenient to accomplish a decomposition of the incident fields into their transversal electric (TE) and transversal magnetic (TM) components to independently insert R_{TE} and R_{TM} in the evaluation of the equivalent current densities:

$$E^{inc} = E^{inc}_{TE} + E^{inc}_{TM} = \mathrm{E}^{\mathrm{inc}}_{\mathrm{TE}}\hat{e}_{TE} + \mathrm{E}^{\mathrm{inc}}_{\mathrm{TM}}\hat{e}_{TM} \tag{4}$$

where $\hat{e}_{TE}$ and $\hat{e}_{TM}$ are the unit vectors in the direction of the TE and TM components of the incident electric field respectively. Likewise, the incident magnetic field can be written in an analogous way.
The Modified Equivalent Current Approximation (MECA) method (Meana et al., 2010) is a new high frequency technique based on the evaluation of a set of equivalent currents to calculate electromagnetic field levels at any observation point. The most important features of this method are:

- MECA deals with electrically large scenarios consisting of dielectric and lossy surfaces due to both electric and magnetic equivalent currents, J^{MECA} and M^{MECA}, are taken into account.
- MECA reduces to the PO expressions when considering PEC obstacles. Therefore, it can be seen as an extension and an improvement of the classic Physical Optics approximation.
- Reflection coefficients are calculated in a rigorous manner.
- A constant amplitude and linear phase current density distribution on the surface of the facets is adopted to analytically solve the radiation integral.

3.1 MECA formulation

Let us suppose a skew plane wave impinging on the interface S with an angle of incidence denoted by θ^{inc} and a unit incident vector $\hat{k}^{inc}$. The two media are now characterised by their respective permittivities (ε_1, ε_2), permeabilities (μ_1, μ_2) and conductivities (σ_1, σ_2). The analysis of the TE and TM components will be presented separately.

3.1.1 TE component

For the incident wave, the coordinate system $\left[\hat{k}^{inc}, \hat{e}_{TM}, \hat{e}_{TE}\right]$ where

$$\hat{e}_{TE} = \frac{\hat{k}^{inc} \times \hat{n}}{\left|\hat{k}^{inc} \times \hat{n}\right|} \qquad (5)$$
$$\hat{e}_{TM} = \hat{e}_{TE} \times \hat{k}^{inc}$$

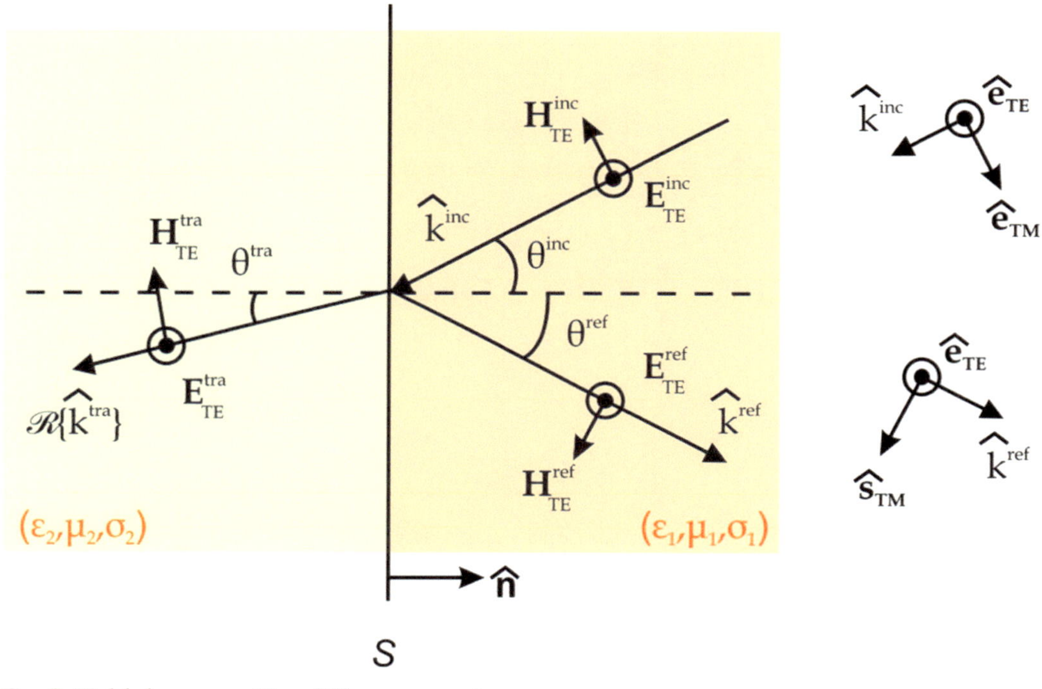

Fig. 2. Field decomposition. TE component.

has been chosen (see Fig. 2). Substituting $\hat{k}^{inc}$ by the unit reflection vector $\hat{k}^{ref}$, the coordinate system $\left[\hat{s}_{TM}, \hat{k}^{ref}, \hat{s}_{TE} = \hat{e}_{TE}\right]$ is defined. Then, the incident electric and magnetic fields are rewritten in terms of these unit vectors:

$$E_{TE}^{inc} = \mathrm{E}_{\mathrm{TE}}^{\mathrm{inc}} \hat{e}_{TE}$$
$$H_{TE}^{inc} = -\mathrm{H}_{\mathrm{TE}}^{\mathrm{inc}} \hat{e}_{TM} = \frac{1}{\eta_1} \mathrm{E}_{\mathrm{TE}}^{\mathrm{inc}} \left(\hat{k}^{inc} \times \hat{e}_{TE}\right) \qquad (6)$$

where the intrinsic impedance of the first medium η_1 is $\eta_{\mathrm{i}} = \sqrt{\dfrac{\mathrm{j}\omega\mu_{\mathrm{i}}}{\mathrm{j}\omega\varepsilon_{\mathrm{i}} + \sigma_{\mathrm{i}}}}$, $\mathrm{i} = 1,2$.

From the Snell law, the angle of reflection θ^{ref} is equal to θ^{inc}. Therefore, $\hat{k}^{ref}$ can be expressed as:

$$\hat{k}^{ref} = \hat{k}^{inc} - 2\hat{n}\left(\hat{k}^{inc} \cdot \hat{n}\right) \tag{7}$$

Inserting the TE reflection coefficient R_{TE}, the electric E_{TE}^{ref} and magnetic H_{TE}^{ref} reflected fields are:

$$\begin{aligned} E_{TE}^{ref} &= \mathrm{R_{TE}} \mathrm{E_{TE}^{ref}} \hat{e}_{TE} \\ H_{TE}^{ref} &== \frac{1}{\eta_1} \mathrm{R_{TE}} \mathrm{E_{TE}^{inc}} \hat{s}_{TM} = \frac{1}{\eta_1} \mathrm{R_{TE}} \mathrm{E_{TE}^{inc}} \left(\hat{k}_{ref} \times \hat{e}_{TE}\right) \end{aligned} \tag{8}$$

Considering that

$$\begin{aligned} E_{TE,TM}^{tot} &= E_{TE,TM}^{inc} + E_{TE,TM}^{ref} \\ H_{TE,TM}^{tot} &= H_{TE,TM}^{inc} + H_{TE,TM}^{ref} \end{aligned} \tag{9}$$

and introducing Eq. (8) in Eq. (9), the total fields for the TE component are:

$$\begin{aligned} E_{TE}^{tot} &= \mathrm{E_{TE}^{inc}} \left(1 + \mathrm{R_{TE}}\right) \hat{e}_{TE} \\ H_{TE}^{tot} &= \frac{1}{\eta_1} \mathrm{E_{TE}^{inc}} \left[\left(\hat{k}^{inc} \times \hat{e}_{TE}\right) + \mathrm{R_{TE}} \left(\hat{k}^{ref} \times \hat{e}_{TE}\right)\right] \end{aligned} \tag{10}$$

3.1.2 TM component

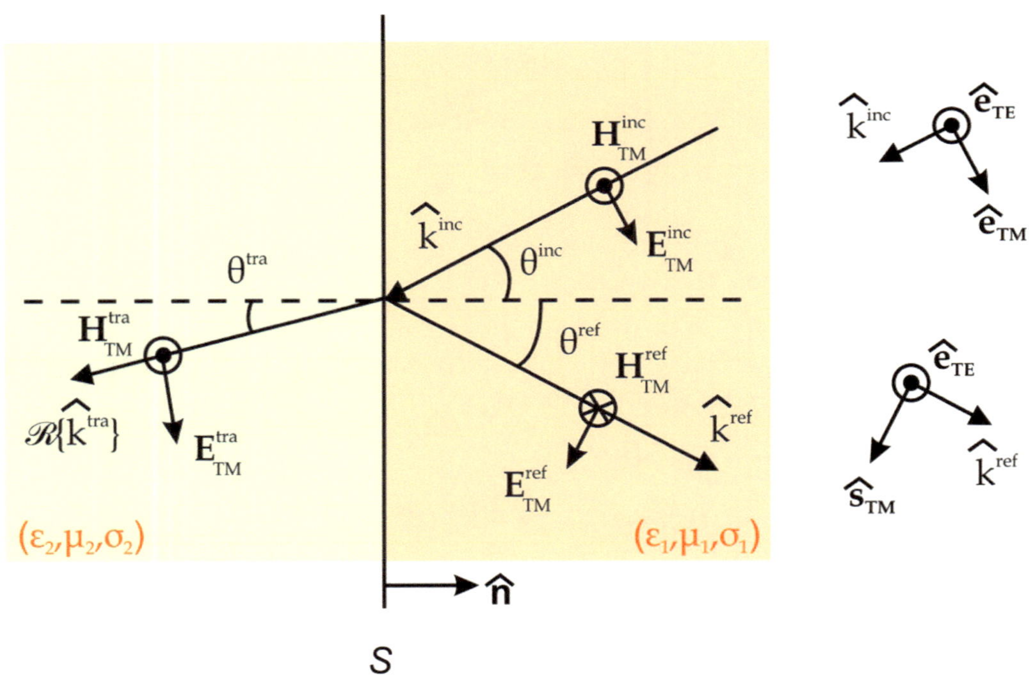

Fig. 3. Field decomposition. TM component.

In Fig. 3 the same directions of incidence, reflection and transmission as well as the boundary S are again depicted. In addition, the TM components of the fields are plotted. In accordance with the reasoning in the previous section, the expressions for the total fields can be obtained, leading to:

$$\begin{aligned} E_{TM}^{tot} &= \mathrm{E}_{\mathrm{TM}}^{\mathrm{inc}}\left[\hat{e}_{TM} + \mathrm{R}_{\mathrm{TM}}\left(\hat{k}^{ref} \times \hat{e}_{TE}\right)\right] \\ H_{TM}^{tot} &= \frac{1}{\eta_1}\mathrm{E}_{\mathrm{TM}}^{\mathrm{inc}}\left(1 - \mathrm{R}_{\mathrm{TM}}\right)\hat{e}_{TE} \end{aligned} \tag{11}$$

3.1.3 MECA equivalent currents

Adding the contributions of the TE and TM components in Eqs. (10) and (11), the total electric and magnetic fields can be written as a function of the incident electric field, the reflection coefficients, the intrinsic impedance of the first medium, the propagation vectors and the unit vector $\hat{e}_{TE}$:

$$\begin{aligned} E^{tot} &= E^{inc} + \mathrm{R}_{\mathrm{TE}}\mathrm{E}_{\mathrm{TE}}^{\mathrm{inc}}\hat{e}_{TE} + \mathrm{R}_{\mathrm{TM}}\mathrm{E}_{\mathrm{TM}}^{\mathrm{inc}}\left(\hat{k}^{ref} \times \hat{e}_{TE}\right) \\ H^{tot} &= \frac{1}{\eta_1}\hat{k}^{inc} \times E^{inc} + \frac{1}{\eta_1}\left[\mathrm{R}_{\mathrm{TE}}\mathrm{E}_{\mathrm{TE}}^{\mathrm{inc}}\left(\hat{k}^{ref} \times \hat{e}_{TE}\right) - \mathrm{R}_{\mathrm{TM}}\mathrm{E}_{\mathrm{TM}}^{\mathrm{inc}}\hat{e}_{TE}\right] \end{aligned} \tag{12}$$

Going back to the boundary conditions, the MECA equivalent current densities are reckoned from the total fields at S in Eq. (12), being the magnetic ones no longer zero:

$$\begin{aligned} J^{MECA} &= \hat{n} \times H^{tot}\Big|_{\mathrm{S}} = \\ &= \frac{1}{\eta_1}\left\{\mathrm{E}_{\mathrm{TE}}^{\mathrm{inc}}\cos\theta^{\mathrm{inc}}\left(1 - \mathrm{R}_{\mathrm{TE}}\right)\hat{e}_{TE} + \mathrm{E}_{\mathrm{TM}}^{\mathrm{inc}}\left(1 - \mathrm{R}_{\mathrm{TM}}\right)\left(\hat{n} \times \hat{e}_{TE}\right)\right\}\Big|_{\mathrm{S}} \\ M^{MECA} &= -\hat{n} \times E^{tot}\Big|_{\mathrm{S}} = \left\{\mathrm{E}_{\mathrm{TE}}^{\mathrm{inc}}\left(1 + \mathrm{R}_{\mathrm{TE}}\right)\left(\hat{e}_{TE} \times \hat{n}\right) + \mathrm{E}_{\mathrm{TM}}^{\mathrm{inc}}\cos\theta^{\mathrm{inc}}\left(1 + \mathrm{R}_{\mathrm{TM}}\right)\hat{e}_{TE}\right\}\Big|_{\mathrm{S}} \end{aligned} \tag{13}$$

As it can be seen and remarked in the previous expressions, the amplitude and phase of the current densities depend on the polarization and also on the angle of incidence θ^{inc} of the incident wave.

3.1.4 TE and TM reflection coefficients

In order to complete the evaluation of the MECA current densities, R_{TE} and R_{TM} have to be determined before moving to the calculation of the electromagnetic fields at the observation points. For this task, the boundary condition for the continuity of the tangential components of the electric and magnetic fields at the boundary S is utilised.

The general dispersion relation applied to the second isotropic medium allows the evaluation of the transmission propagation vector $\hat{k}^{tra}$ (its real part is depicted in Fig. 2 and Fig. 3) due to

$$k \cdot k = \omega^2 \mu_{\mathrm{i}}\left(\varepsilon_{\mathrm{i}} - \frac{\mathrm{j}\sigma_{\mathrm{i}}}{\omega}\right) \tag{14}$$

where $k = [\mathrm{k}_1 \quad \mathrm{k}_2 \quad \mathrm{k}_3]$ denotes any propagation vector, the subscript i refers to the first $(\mathrm{i}=1)$ or second medium $(\mathrm{i}=2)$ and $\omega = 2\pi\mathrm{f}$, with f being the working frequency. If $\sigma_{\mathrm{i}} \neq 0$, the right hand side of Eq. (14) is a complex number and the analysis is better performed introducing the real and imaginary parts of the vector k, denoted by $\Re\{k\}$ and $\Im\{k\}$ respectively. Particularising for the reflection and transmission propagation vectors:

$$\begin{cases} \left|\Re\{k^{ref}\}\right|^2 - \left|\Im\{k^{ref}\}\right|^2 = \omega^2\mu_1\varepsilon_1 \\ 2\Re\{k^{ref}\}\cdot\Im\{k^{ref}\} = 0 \\ \\ \left|\Re\{k^{tra}\}\right|^2 - \left|\Im\{k^{tra}\}\right|^2 = \omega^2\mu_2\varepsilon_2 \\ 2\Re\{k^{tra}\}\cdot\Im\{k^{tra}\} = -\omega\mu_2\sigma_2 \end{cases} \tag{15}$$

The continuity condition at S for the first component of the propagation vector means that

$$\Re\{\mathrm{k}_1^{\mathrm{inc}}\} = \Re\{\mathrm{k}_1^{\mathrm{ref}}\} = \Re\{\mathrm{k}_1^{\mathrm{tra}}\} = \frac{2\pi}{\lambda_1}\sin\theta^{\mathrm{inc}} \tag{16}$$

where λ_1 is the wavelength in the first medium and

$$\Im\{\mathrm{k}_1^{\mathrm{inc}}\} = \Im\{\mathrm{k}_1^{\mathrm{ref}}\} = \Im\{\mathrm{k}_1^{\mathrm{tra}}\} = 0 \tag{17}$$

Inserting Eqs. (16) and (17) in Eq. (15), the real and imaginary parts of the third component of the transmission propagation vector are obtained as:

$$\begin{aligned} \Re\{\mathrm{k}_3^{\mathrm{tra}}\} &= \frac{1}{\sqrt{2}}\left[\Re\{k^{tra2}\} - \left(\frac{2\pi}{\lambda_1}\right)^2\sin^2\theta^{\mathrm{inc}} + \sqrt{\Re\{k^{tra2}\} - \left(\frac{2\pi}{\lambda_1}\right)^2\sin^2\theta^{\mathrm{inc}} + \Im\{k^{tra2}\}}\right]^{1/2} \\ \Im\{\mathrm{k}_3^{\mathrm{tra}}\} &= \frac{\Im\{k^{tra2}\}}{2\Re\{\mathrm{k}_3^{\mathrm{tra}}\}} \end{aligned} \tag{18}$$

Once $\mathrm{k}_3^{\mathrm{tra}} = \Re\{\mathrm{k}_3^{\mathrm{tra}}\} + \mathrm{j}\Im\{\mathrm{k}_3^{\mathrm{tra}}\}$ is determined, R_{TE} is (Staelin et al., 1993)

$$\mathrm{R}_{\mathrm{TE}} = \frac{\mu_2 \, {}^{2\pi}\!/_{\lambda_1}\cos\theta^{\mathrm{inc}} - \mu_1\mathrm{k}_3^{\mathrm{tra}}}{\mu_2 \, {}^{2\pi}\!/_{\lambda_1}\cos\theta^{\mathrm{inc}} + \mu_1\mathrm{k}_3^{\mathrm{tra}}} \tag{19}$$

and the corresponding R_{TM} is

$$\mathrm{R}_{\mathrm{TM}} = -\frac{\left(\varepsilon_2 - \mathrm{j}\,{}^{\sigma_2}\!/_{\omega}\right){}^{2\pi}\!/_{\lambda_1}\cos\theta^{\mathrm{inc}} - \varepsilon_1\mathrm{k}_3^{\mathrm{tra}}}{\left(\varepsilon_2 - \mathrm{j}\,{}^{\sigma_2}\!/_{\omega}\right){}^{2\pi}\!/_{\lambda_1}\cos\theta^{\mathrm{inc}} + \varepsilon_1\mathrm{k}_3^{\mathrm{tra}}} \tag{20}$$

In case of working with non-metallic obstacles, the reflection coefficients become the well-known and simpler Snell's expressions where the calculation of R_{TE} and R_{TM} is straightforward.
When the second medium is a PEC ($\sigma_2 \to \infty$), the absolute value of the third component of k^{tra} tends to infinite and consequently $\mathrm{R_{TE}} = -1$ as well as $\mathrm{R_{TM}} = -1$. Substituting these values in Eq. (12), the total magnetic field equals $2H^{inc}$ at the boundary S. Therefore, J^{MECA} and M^{MECA} in Eq. (13) reduces to J^{PO} and M^{PO} in Eq. (2).

3.2 Electromagnetic field levels

Once J^{MECA} and M^{MECA} can be calculated with the inclusion of the reflection coefficients in the previous section, the evaluation of the scattered fields due to this set of equivalent current densities on the surface of an obstacle requires solving a radiation integral. For arbitrary distributions and geometries, an analytical solution can not be reached. As a consequence, two assumptions are imposed (Arias et al. 2000; Lorenzo et al., 2005):

- The incident wave is a plane wave.
- The obstacle consists of flat triangular facets.

With the objective of distinguishing between the reflected fields at the boundary and the scattered fields at the observation points, the latter are denoted by E^s and H^s. Note that these do not take into account the contribution of the diffraction phenomenon.

3.2.1 Incident plane wave approximation

The position vector $\vec{r}\,'$ of a point on the surface of a flat triangular facet can be written as:

$$\vec{r}\,' = \vec{r}_i + \vec{r}\,'' \tag{21}$$

where $\vec{r}_i$ is the position vector of the barycentre of the i-th facet and $\vec{r}\,''$ is the vector from the barycentre to the source point (see Fig. 4).
The current densities are particularised for the i-th facet, J_i^{MECA} and M_i^{MECA}, having constant amplitudes , $J_{i0}^{MECA} = J_i^{MECA}\big|_{\vec{r}=\vec{r}_i}$ and $M_{i0}^{MECA} = M_i^{MECA}\big|_{\vec{r}=\vec{r}i}$ and a linear phase distribution which varies with the unit direction of the Poynting vector as:

$$\hat{p}_i = \frac{\Re\left\{E_i^{inc} \times \left(H_i^{inc}\right)^*\right\}}{\left|\Re\left\{E_i^{inc} \times \left(H_i^{inc}\right)^*\right\}\right|} \tag{22}$$

where E_i^{inc} and H_i^{inc} are the incident electric and magnetic fields at $\vec{r}_i$ respectively, and * denotes the complex conjugate value of the complex vector in brackets. Therefore, in a medium where the working wavelength is λ_1,

$$\begin{aligned} J_i^{MECA} &= J_{i0}^{MECA} \exp\left\{\mathrm{j}2\pi/_{\lambda_1}\, \hat{p}_i \cdot \vec{r}\,''\right\} \\ M_i^{MECA} &= M_{i0}^{MECA} \exp\left\{\mathrm{j}2\pi/_{\lambda_1}\, \hat{p}_i \cdot \vec{r}\,''\right\} \end{aligned} \tag{23}$$

The reason to introduce the linear phase variation is that this technique allows employing larger facets than when assuming a constant phase distribution.

3.2.2 Application to flat triangular facets

In order to analytically evaluate the radiation integral, the observation points $\vec{r}$ are supposed to be in the far field of the radiating facet - but not necessarily in the far field of the whole scenario. Most of times this constraint is not an additional restriction because applications, such us radar cross section, directly impose huge distances between the object under test and the coordinates where the fields are determined.

Mathematically, the magnitude of the vector between the source point $\vec{r}\,'$ and the observation point $\vec{r}$ is approximated (Balanis, 1989) by the magnitude of the latter $\left|\vec{R}\right| = \left|\vec{r} - \vec{r}\,'\right| \approx \left|\vec{r}\right| \equiv \mathrm{r}$. The phase is taken as $\angle\vec{R} = \angle\vec{r} - \vec{r}\,' \approx \angle \mathrm{r}\text{-}\hat{r}\cdot\vec{r}\,'$, where $\hat{r} = \vec{r}/\left|\vec{r}\right|$.

The formulation to compute the scattered fields can be derived from the Maxwell's equations, e.g. using the magnetic vector potential A and the electric potential vector F separately and combining both solutions. The resulting expressions are written in terms of the Green's function in unbounded media G and its gradient $\nabla\mathrm{G}$. The far field approximation allows expressing these functions as:

$$\begin{aligned} \mathrm{G} &= \frac{1}{4\pi}\frac{\exp\left\{-\mathrm{j}{}^{2\pi}\!/_{\lambda_1}\left|\vec{R}\right|\right\}}{\left|\vec{R}\right|} \approx \frac{1}{4\pi}\frac{\exp\left\{-\mathrm{j}{}^{2\pi}\!/_{\lambda_1}\mathrm{r}\right\}}{\mathrm{r}}\exp\left\{\mathrm{j}{}^{2\pi}\!/_{\lambda_1}\hat{r}\cdot\vec{r}\,'\right\} \\ \nabla\mathrm{G} &= -\mathrm{j}{}^{2\pi}\!/_{\lambda_1}\,\mathrm{G}\,\hat{R}\left(1+\frac{1}{\mathrm{j}{}^{2\pi}\!/_{\lambda_1}\mathrm{R}}\right) \approx -\mathrm{j}{}^{2\pi}\!/_{\lambda_1}\,\mathrm{G}\hat{r} \end{aligned} \tag{24}$$

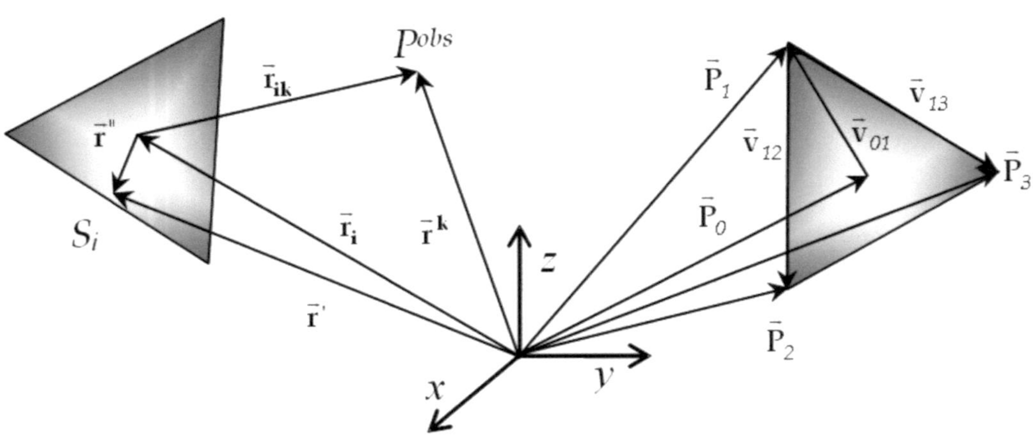

Fig. 4. Coordinate system and nomenclature for the triangular facets: on the left, for section 3.2.1 and on the right, for section 3.2.2.

Consequently, the scattered fields are:

$$E^s \approx \frac{j}{2\lambda_1}\frac{\exp\left\{-j2\pi/\lambda_1\, r\right\}}{r}\int_S\left[\hat{r}\times M^{MECA}-\left(\hat{r}\times\eta_1 J^{MECA}\times\hat{r}\right)\right]\exp\left\{j2\pi/\lambda_1\,\hat{r}\cdot\vec{r}\,'\right\}\partial s'$$
$$H^s \approx -\frac{j}{2\lambda_1}\frac{\exp\left\{-j2\pi/\lambda_1\, r\right\}}{r}\int_S\left[\hat{r}\times J^{MECA}-\left(\hat{r}\times\hat{r}\times\frac{1}{\eta_1}M^{MECA}\right)\right]\exp\left\{j2\pi/\lambda_1\,\hat{r}\cdot\vec{r}\,'\right\}\partial s' \quad (25)$$

and they can be simplified inserting the auxiliary vectors E^a and H^a

$$E^s \approx \frac{j}{2\lambda_1}\frac{\exp\left\{-j2\pi/\lambda_1\, r\right\}}{r}\left[E^a-\eta_1 H^a\times\hat{r}\right]$$
$$H^s \approx -\frac{j}{2\lambda_1}\frac{\exp\left\{-j2\pi/\lambda_1\, r\right\}}{r}\left[H^a-\frac{1}{\eta_1}\hat{r}\times E^a\right] \quad (26)$$

with

$$E^a = \int_S \hat{r}\times M^{MECA}\exp\left\{j2\pi/\lambda_1\,\hat{r}\cdot\vec{r}\,'\right\}\partial s'$$
$$H^a = \int_S \hat{r}\times J^{MECA}\exp\left\{j2\pi/\lambda_1\,\hat{r}\cdot\vec{r}\,'\right\}\partial s' \quad (27)$$

Considering that the expressions of E^a and H^a are quite similar, the following reasoning is only performed for the magnetic fields, so the contribution due to the electric current density J^{MECA} on the i-th facet at the observation point $\vec{r}^{\,k}$ is given by:

$$H^a_{ik} = \exp\left\{j2\pi/\lambda_1\,\hat{r}_{ik}\cdot\vec{r}_i\right\}\int_{S_i}\left(\hat{r}_{ik}\times J_i^{MECA}\right)\exp\left\{j2\pi/\lambda_1\,\hat{r}_{ik}\cdot\vec{r}\,''\right\}\partial s_i' \quad (28)$$

where S_i refers to the surface of the i-th facet and $\hat{r}_{ik}$ is the unit vector from the barycentre of that facet to the observation point k. Substituting J_i^{MECA} by its expression in Eq. (23), where the constant amplitude and linear phase variation distributions were inserted and extracting the terms from the integrand, Eq. (28) is rewritten as:

$$H^a_{ik} = \exp\left\{j2\pi/\lambda_1\,\hat{r}_{ik}\cdot\vec{r}_i\right\}\left(\hat{r}_{ik}\times J_{i0}^{MECA}\right)\mathrm{I}(\hat{r}_{ik})$$
$$\mathrm{I}(\hat{r}_{ik}) = \int_{S_i}\exp\left\{j2\pi/\lambda_1\,(\hat{r}-\hat{p}_i)\cdot\vec{r}\,''\right\}\partial s_i' \quad (29)$$

In case of dealing with flat triangular facets, $\vec{r}\,''$ can be expressed as a function of the vectors $\vec{v}_{mn}$ in Fig. 4:

$$\vec{r}\,'' = \vec{v}_{01} + f\vec{v}_{12} + g\vec{v}_{13} \quad (30)$$

with $\vec{v}_{mn} = \vec{P}_{\mathrm{n}} - \vec{P}_{\mathrm{m}}$, being $\vec{P}_{\mathrm{l}}$ the position vector of the vertices $\mathrm{l} = 1,2,3$ of the facet and $\vec{P}_0$, the position vector of the barycentre. f and g are real coefficients. Taking into account this transformation, the integral in Eq. (29) is converted into:

$$\mathrm{I}(\hat{r}) = 2\mathrm{A}\exp\left\{-\mathrm{j}\frac{\mathrm{a}+\mathrm{b}}{3}\right\}\int_0^1\int_0^{1-\mathrm{f}} \exp\{\mathrm{j}(\mathrm{af}+\mathrm{bg})\}\partial\mathrm{g}\,\partial\mathrm{f} \qquad (31)$$

where A is the area of the facet and the values of a and b are defined as:

$$\mathrm{a} = \frac{2\pi}{\lambda_1}\vec{v}_{12}\cdot(\hat{r}-\hat{p}_i)$$
$$\mathrm{b} = \frac{2\pi}{\lambda_1}\vec{v}_{13}\cdot(\hat{r}-\hat{p}_i) \qquad (32)$$

The solution to the Eq. (31) is obtained by parts, but in order to overcome some singularities, five different cases are detailed (Arias et al. 2000) in Table 1.

a	b	$\mathrm{I}(\hat{r})$
Any	Any	$2\mathrm{A}\exp\left\{-\mathrm{j}\frac{\mathrm{a}+\mathrm{b}}{3}\right\}\left[\frac{\mathrm{a}\exp\{\mathrm{jb}\}-\mathrm{b}\exp\{\mathrm{ja}\}+\mathrm{b}-\mathrm{a}}{(\mathrm{a}\text{-}\mathrm{b})\mathrm{ab}}\right]$
0	Any	$2\mathrm{A}\exp\left\{-\mathrm{j}\frac{\mathrm{b}}{3}\right\}\left[\frac{1+\mathrm{jb}-\exp\{\mathrm{jb}\}}{\mathrm{b}^2}\right]$
Any	0	$2\mathrm{A}\exp\left\{-\mathrm{j}\frac{\mathrm{a}}{3}\right\}\left[\frac{1+\mathrm{ja}-\exp\{\mathrm{ja}\}}{\mathrm{a}^2}\right]$
0	0	A
Any	a	$2\mathrm{A}\exp\left\{-\mathrm{j}\frac{2\mathrm{a}}{3}\right\}\left[\frac{\exp\{\mathrm{ja}\}(1-\mathrm{ja})-1}{\mathrm{a}^2}\right]$

Table 1. Solution to the integral in Eq. (31). Different cases are presented to overcome the singularities in the original expression.

The formulation of the Modified Equivalent Current Approximation method is completed once all the intermediate results that have been described in this section are combined to evaluate the scattered fields at a specific observation point because of one radiating facet. Until this point, the steps to follow are summarised in:

- Decompose the incident electric field into its TE and TM components.
- Obtain the reflection coefficients in Eqs. (19) and (20).
- Calculate the MECA electric and magnetic current densities in Eq. (13).
- Particularise the expressions for the barycentre of every radiating facet.
- Compute the integral $\mathrm{I}(\hat{r}_{ik})$ with the help of Table 1.

- Evaluate the auxiliary fields E^a and H^a introducing the expressions of the MECA current densities and the result of the radiation integral.
- Calculate the scattered fields E^s and H^s

Afterwards, the superposition theorem is employed to add the contribution due to all the radiating facets in the geometry for the observation point *k*:

$$H_k^s = \sum_i H_{ik}^s \tag{33}$$

If an evaluation of the line of sight between the radiating sources and the obstacle were necessary because of its shape, this task would be faced at the beginning of the whole.

3.3 Validation schemes

In this section, two canonical examples for different constitutive parameters and geometries are presented: one assures the good behaviour in high frequency and the other consists in the analysis of electrically large surfaces. In the first scenario, the validation is accomplished in terms of the Radar Cross Section (RCS) which is calculated as:

$$\mathrm{RCS} = 4\pi \lim_{r\to\infty} \mathrm{r}^2 \frac{|E|^2}{|E^{inc}|^2} \tag{34}$$

Since the field levels due to the diffraction contribution are not significant in the selected examples, this effect can be neglected in the computation of the RCS. Consequently, the term in the numerator E only takes into account the electric field levels owing to reflections on the surface.

The methods to contrast the results provided by MECA are an analytical solution taken from the references and the full-wave technique the Method of Moments (MoM) (Medgyesi-Mitschang et al., 1994).

3.3.1 High frequency behaviour

As a high frequency technique, MECA has to show an accurate behaviour when the dimensions of the obstacle are large in comparison with the working wavelength. A good example to test this is a sphere, whose monostatic RCS can be obtained theoretically for both PEC (Balanis, 1989) and non-metallic (Van-Bladel, 2007) characterisations. A frequency sweep is performed for a sphere of radius a with relative permittivity and permeability of $\varepsilon_\mathrm{r} = 2.2$, $\mu_\mathrm{r} = 1.1$ respectively and conductivity of $\sigma = 7\omega\varepsilon_0$, being $\varepsilon_0 = 8.854 \cdot 10^{-12}$ F/m, for the lossy case. The region in Fig. 5 for $2\pi/\lambda\, \mathrm{r} > 10$ corresponds to the high frequency zone where a great coincidence in the results can be observed.

3.3.2 Electrically large surfaces

In this setup, a resonant horizontal dipole is placed at a fixed position whose height is of *h=1.5* metres above an electrically large flat surface as depicted in Fig. 6. The position of the observation point $\mathrm{P^{obs}}$ is determined by two correlated variables: the horizontal distance from the source to $\mathrm{P^{obs}}$, denoted by d and the height of $\mathrm{P^{obs}}$, denoted by H. The angle υ is defined with the purpose of representing the results.

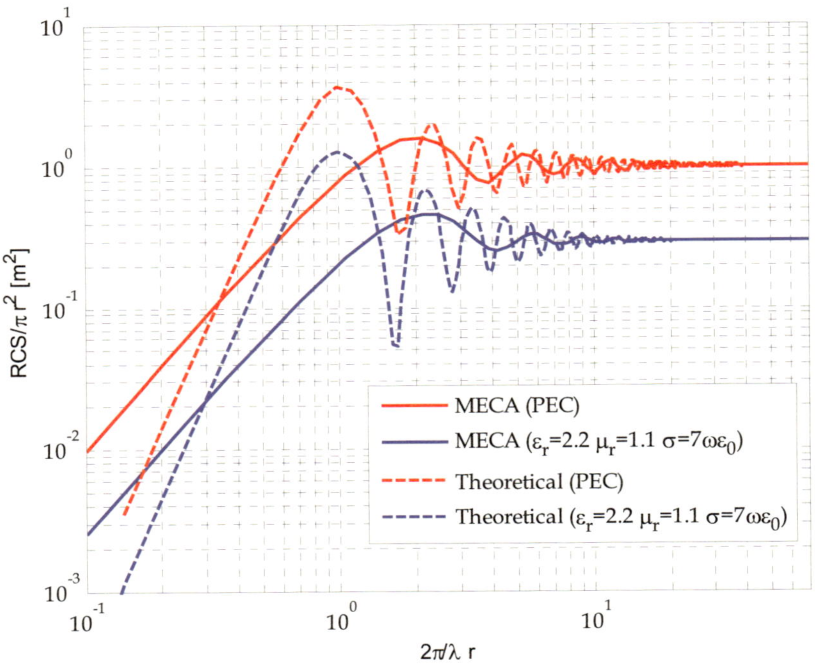

Fig. 5. Monostatic RCS of a sphere. PEC and lossy characterization ($\varepsilon_r = 2.2$, $\mu_r = 1.1$, $\sigma = 7\omega\varepsilon_0$).

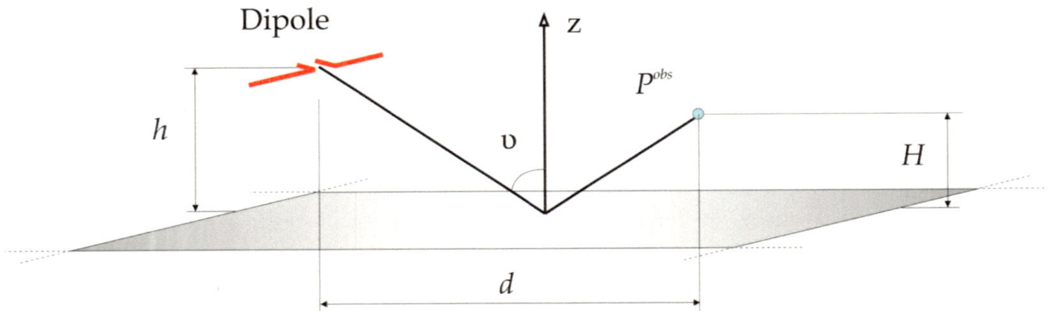

Fig. 6. Scheme for testing electrically large surfaces. The dipole is placed horizontally at a height *h*. The position of the observation points varies with the distance from the dipole *d*. For the sake of clarity in the representation of the results, the angle υ has been inserted.

For the working frequency of 1800 MHz and the values of H ranging from 1 to 35 metres, the magnitude of the total electric field $|E^{tot}|$, including the direct illumination from the source, is evaluated. In Fig. 6 MECA results are compared with those by MoM (Medgyesi-Mitschang, 1994) for the following constitutive parameters: $\varepsilon_r = 3.5$, $\mu_r = 1$ and $\sigma = 0.5\,\text{S/m}$. On the left, $|E^{tot}|$ is plotted as a function of υ, while on the right the variable is the logarithm of the distance d in wavelengths, d_λ, so the small deviation –less than 2dB in the worst case- can be appreciated. Consequently, a high degree of overlapping in the curves of MoM and MECA clearly demonstrates the accuracy of the high frequency technique.

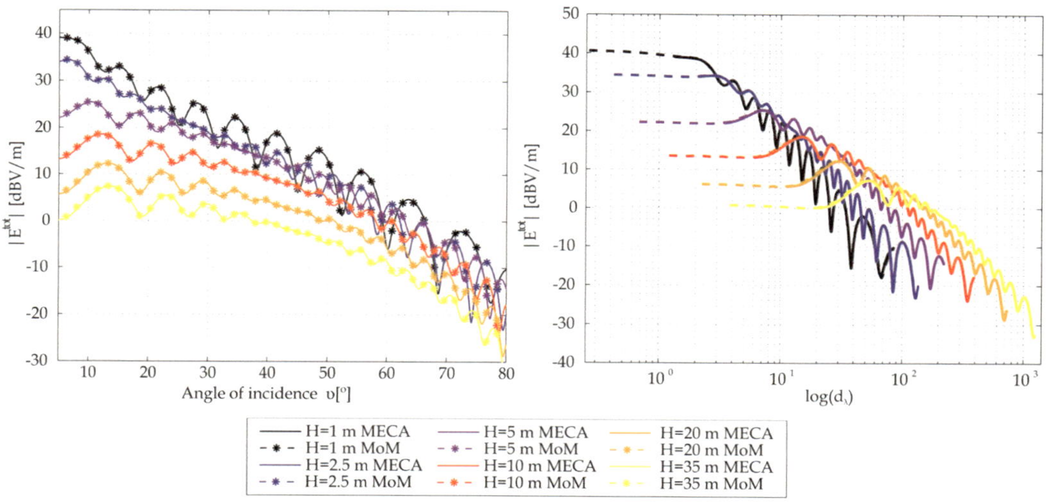

Fig. 7. Magnitude of the total electric field for the electrically large surface configuration. On the left, the results have been represented as a function of the angle υ and on the right, as a function of the logarithm of the distance in wavelengths. MECA and MoM show a great coincidence.

4. Fast visibility algorithm for solving the visibility problem

Because MECA (also PO) predicts null equivalent current densities in shadowed regions and non-null equivalent current densities in those regions with line of sight from the source point, a distinction between non-illuminated and illuminated surfaces is mandatory prior to the evaluation of electromagnetic levels at the observation points. This distinction is known as the visibility problem and it is widely found in computer graphics (Dewey, 1988; Foley, 1992; Bittner & Wonka, 2003), e.g. virtual reality or games, but also in the context of electromagnetic problems such as radio electric coverage evaluation or tracking applications in radar.

Within the classic algorithms, the Z-buffer and the Painter's algorithm can be cited. Assuming the screen in the XY plane, both are based on the identification of the z-coordinate of the elements in the geometry to finally project the nearest parts of the objects. Another method is the Binary Space Partitioning (BSP) (Fuch et al. 1980; Gordon & Chen, 1991) which recursively divides the space into front and back semi-spaces, creating a binary tree at the same time. Afterwards, and in accordance with the target direction, the tree is walked and a priority list of the facets can be built.

Most of the techniques to solve the visibility problem are thought to project an image onto a display unit, so the algorithm itself has not better resolution than the pixel size. Therefore, an error is introduced, although almost imperceptible to the human eye. In order to overcome it, solutions where no approximations are made can be implemented. For example, in the Trimming method (Meana et al., 2009) when a piece of surface is partially

occluded by other one, only the region in shadow is trimmed and removed from the original geometry. At the end, the remaining surface is the exact part that can be seen from a specific point of view. This type of algorithms is very time consuming but its precision counteracts this disadvantage.
To speed up the computation of the existing techniques, some modifications in the phase of preprocessing can be inserted. One option to carry out this is to split the scenario into parallelepipedic or conic macrodomains. Then the facets are classified in one or other macrodomain depending on their position. As a consequence, a preliminary discrimination is accomplished by blocks, so the facets are discarded faster. In this category, the Angular Z-buffer (AZB) (Cátedra et al., 1998) or the Space Volumetric Partitioning (SVP) (Cátedra & Arriaga, 1999) is included.
Additional acceleration can be achieved when the same algorithms are developed in Graphic Processing Units (GPUs) instead of Central Process Units (CPUs) (Ricks & Kuhlen, 2010). Fortunately, this is not restricted to visibility algorithms but it is suitable for all kind of algorithms consisting of loops, matrix operations, etc. As a consequence, generic GPUs are being extensively utilised to run the codes with smaller costs in comparison with CPU parallelisation. Even more, thanks to the use of libraries, e.g. DirectX or OpenGL (Shreiner, 2004) or the interaction between Matlab® and the graphic cards by means of the free plug-in by Nvidia® or Jacket by AccelerEyes®, the implementation of the routines turns into less complex programming.
In the following section the Pyramid method (Meana et al., 2009) is described due to its suitability to deal with flat triangular facets. Briefly, a right pyramid is built so that its vertex coincides with the source point. The i-th wall is the plane containing the source point and the i-th edge of the facet under test. Thus, a point behind this facet and inside the walls of the pyramid is occluded. This is a fast operation that is accomplished by substituting the coordinates of the observation point in the equation of the plane for all the walls.

4.1 Pyramid method

The scenario consists of N facets which have been sorted by their distance from the source point $\vec{s}$ to their barycentres, so the closest facet, denoted by F_1, is always seen. Additionally, the origin of the coordinate system has been moved to the coordinates of $\vec{s}$.
In order to know whether a facet F_j occludes a generic point in the geometry $\vec{P}$, the plane in which F_j is contained is defined:

$$\hat{n}^j \cdot \left(\begin{bmatrix} \mathrm{x} & \mathrm{y} & \mathrm{z} \end{bmatrix} - \vec{P}_0^j \right) = 0 \tag{35}$$

where $\hat{n}^j$ is the unit outward normal vector and $\vec{P}_0^j$ is the barycentre of F_j. Similarly, the three planes containing one of the edges of the F_j and $\vec{s}$ are written as:

$$\frac{\vec{V}_2^j \times \vec{V}_1^j}{\left|\vec{V}_2^j \times \vec{V}_1^j\right|} \cdot \vec{P} = 0, \frac{\vec{V}_3^j \times \vec{V}_2^j}{\left|\vec{V}_3^j \times \vec{V}_2^j\right|} \cdot \vec{P} = 0, \frac{\vec{V}_1^j \times \vec{V}_3^j}{\left|\vec{V}_1^j \times \vec{V}_3^j\right|} \cdot \vec{P} = 0 \tag{36}$$

with $\vec{V}_i^j$ being the position vector of the i-th vertex ($\mathrm{i} = 1,2,3$) of F_j. These planes constitute the walls of the pyramid whose base is the facet under test. There are two conditions to conclude that $\vec{P}$ can not be seen from the source point:

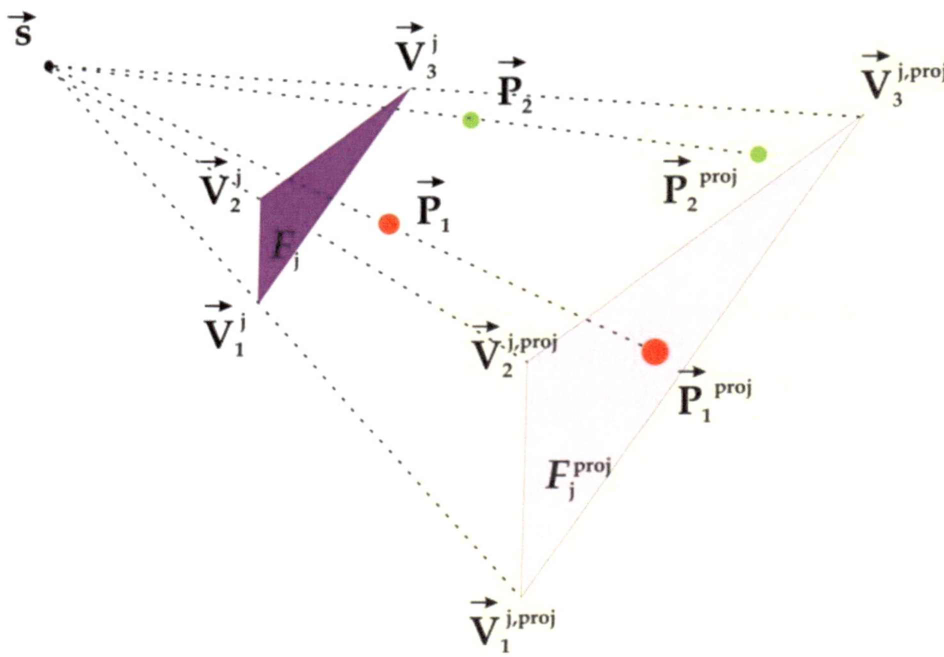

Fig. 8. Pyramid method. In the image, a projection plane in light yellow has been selected in order to clearly remark that the facet F_j occludes the observation point $\vec{P}_1$. On the other hand, $\vec{P}_2$ has line of sight with the source $\vec{s}$.

- $\vec{P}$ is behind the plane in Eq. (35)

$$\hat{n}^j \cdot \left(\vec{P} - \vec{P}_0^j\right) < 0 \tag{37}$$

- $\vec{P}$ is inside the three walls. An easy way to compute this is:

$$\sum_{i=1}^{3} \operatorname{sign}\left\{\left|\frac{\vec{V}_{i+1}^j \times \vec{V}_i^j}{\left|\vec{V}_{i+1}^j \times \vec{V}_i^j\right|} \cdot \vec{P}\right|\right\} = 3 \tag{38}$$

where $\vec{V}_4^j = \vec{V}_1^j$ and the vertices are supposed to be in clockwise or counterclockwise order.

The algorithm determines the occlusion of points by a triangular facet as depicted in Fig. 8 , but the Pyramid method can be applied to determine the existence of line of sight among facets. The process to perform this is based on simplifying the representation of the triangles choosing some important points like the barycentre, vertices, inner points and executing the algorithm for them.

Even though the expressions have been particularised for the case of triangular facets, the implementation can be accomplished for any polygon by including additional walls in the pyramid and substituting the upper limit of the summatory in Eq. (38).

A quite similar algorithm to the Pyramid method is the Cone method (Meana et al., 2009), but a right cone is defined instead of a right pyramid. Its radius is a mean value of the distances from the barycentre of the facet under test to each of its vertices. Analogically, any point behind that facet and inside the cone is occluded. This is a faster algorithm because only one operation comparing the cone angle and the angle between the observation point and the axis of the cone is carried out. On the other hand, the exactness diminishes.

5. Application examples

The scope of application of the Modified Equivalent Current Approximation comprises all the problems which have been analysed by means of the Physical Optics traditionally, but also the extension to dielectric and lossy materials. In the following paragraphs some of these fields where MECA could be employed are summarised.

The RCS computation is one of the most cited topics in the literature: from canonical geometries (e.g. spheres, flat plates and dihedrals) (Griesser & Balanis, 1987; Ross, 1966) for contrasting and validating results to electrically large random surfaces and complex targets (ships or airplanes) (Adana et al., 2000; Uluisik et al., 2008) with a remarkable decrease in the computational time in comparison to full wave techniques. The consideration of absorbing materials allows studying the reduction in the radar signature of aircrafts or missiles.

In order to deal with more realistic scenarios in the RCS computation, the analysis of open-ended cavities including reflections and resonances (Burkholder & Lundin, 2005) is also a line of investigation MECA can face. Likewise, MECA has proven a good accuracy when dealing with electrically large rough surfaces (Meana et al., 2010) that can model the sea surface or the orography of a rural environment. Therefore, the evaluation of the electromagnetic levels for RCS or other applications can consider not only the target itself but also its surroundings.

Another field of application is the design of reflector antennas (Boag & Letrou, 2003; Lorenzo et al., 2005) with or without dielectric radomes because of their electric size. They are usually fed by an array of antennas whose radiation pattern is determined to compute the equivalent current densities and to calculate the fields at distances much larger than the wavelength afterwards.

On the other hand, the Physical Optics approximation and its extensions can also provide some reference values to validate new algorithms or the results by some imaging or shape reconstruction techniques (Saeedfar & Barkeshli, 2006). Due to the fact that the illuminating sources are known a priori, the simulation of a set of constitutive parameters and the shape of the objects can be tackled in a fast manner.

Although the high frequency techniques have lost relevance in the recent years due to the Fast Multipole Method (FMM) -developed by Rohklin in the field of acoustic dispersion (Rohklin, 1985; Rohklin, 1990) and then extended to electromagnetic dispersion (Engheta et al., 1992) - in joint with multilevel schemes and other acceleration methodologies, the advances in the terahertz band will become notable again in the foreseeable future. This means similar applications but working at higher frequencies where the wavelength is smaller than 1 mm and, as a consequence, most of the obstacles and objects is electrically large. Some papers have already proven the validity of the asymptotic approximations at these frequencies.

In addition to the previously enumerated fields of application of the MECA method and in spite of the typical employment of Ray Tracing and Launching in joint with empirical models to deal with the radioelectric coverage evaluation (Papkelis et al., 2007), this problem must be added to that list. The reason is that it fulfils all the requisites detailed through this chapter. In the next section an application example will be shown.

Fig. 9. Rural environment for evaluating the radioelectric coverage. The scenario is 17 km long by 10 km width and the 13 base stations have been depicted in red at their real emplacements.

5.1 Radioelectric coverage

Network planning and optimization in rural and urban environments can be studied for different radiocommunication systems (*Global System for Mobile Communications* -GSM -, *Universal Mobile Telecommunications System* -UMTS - or radio and television broadcasting). This example focuses on the radioelectric coverage evaluation of the *General Packet Radio Service* -GPRS - at the working frequency of 1800 MHz, where the scenario consists of a rural terrain of 17 km long by 10 km wide with a uniform soil characterization, $\varepsilon_r = 3$, $\mu_r = 1$, $\sigma = 0.001\,\mathrm{S/m}$. Some relevant information about the scenario is that is mainly compound of mountains and valleys where there are not big villages. Thus, regions with no coverage are expected there while, on the other hand, a town on the right and a freeway at the bottom of the map suggest a better coverage in these areas.

The radio electric stations are located at their real emplacements (see Fig. 10) simulating their electromagnetic behaviour based on the available data, such as configuration parameters (e.g. power, gain), provided by the telecommunications company. As a consequence, the radiation patterns were synthesised assuming a dipole array with a different number of elements for every different antenna. Once the illumination from the sources is explicitly determined, the MECA equivalent current densities are computed and the electromagnetic field levels are calculated.

The power density at the observation points situated at 1.5 metres above the terrain is represented in Fig. 10. A threshold has been obtained by taking into account the sensibility of the receivers, so that the minimum in the colorbar has been fixed in a lower value. This means that any region in dark blue can not initiate or maintain a communication with the base station. In order to improve the coverage in specific zones, new base stations could be added and a fast evaluation can be accomplished by computing only the direct illumination due to those stations.

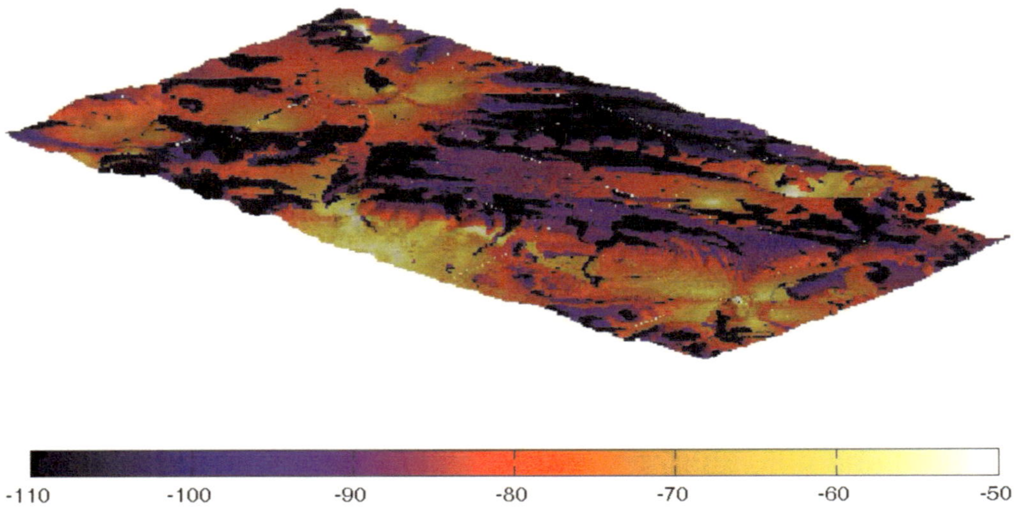

Fig. 10. Power density in dBW/m[2] for the rural scenario with MECA. The electromagnetic field levels have been evaluated for a penetrable characterization of the terrain given by $\varepsilon_r = 3$, $\mu_r = 1$, $\sigma = 0.001$ S/m .

6. Conclusion

In this chapter "*High frequency techniques: the Physical Optics approximation and the Modified Equivalent Current Approximation (MECA)*" the whole process to compute electromagnetic field levels based on the high frequency technique Modified Equivalent Current Approximation has been presented. MECA is an extension of the Physical Optics formulation for penetrable and non-metallic objects based on the equivalence principle, so not only electric but also magnetic equivalent current densities are taken into account for dielectric and lossy materials.

One of the most relevant features of MECA is that the reflection coefficients are calculated from the general dispersion relation for the transversal electric and transversal magnetic components independently. Therefore any conductivity and relative permittivity and permeability are allowed. Then, they are inserted in the corresponding TE/TM analysis which was shown in Fig. 2 and Fig. 3, to be finally combined in order to obtain the MECA equivalent current densities. With the objective of determining the electromagnetic fields at the observation points analytically, the incident wave is supposed to be a plane wave which impinges on the surface and generates a current density distribution with constant amplitude and linear phase variation. Assuming a flat triangular facet, the radiation integral can be solved by parts.

The good behaviour was proven in the validation examples, where the results from the frequency sweep in the high frequency region agreed the theoretical values. Likewise, an excellent overlapping was obtained for different angles of incidence when dealing with a non-PEC electrically large surface.

Because one of the constraints to employ PO and MECA is the determination of line of sight between the source and the observation points, some algorithms to solve the visibility problem were described. The classic methods are complemented by acceleration techniques and then, they are translated into the GPU programming languages. The Pyramid method was explained as an example of fast algorithm which was specifically developed for evaluating the occlusion by flat facets. Undoubtedly, this can be employed in joint with the MECA formulation, but the Pyramid method can also be helpful in other disciplines of engineering.

Throughout the section "*Application examples*", the way MECA becomes a powerful and efficient method to tackle different scattering problems for electrically large scenarios was satisfactorily demonstrated by means of the example consisting in the evaluation of the radio electric coverage in a rural environment. In addition to this, other fields of application were suggested from the RCS computation to imaging techniques, covering a wide range of electromagnetic problems.

7. References

Arias, A.M., Rubiños, J.O., Cuiñas, I., & Pino, A.G. (2000). Electromagnetic Scattering of Reflector Antennas by Fast Physical Optics Algorithms, *Recent Res. Devel. Magnetics*, Vol. 1, No. 1, pp. 43-63

Adana, F.S. de, Lozano, P., Gisbert, F., Sudupe, I., Pérez, J., & Cátedra, M.F. (2000) Application of the PO to the Computation of the Monostatic RCS of Arbitrary Bodies Modeled by Plane Facets of Dielectric and Magnetic Material, *Proceedings of 2000 USCN/URSI National Radio Science Meeting*, Salt Lake City, Utah, USA, 2000

Balanis, C. A. (1989). *Advanced Engineering Electromagnetics*, John Wiley & Sons, ISBN 978-0471621942, New York, USA

Bittner, J., & Wonka, P. (2003). Visibility in Computer Graphics, *Environment and Planning B: Planning and Design*, Vol. 30, No. 5,

Boag, A., & Letrou, C. (2003). Fast Radiation Pattern Evaluation for Lens and Reflection Antennas. *IEEE Transactions on Antennas and Propagation*, Vol. 51, No. 5, pp. 1063-1068

Burkholder, R.J. & Lundin, T. (2005). Forward-Backward Iterative Physical Optics Algorithm for Computing the RCS of Open-Ended Cavities. *IEEE Transactions on Antennas and Propagation*, Vol.53, No.2, pp. 793- 799

Cátedra, M. F., Pérez, J., Sáez de Adana, F., & Gutiérrez O., (1998). Efficient Ray-Tracing Techniques for Three Dimensional Analyses of Propagation in Mobile Communications: Application to Picocell and Microcell Scenarios. *IEEE Antennas and Propagation Magazine*, Vol. 40, No. 2, pp. 15-28

Cátedra, M.F., & Arriaga, J.P. (1999). *Cell Planning for Wireless Communications*, Artech House, ISBN 978-0890066010, Boston, USA

Dewey, B.R. (1988). *Computer graphics for Engineers*, Harpercollins College Div., ISBN 978-0060416706, USA

Engheta, N., Murphy, W.D., Rokhlin, V., & Vassiliou, M.S. (1992). The Fast Multipole Method (FMM) for Electromagnetic Scattering, *IEEE Transactions on Antennas and Propagation*, Vol. 40, No. 6, pp. 634-641

Foley, J.D. (1992). *Computer Graphics: Principles and Practice in C* (2nd edition), ISBN 0201848406, Addison-Wesley, USA

Fuch, H., Kedem, Z.M., & Naylor, B.F. (1980). On Visible Surface Generation by Priori Tree Structures. *ACM SIGGRAPH Computer Graphics*, Vol. 14, No. 3, pp. 124-133

Glassner, A.S. (1989). *An Introduction to Ray Tracing*, Academic Press, ISBN 978-0-12-286160-4, San Diego, USA

Gordon, D., & Chen, S. (1991). Front-to-Back Display of BSP Trees. *IEEE Computer Graphics and Applications*, Vol. 11, No. 5, pp. 79-85

Griesser, T., & Balanis, C. (1987). Backscatter Analysis of Dihedral Corner Reflectors Using Physcal Optics and the Physical Theory of Diffraction, *IEEE Antennas and Propagation Magazine*, Vol. 35, No. 10, pp. 1137-1147

Harrington, R.F. (2001). *Time-Harmonic Electromagnetic Fields* (2nd edition), McGraw Hill, ISBN 978-0471208068, USA

Hodges, R.E., & Rahmat-Samii, Y. (1993). Evaluation of Dielectric Physical Optics in Electromagnetic Scattering, *Proceedings 1993 Antennas and Propagation Society InternationalSymposium*, USA, 1993

Keller, J.B. (1962). Geometrical Theory of Diffraction. *Journal of the Optical Society of America*, Vol. 52, No. 3, pp. 116-130

Kempel, L.C., Chatterjee, A., & Volakis, J.L. (1998). *Finite Element Method Electromagnetics* (1st edition), IEEE, USA

Lorenzo, J. A. M., Pino, A. G., Vega, I., Arias, M., & Rubiños, O. (2005). ICARA: Induced-Current Analysis of Reflector Antennas. *IEEE Antennas and Propagation Magazine*, Vol.47, No.2, pp. 92-100

Meana, J. G., Las-Heras, F., & Martínez-Lorenzo, J. Á. (2009). A Comparison Among Fast Visibility Algorithms Applied to Computational Electromagnetics. *Applied Computational Electromagnetics Society Journal*, Vol.24, No.3, pp. 268-280

Meana, J. G., Martínez-Lorenzo, J. Á., Las-Heras, F., & Rappaport, C. (2010). Wave Scattering by Dielectric and Lossy Materials Using the Modified Equivalent Current Approximation (MECA). *IEEE Transactions on Antennas and Propagation*, Vol. 58, No. 11, pp. 3757-3761

Medgyesi-Mitschang, L.N., Putnam, J.M., & Gedera, M.B. (1994). Generalized Method of Moments for Three-Dimensional Penetrable Scatterers, *Journal of the Optical Society of America A*, Vol. 11, No. 4, pp. 1383-1398

Pathak, P.H., & Kouyoumjian, R.G. (1974). A Uniform Geometrical Theory of Diffraction for an Edge in a Perfectly Conducting Surface. *Proceedings of the IEEE*, Vol. 62, No. 11, pp. 1448-1461

Papkelis, E.G., Psarros, I., Ouranos, I.C., Moschovitis, C.G., Karakatselos, K.T., Vagenas, E., Anastassiu, H.T., & Frangos, P.V. (2007). A Radio-Coverage Prediction Model in Wireless Communication Systems Based on Physical Optics and the Physical Theory of Diffraction. *IEEE Antennas and Propagation Magazine*, Vol. 49, No. 2, pp. 156-165

Rappaport, C.M., & McCartin, B.J. (1991). FDFD Analysis of Electromagnetic Scattering in Anisotropic Media Using Unconstrained Triangular Meshes. *IEEE Transactions on Antennas and Propagation*, Vol. 39, No. 3, pp. 345-349

Rengarajan, S.R., & Gillespie, E.S. (1988). Asymptotic Approximations in Radome Analysis. *IEEE Transactions on Antennas and Propagation*, Vol. 36, No. 3, pp. 405-414

Ricks, T., & Kuhlen, T. (2010). Accelerating Radio Wave Propagation Algorithms by Implementation on Graphics Hardware, In: *Wave Propagation in Materials for Modern Applications*, Andrey Petrin, pp. 103-122, Intech, Vienna, Austria

Rokhlin, V. (1985). Rapid Solution of Integral Equations of Classical Potential Theory. *Journal of Computational Physics*, Vol. 60, No. 9, pp. 187-207

Rokhlin, V. (1990). Rapid Solution of Integral Equations of Scattering Theory in Two Dimensions. *Journal of Computational Physics*, Vol. 86, No. 2, pp. 414-439

Ross, R.A. (1966). Radar Cross Section of Rectangular Flat Plates as a Function of Aspect Angle. *IEEE Transactions on Antennas and Propagation*, Vol. 14, No. 8, pp. 329-335

Rossi, J.P., & Gabillet, Y. (2002). A Mixed Launching/Tracing Method for Full 3-D UHF Propagation Modeling and Comparison with Wide-Band Measurements. *IEEE Transactions on Antennas and Propagation*, Vol. 50, No. 4, pp. 517-523

Saeedfar, A., & Barkeshli, K. (2006). Shape Reconstruction of Three-Dimensional Conducting Curved Plates Using Physical Optics, NURBS Modeling, and Genetic Algorithm. *IEEE Transactions on Antennas and Propagation*, Vol.54, No.9, pp. 2497-2507

Sáez de Adana, F., González, I., Gutiérrez, O., Lozano, P., & Cátedra, M.F. (2004) Method Based on Physical Optics for the Computation of the Radar Cross Section Including Diffraction and Double Effects of Metallic and Absorbing Bodies Modeled with Parametric Surfaces. *IEEE Transactions on Antennas and Propagation*, Vol. 52, No. 12, pp. 3295-3303

Shreiner, D. (2004). *OpenGL Reference Manual: the Official Reference Document to OpenGL, Version 1.4*, Addison-Wesley, ISBN 978-0321173836, London, UK

Staelin, D.H., Morgenthaler, A.W., & Kong, J.A. (1993). *Electromagnetic Waves*, Prentice Hall, ISBN 978-0132258715, USA

Taflove, A., & Umashankar, K.R. (1987). The Finite Difference Time Domain FD-TD Method for Electromagnetic Scattering and Interaction Problems. *Journal of Electromagnetic Waves and Applications*, Vol. 1, No. 4, pp. 363-387

Ufimtsev, P.Y. (1962). Method of Edge Waves in the Physical Theory of Diffraction, *Izd-vo Sov. Radio*, pp. 1-243 (translated by U.S. Air Force Foreign Technology Division, Wright-Patterson AFB, OH)

Uluisik, C., Cakis, G., Cakis, M., & Sevgi, L. (2008). Radar Cross Section (RCS) Modeling and Simulation, Part 1: A Tutorial Review of Definitions, Strategies, and Canonical Examples. *IEEE Antennas and Propagation Magazine*, Vol. 50, No. 1, pp. 115-126

Van-Bladel, J. (2007). *Electromagnetic Fields* (2nd edition), IEEE Press,Wiley-Interscience, ISBN 978-0471263883, USA

Part 4

Propagation in Guided Media

9

Electrodynamics of Multiconductor Transmission-line Theory with Antenna Mode

Hiroshi Toki and Kenji Sato
Research Center for Nuclear Physics (RCNP), Osaka University and National Institute of Radiological Sciences (NIRS)
Japan

1. Introduction

In the modern life we depend completely on the electricity as the most useful form of energy. The technology on the use of electricity has been developed in all directions and also in very sophisticated manner. All the electric devices have to use electric power (energy) and they use both direct current (DC) and alternating current (AC). Today a powerful technology of manipulation of frequency and power becomes available due to the development of chopping devices as IGBT and other methods. This technology of manipulating electric current and voltage, however, unavoidably produces electromagnetic noise with high frequency. We are now filled with electromagnetic noise in our circumstance.

This situation seems to be caused by the fact that we do not have a theory to describe the electromagnetic noise and to take into account the effect of the circumstance in the design of electric circuit. We have worked out such a theory in one of our papers as "Three-conductor transmission-line theory and origin of electromagnetic radiation and noise" (Toki & Sato (2009)). In addition to the standard two-conductor transmission-line system, we ought to introduce one more transmission object to treat the circumstance. As the most simple object, we introduce one more line to take care of the effect of the circumstance. This third transmission-line is the place where the electromagnetic noise (electromagnetic wave) goes through and influences the performance of the two major transmission-lines. If we are able to work out the three-conductor transmission-line theory by taking care of unwanted electromagnetic wave going through the third line, we understand how we produce and receive electromagnetic noise and how to avoid its influence.

To this end, we had to introduce the coefficient of potential instead of the coefficient of capacity, which is used in all the standard multi-conductor transmission line theories (Paul (2008)). We are then able to introduce the normal mode voltage and current, which are usually considered in ordinary calculations, and at the same time the common mode voltage and current, which are not considered at all so far and are the sources of the electromagnetic noise (Sato & Toki (2007)). We are then able to provide the fundamental coupled differential equations for the TEM mode of the three-conductor transmission-line theory and solve the coupled equations analytically. As the most important consequence we obtain that the main two transmission-lines should have the same qualities and same geometrical shapes and their distances to the third line should be the same in order to decouple the normal mode from the common mode. The symmetrization is the key word to minimize the influence of the circumstance and hence the electromagnetic noise to the electric circuit. The symmetrization makes the normal mode decouple from the common mode and hence we are able to avoid the influence of the common mode noise in the use of the normal mode (Toki & Sato (2009)). The symmetrization has been carried out at HIMAC (Heavy Ion Medical Accelerator in Chiba) (Kumada (1994)) one and half decade ago and at Main Ring of J-PARC recently (Kobayashi (2009)). Both synchrotrons are working well at very low noise level.

As the next step, we went on to develop a theory to couple the electric circuit theory with the antenna theory (Toki & Sato (2011)). This work is motivated by the fact that when the electromagnetic noise is present in an electric circuit, we observe electromagnetic radiation in the circumstance. In order to complete the noise problem we ought to couple the performance of electric circuit with the emission and absorption of electromagnetic radiation in the circuit. To this end, we introduce the Ohm's law as one of the properties of the charge and current under the influence of the electromagnetic fields outside of a thick wire. As a consequence of the new multi-conductor transmission-line theory with the antenna mode, we again find that the symmetrization is the key technology to decouple the performance of the normal mode from the common and antenna modes (Toki & Sato (2011)).

The Ohm's law is considered as the terminal solution of the equation of motion of massive amount of electrons in a transmission-line of a thick wire with resistance, where the collisions of electrons with other electrons and nuclei take place. This consideration is able to put the electrodynamics of electromagnetic fields and dynamics of electrons in the field theory. We are also able to discuss the skin effect of the TEM mode in transmission-lines on the same footing.

In this paper, we would like to formulate the multi-conductor transmission-line theory on the basis of electrodynamics, which includes naturally the Maxwell equations and the Lorentz force.

This paper is arranged as follows. In Sect.2, we introduce the field theory on electrodynamics and derive the Maxwell equation and the Lorentz force. In Sect.3, we develop the multiconductor transmission-line (MTL) equations for the TEM mode. We naturally include the antenna mode by taking the retardation potentials. In Sect.4, we provide a solution of one antenna system for emission and absorption of radiation. In Sect.5, we discuss a three-conductor transmission-line system and show the symmetrization for the decoupling of the normal mode from the common and antenna modes. In Sect.6, we introduce a recommended electric circuit with symmetric arrangement of power supply and electric load for good performance of the electric circuit. Sect.7 is devoted to the conclusion of the present study.

2. Electrodynamics

We would like to work out the multiconductor transmission-line (MTL) equation with electromagnetic emission and absorption. To this end, we should work out fundamental equations for a multiconductor transmission-line system by using the Maxwell equation and the properties of transmission-lines. We shall work out electromagnetic fields outside of multi-conductor transmission-lines produced by the charges and currents in the transmission-lines. In this way, we are able to describe electromagnetic fields far outside of the transmission-line system so that we can include the emission and absorption of electromagnetic wave. For this purpose, we take the electrodynamics field theory, since a multiconductor transmission-line system is a coupled system of charged particles and electromagnetic fields. In this way, we are motivated to treat the scalar potential in the same way as the vector potential and find it natural to use the coefficients of potential instead of the coefficients of capacity as the case of the coefficients of inductance.

We discuss here the dynamics of charged particles with electromagnetic fields in terms of the modern electrodynamics field theory. For those who are not familiar to this theory, you can skip this paragraph and start with the equations (6) and (7). In the electrodynamics, we have the gauge theory Lagrangian, where the interaction of charge and current of a Fermion (electron) field ψ with the electromagnetic field A_μ is determined by the following Lagrangian,

$$L = \frac{1}{4} F_{\mu\nu}(x) F^{\mu\nu}(x) + \bar{\psi}(i\gamma_\mu D^\mu - m)\psi \,. \tag{1}$$

with $D^\mu = \partial^\mu - ieA^\mu$, where A^μ is the electromagnetic potential. Here, $F_{\mu\nu}(x) = \partial_\mu A_\nu(x) - \partial_\nu A_\mu(x)$ is the anti-symmetric tensor with the four-derivative defined as $\partial_\mu = \frac{\partial}{\partial x^\mu} = (\frac{\partial}{c\partial t}, \nabla)$ and the four-coordinate as $x^\mu = (ct, \mathbf{x})$. Here, electrons are expressed by the Dirac field ψ, which possesses spin as the source of the permanent magnet and therefore we do not have to introduce the notion of the perfect conductor anymore (Maxwell (1876)). The vector current is written by using the charged field as $j^\mu = \bar{\psi}\gamma^\mu\psi$. The variation of the above Lagrangian with respect to A^μ provides the Maxwell equation with a source term expressed in the covariant form (Maxwell (1876)).

$$\partial^\mu F_{\mu\nu}(x) = ej_\nu(x) \tag{2}$$

They are Maxwell equations, which become clear by writing explicitly the anti-symmetric tensor in terms of the electric field **E** and magnetic field **B**.

$$F_{\mu\nu} = \begin{pmatrix} 0 & \frac{1}{c}E_x & \frac{1}{c}E_y & \frac{1}{c}E_z \\ -\frac{1}{c}E_x & 0 & -B_z & B_y \\ -\frac{1}{c}E_y & B_z & 0 & -B_x \\ -\frac{1}{c}E_z & -B_y & B_x & 0 \end{pmatrix} \tag{3}$$

Here, $\mathbf{E} = -\nabla V - \frac{\partial \mathbf{A}}{\partial t}$ and $\mathbf{B} = \nabla \times \mathbf{A}$. The two more equations are explicitly written as $\nabla \cdot \mathbf{E} = \frac{1}{\varepsilon} q$ and $\nabla \times \mathbf{B} - \frac{1}{c^2}\frac{\partial \mathbf{E}}{\partial t} = \mu \mathbf{j}$ by using the above equation of motion (2).
It is convenient to write the Maxwell equation in the covariant form for the symmetry of the relevant quantities without worrying about the factors as c, μ and ε. The four-vector potential is written by the scalar and vector potentials as $A^\mu(x) = (V(x)/c, \mathbf{A}(x))$ and the four-current, which is a source term of the potentials, is given as $ej^\mu = \mu(cq, \mathbf{j})$. Here, the charge q and current $\mathbf{j}$ are both charge and current densities. The contra-variant four vector x^μ is related with the co-variant four vector x_μ as $x^\mu = g^{\mu\nu}x_\nu$. Here, the metric is $g^{\mu\nu} = 1$ for $\mu = \nu = 0$ and $g^{\mu\nu} = -1$ for $\mu = \nu = 1, 2, 3$ and zero otherwise (Bjorken (1970)). The Maxwell equation (2) gives the following differential equation (Maxwell (1876)).

$$\partial^\mu \partial_\mu A_\nu(x) - \partial^\mu \partial_\nu A_\mu(x) = ej_\nu(x) \tag{4}$$

In order to simplify the differential equation and also to keep the symmetry among the scalar and vector potentials, we take the Lorenz gauge $\partial^\mu A_\mu(x) = 0$ (Lorenz (1867); Jackson (1998)). In this case, we get a simple covariant equation for the potential with the source current.

$$\partial^\mu \partial_\mu A_\nu(x) = ej_\nu(x) \tag{5}$$

This expression based on the field theory shows the fact that the dynamics of the four-vector potential A_ν is purely given by the corresponding source current j_ν. This fact should be contrasted with the standard notion that the time-dependent electric and magnetic fields are the sources from each other through the Ampere-Maxwell's law and the Faraday's law in the Maxwell equation. When there is no source term $j_\nu = 0$ in the space outside of the conductors, the four-vector potential satisfies the wave equation with the light velocity. In

the electrodynamics, the propagation of electromagnetic wave with the velocity of light is the property of a vector particle with zero mass.
We express now the four-vectors in the standard three-vector form. The scalar potential $V(\mathbf{x},t)$ and the vector potential $\mathbf{A}(\mathbf{x},t)$ should satisfy the following equations with sources in the Lorenz gauge.

$$\left(\frac{\partial^2}{c^2\partial t^2} - \nabla^2\right) V(\mathbf{x},t) = \frac{1}{\varepsilon} q(\mathbf{x},t) \tag{6}$$

$$\left(\frac{\partial^2}{c^2\partial t^2} - \nabla^2\right) \mathbf{A}(\mathbf{x},t) = \mu \mathbf{j}(\mathbf{x},t) \tag{7}$$

These two second-order differential equations (6) and (7) clearly show that the charge and current are the sources of electromagnetic fields. For the propagation of electromagnetic power through a MTL system, we are interested in the electromagnetic fields outside of thick electric wires with resistance. In this case, we are able to solve the differential equations by using retardation charge and current (Lorenz (1867); Rieman (1867); Jackson (1998)).

$$V(\mathbf{x},t) = \frac{1}{4\pi\varepsilon}\int d\mathbf{x}' \frac{q(\mathbf{x}', t - \frac{|\mathbf{x}-\mathbf{x}'|}{c})}{|\mathbf{x}-\mathbf{x}'|} \tag{8}$$

$$\mathbf{A}(\mathbf{x},t) = \frac{\mu}{4\pi}\int d\mathbf{x}' \frac{\mathbf{j}(\mathbf{x}', t - \frac{|\mathbf{x}-\mathbf{x}'|}{c})}{|\mathbf{x}-\mathbf{x}'|} \tag{9}$$

These expressions are valid for the scalar and vector potentials outside of the transmission-lines. The presence of the retardation effect in the time coordinate in the integrand is important for the production of electromagnetic radiation. The retardation terms generate a finite Poynting vector going out of a surface surrounding the MTL system not only at a far distance but also at a boundary.
This part is related with the derivation of the Lorentz force from the field theory. You may skip this part and directly move to the next section. It is important to derive the current conservation equation of the field theory, which is related with the behavior of charged particles. The current conservation is derived by writing an equation of motion for ψ using the above Lagrangian as

$$(i\gamma_\mu\partial^\mu + e\gamma_\mu A^\mu - m)\psi(x) = 0\,. \tag{10}$$

Using this Dirac equation together with the complex-conjugate Dirac equation, we obtain

$$\partial_\mu j^\mu(x) = 0\,, \tag{11}$$

which is the charge conservation law of the field theory. The electromagnetic potential for a charged particle is given from the above equation as $ej_\mu A^\mu$. From this expression, we are able to derive an electromagnetic force exerted on a charged particle. To write it explicitly, we ought to use a Lagrangian of a point particle with the electromagnetic potential $ej_\mu A^\mu$, where $j^\mu = (c, \mathbf{v})$.

$$L = \frac{1}{2}m(\frac{d\mathbf{x}}{dt})^2 - eV(x) + e\mathbf{v}\cdot\mathbf{A}(x) \tag{12}$$

We use the Euler equation $-\frac{\partial L}{\partial \mathbf{x}} + \frac{d}{dt}\frac{\partial L}{\partial \mathbf{v}} = 0$, we get

$$m\frac{d^2\mathbf{x}}{dt^2} + e\nabla V(x) - e\nabla(\mathbf{A}(x)\cdot\mathbf{v}) + e\frac{d\mathbf{A}(x)}{dt} = 0\,. \tag{13}$$

Here, $\mathbf{v} = \frac{d\mathbf{x}}{dt}$ is used. We have the relations

$$\frac{d\mathbf{A}(x)}{dt} = \frac{\partial \mathbf{A}(x)}{\partial t} + (\mathbf{v}\cdot\nabla)\mathbf{A}(x)\,, \tag{14}$$

and

$$\mathbf{B}\times\mathbf{v} = (\nabla\times\mathbf{A})\times\mathbf{v} = (\mathbf{v}\cdot\nabla)\mathbf{A}(x) - \nabla(\mathbf{A}(x)\cdot\mathbf{v})\,. \tag{15}$$

Hence, the Lorentz force is written as

$$\mathbf{F}_L = e\mathbf{E}(x) + e\mathbf{v}\times\mathbf{B}(x)\,. \tag{16}$$

with $\mathbf{E}(x) = -\nabla V(x) - \frac{\partial \mathbf{A}(x)}{\partial t}$. Charged particles are influenced by the electromagnetic field through the Lorentz force given above. In the present discussion, we use the phenomenological relation in terms of the Ohm's law for the relation of the current with the electromagnetic field. Because the total energy should be conserved, the summation of electromagnetic power of circuit, energy of emission and absorption of electromagnetic wave, and Joule's heat energy is kept constant in a multiconductor transmission-line system.

3. Multiconductor transmission-line theory with radiation

We start with the properties of transmission-lines, where the charge and current are present and they oscillate in space and time for the propagation of electromagnetic energy through the transmission-lines. We introduce N parallel lines numbered by i $(= 1,...,N)$ and its direction x with a round cross section of a thick wire with resistance. First of all, we have the charge conservation equation (11) of the field theory, which indicates the conservation of charge $\partial q/\partial t + \nabla \mathbf{j} = 0$ and the continuity equation of the standard electromagnetism. We introduce i-th current and i-th charge by integrating $\mathbf{j}$ and q over the cross section of each transmission-line at a space-time position x, t taking into account the skin effect in the transmission-line, $I_i(x,t) = \int ds j_i^x(x,y,z,t)$ and $Q_i(x,t) = \int ds q_i(x,y,z,t)$, where $ds = dydz$.

$$\frac{\partial I_i(x,t)}{\partial x} = -\frac{\partial Q_i(x,t)}{\partial t} \tag{17}$$

Here, $I_i(x,t)$ and $Q_i(x,t)$ denote the conduction current and true electric charge of the i-th transmission-line at a position x and a time t. The subcript i indicates the charge and current of the i-th transmission-line. This equation indicates that a current goes through a transmission line while satisfying the continuity equation. Hence, the next natural equation for a transmission-line is the Ohm's law for a current due to an electric field. The Ohm's law relates the electric field $E_i^x(x,t)$ at the inner surface of the resistive conductor in the direction of the current through the resistance R_i with the current $I_i(x,t)$.

$$R_i I_i(x,t) = E_i^x(x,t) \tag{18}$$

Here, the superscript x denotes the x component of the electric field of the i-th transmission-line. We note that the resistance R_i should depend on the wave-length of the electromagnetic wave going through each transmission-line due to the skin effect (Takeyama (1983)). With a finite $E^{\parallel}$ at the surfaces of transmission-lines together with B^{θ} perpendicular to both $E^{\perp}$ and $E^{\parallel}$, we have an electromagnetic wave in far distance. The boundary condition in the direction of the current even for the resistive conductor is identical to that for the perfect conductor so that $E^{\parallel}$ is equivalent to $E_i^x(x,t)$ as

$$E_i^{\parallel}(x,t) = E_i^x(x,t)\ . \tag{19}$$

The electric field $E_i^{\parallel}(x,t)$ is expressed in terms of the scalar potential $V_i(x,t)$ and the vector potential $A_i(x,t)$ in the direction of the current $I_i(x,t)$.

$$E_i^{\parallel}(x,t) = -\frac{\partial V_i(x,t)}{\partial x} - \frac{\partial A_i(x,t)}{\partial t} \tag{20}$$

Hence, from Eqs. (18), (19) and (20) we have the following relation.

$$-\frac{\partial V_i(x,t)}{\partial x} - \frac{\partial A_i(x,t)}{\partial t} = R_i I_i(x,t) \tag{21}$$

It is very interesting to point out that this equation with $R_i = 0$ corresponds to the expression of the electromagnetic potentials at the surface of the transmission-line for the TEM mode, which is worked out for the transverse electric and magnetic fields around the i-th conductor-line (Toki & Sato (2009); Paul (2008)). In this sense, we want to note again that the scalar and vector potentials here are those at the surface of the i-th conductor-line so that the TEM mode fields are obtained by using the Maxwell equation at the boundary and the outside of the conductor-line. The TEM mode fields are produced by the current and the charge in thick wires and the Ohm's law provides the effect of the TEM mode fields on these currents. We ought to solve the resulting coupled equations for the propagation of the TEM mode through a multiconductor transmission line system.

We consider now a MTL system consisting of many nearby parallel lines with circular cross sections numbered by $i = 1,..N$. We relate then the scalar and vector potentials at the surface of each line with charges and currents in all the lines. The charges and currents are present in the transmission-lines and we express them as $Q_i(x,t)$ and $I_i(x,t)$. The relations of the charge and current with the scalar and vector potentials have been worked out above as the properties of each transmission-line. We take the direction of the current in the x direction and the integral over $\mathbf{x}'$ is replaced by summation over parallel lines over j and integral in the direction x' of the parallel lines. Because the distance $|\mathbf{x} - \mathbf{x}'|$ is given as $((x-x')^2 + d_{ij}^2)^{1/2}$ where d_{ij} is a distance between two parallel ij lines, we can write the scalar and vector potentials at the surface of the i-th line as

$$V_i(x,t) = \frac{1}{4\pi\varepsilon}\sum_{j=1}^{N}\int_0^l dx' \frac{Q_j(x', t - \sqrt{(x-x')^2 + d_{ij}^2}/c)}{\sqrt{(x-x')^2 + d_{ij}^2}}\ , \tag{22}$$

$$A_i(x,t) = \frac{\mu}{4\pi}\sum_{j=1}^{N}\int_0^l dx' \frac{I_j(x', t - \sqrt{(x-x')^2 + d_{ij}^2}/c)}{\sqrt{(x-x')^2 + d_{ij}^2}}\ . \tag{23}$$

The denominators of the above two equations indicate the distance of two points in two lines denoted by ij. For the diagonal case $i = j$, $d_{ij} = 0$ and as will be discussed the finite size effect

of each transmission-line is to be taken care by using the geometrical mean distance (GMD) in the same manner as the Neumann's formula (Takeyama (1983)). We also mention here that we define the scalar and vector potentials at the surface of each transmission-line. Hence, we consider that d_{ij} is of the order of the radius of each thick wire. We assume that the length of the wire l is much larger than the radius of each wire and the distance between two lines, $l \gg d_{ij}$. Hence, we consider the case where all the transmission-lines are packed together.
These four equations; the continuity equation (17), the combined equation (21) of the Ohm's law (18) and the boundary condition (19), the scalar potential (22) and the vector potential (23), are the fundamental equations of the MTL system. We are able to know the performance of a MTL system by solving these four equations, which are now coupled integro-differential equations. Here, it is important to comment that the expressions for the scalar potential (22) and the vector potential (23) provide the electromagnetic fields outside of the wires and even at far distance if we introduce other coordinates y, z in addition to x to express the entire space. We further comment that the electromotive force (EMF) method for the input impedance of an antenna uses back the entire radiation energy to calculate the electromagnetic field at the surface of a wire (Stratton (1941)). Hence, these four equations are able to provide the behavior of electromagnetic wave even far outside of the MTL system. Therefore, when we solve these four equations we know not only the behavior of the MTL system but also the electromagnetic fields in the entire space outside of the thick wires. We are then able to include naturally emission and absorption of the EM waves. We comment here that the retardation charge and current in Eqs. (22) and (23) are responsible for a Poynting vector going out at far distance.
We treat the retardation effect in the integral by considering that the coupled differential equations are linear and all the quantities have the time dependence as $Q_i(x,t) = Q_i(x)e^{-j\omega t}$ and $I_i(x,t) = I_i(x)e^{-j\omega t}$. Inserting these expressions to the above equations, we get

$$V_i(x,t) = \frac{1}{4\pi\varepsilon}\sum_{j=1}^{N}\int_0^l dx' \frac{Q_j(x',t)e^{j\omega\sqrt{(x-x')^2+d_{ij}^2}/c}}{\sqrt{(x-x')^2+d_{ij}^2}}, \tag{24}$$

$$A_i(x,t) = \frac{\mu}{4\pi}\sum_{j=1}^{N}\int_0^l dx' \frac{I_j(x',t)e^{j\omega\sqrt{(x-x')^2+d_{ij}^2}/c}}{\sqrt{(x-x')^2+d_{ij}^2}}. \tag{25}$$

These expressions for the scalar and vector potentials provide the right behaviors of electromagnetic fields far outside of the MTL system. Hence, these relations together with the continuity equation and the combined relation of the Ohm's law and the boundary condition provide a proper set of equations of electromagnetic waves with radiation. Since these integro-differential coupled equations are difficult to handle, we shall find an appropriate approximation.
In order to find out an appropriate approximation at a boundary of a thick wire, we study the property of the integrand with the retardation terms. The function $1/\sqrt{(x-x')^2+d_{ij}^2}$ has a strong peak at $x' = x$ and drops rapidly as x' deviates from x. Furthermore, the real part of the factor $e^{j\omega\sqrt{(x-x')^2+d_{ij}^2}/c}$ behaves as $cos(\omega\sqrt{(x-x')^2+d_{ij}^2}/c)$ and provides a further cutoff with $|x-x'|$. Hence, the integral has a dominant contribution in the narrow region close to the position $x' = x$. Hence, it is a good approximation to pull out the charge Q_i and current I_i from the integral by taking their arguments at x. This fact indicates that the electric field has the perpendicular component to the transmission-line and the magnetic field has the axial component produced by the current at the same coordinate. Hence, the

TEM mode propagates through the transmission-lines. We shall call this as the TEM mode approximation. It should be noted here as mentioned before that the real part of the scalar and vector potentials could satisfy the boundary condition of $E^{\perp}$ for the resistive conductor due to the TEM mode approximation. The imaginary part behaves as $sin(\omega\sqrt{(x-x')^2+d_{ij}^2}/c)$ and together with the denominator, the integrant is the zero-th order spherical Bessel function $j_0(\omega\sqrt{(x-x')^2+d_{ij}^2}/c)$ with some factor and drops rapidly with $|x-x'|$ and oscillates at large ω. We may take the TEM mode approximation for the imaginary part as well at large ω, but it seems better to keep the charge and current in the integral. This is particularly the case when the angular velocity ω is small. Hence, we write the scalar and vector potentials in the TEM mode approximation for the real part and keep the integral form for the imaginary part.

$$V_i(x,t) = \frac{1}{4\pi\varepsilon}\sum_{j=1}^{N}\int_0^l dx' \frac{cos(\omega\sqrt{(x-x')^2+d_{ij}^2}/c)}{\sqrt{(x-x')^2+d_{ij}^2}} Q_j(x,t) + j\frac{1}{4\pi\varepsilon}\int_0^l dx' \frac{Q_t(x',t)sin(\omega|x-x'|/c)}{|x-x'|}, \tag{26}$$

$$A_i(x,t) = \frac{\mu}{4\pi}\sum_{j=1}^{N}\int_0^l dx' \frac{cos(\omega\sqrt{(x-x')^2+d_{ij}^2}/c)}{\sqrt{(x-x')^2+d_{ij}^2}} I_j(x,t) + j\frac{\mu}{4\pi}\int_0^l dx' \frac{I_t(x',t)sin(\omega|x-x'|/c)}{|x-x'|}. \tag{27}$$

It is very important to notice that the d_{ij} dependence is negligibly small when $d_{ij} \ll \frac{c}{\omega}$ and we drop the ij dependence in the imaginary part. Hence, we can sum up over the wire number and write the total charge and current as $Q_t(x',t) = \sum_j^N Q_i(x',t)$ and $I_t(x',t) = \sum_j^N I_i(x',t)$. We write therefore the above relations as

$$V_i(x,t) = \sum_j P_{ij}(\omega)Q_j(x,t) + jM_e Q_t^I(l,x,t), \tag{28}$$

$$A_i(x,t) = \sum_j L_{ij}(\omega)I_j(x,t) + jM_m I_t^I(l,x,t).$$

Here, we have defined the integrated charge and current as

$$Q_t^I(l,x,t) = \int_0^l dx' \frac{Q_t(x',t)sin(\omega|x-x'|/c)}{|x-x'|} \tag{29}$$

$$I_t^I(l,x,t) = \int_0^l dx' \frac{I_t(x',t)sin(\omega|x-x'|/c)}{|x-x'|}.$$

with the coefficients defined as $M_e = \frac{1}{4\pi\varepsilon}$ and $M_m = \frac{\mu}{4\pi}$. It is very important to note that the integrals of the charge and current over the wire length generate the parallel component of the electric field at the wire surface. This is important for the radiation of the EM wave in the far distance.

We shall calculate now these coefficients P_{ij} and L_{ij} by using the Neumann's formula (Takeyama (1983)). To this end, we have to take into account the finite size effect of each transmission-line and also the skin effect. We shall study the finite size effect including the skin effect in a future publication (Sato & Toki (2011)). We first write the well known coefficient of inductance L_{ij} given by the Neumann's formula (Takeyama (1983)).

$$L_{ij}(\omega) = \frac{\mu}{4\pi}\int_0^l dx' \frac{cos(\omega\sqrt{(x-x')^2+d_{ij}^2}/c)}{\sqrt{(x-x')^2+d_{ij}^2}} \tag{30}$$

$$\sim \frac{\mu}{4\pi l}\int_0^l dx \int_0^l dx' \frac{cos(\omega\sqrt{(x-x')^2+d_{ij}^2}/c)}{\sqrt{(x-x')^2+d_{ij}^2}}$$

$$= \frac{\mu}{2\pi}\left(\ln\frac{2\tilde{l}(\omega)}{d_{ij}} - 1\right) .$$

The second line of this expression is the approximation of the Neumann's formula. We have to take into account further the finite size effect together with the skin effect. These effects are worked out by introducing the geometrical mean distance (GMD). The GMD is defined as (Takeyama (1983))

$$\ln\tilde{a}_{ij} = \frac{1}{S_i S_j}\int\int \ln r(s_i, s_j : d_{ij}) ds_i ds_j , \tag{31}$$

where $r(s_i, s_j : d_{ij})$ includes the distance between the two lines d_{ij} and the skin effect. Here, $ds_i(ds_j)$ is a small area in a wire $i(j)$ with the total area $S_i(S_j)$. With the GMD, we can finally write the coefficient of inductance as

$$L_{ij}(\omega) = \frac{\mu}{2\pi}\left(\ln\frac{2\tilde{l}(\omega)}{\tilde{a}_{ij}} - 1\right) . \tag{32}$$

We can work out the coefficients of potential P_{ij} exactly in the same way as those of inductance L_{ij} in the TEM mode approximation (Toki & Sato (2009)). This should be the case, because of the continuity equation, which forces the spatial distributions of the charge and current are the same.

$$P_{ij}(\omega) = \frac{1}{2\pi\varepsilon}\left(\ln\frac{2\tilde{l}(\omega)}{\tilde{a}_{ij}} - 1\right) \tag{33}$$

The usual coefficients used are the coefficients of capacity C in the MTL equations (Paul (2008)). This coefficient C_{ij} is the matrix inversion of the coefficients of potential $C = P^{-1}$. We mention that it is an essential feature to write P instead of C in the present derivation, because a capacitance per unit length is no longer an adequate quantity in the MTL theory (Toki & Sato (2009)).

We can work out the relations among the charges and currents and the scalar and vector potentials together with the electric resistances for the TEM mode. In order to use these relations we take time derivatives of the above coupled equations.

$$\frac{\partial V_i(x,t)}{\partial t} = \sum_j^N P_{ij}\frac{\partial Q_j(x,t)}{\partial t} + jM_e\frac{\partial Q_t^I(l,x,t)}{\partial t}\,, \tag{34}$$

$$\frac{\partial A_i(x,t)}{\partial t} = \sum_j^N L_{ij}\frac{\partial I_j(x,t)}{\partial t} + jM_m\frac{\partial I_t^I(l,x,t)}{\partial t}\,. \tag{35}$$

Here, we have dropped writing ω for the coefficients of all the terms for simplicity of writing. By replacing the charge $Q_j(x,t)$ by the current using the continuity equation (17), the above equation (34) provides

$$\frac{\partial V_i(x,t)}{\partial t} = -\sum_j^N P_{ij}\frac{\partial I_j(x,t)}{\partial x} + jM_e\frac{\partial Q_t^I(l,x,t)}{\partial t}\,. \tag{36}$$

We use the combined equation (21) of the Ohm's law (18) and the boundary condition (19) in the above equation (35) in order to write the following equation in terms of the scalar potential as

$$\frac{\partial V_i(x,t)}{\partial x} = -\sum_j^N L_{ij}\frac{\partial I_j(x,t)}{\partial t} - jM_m\frac{\partial I_t^I(l,x,t)}{\partial t} - R_i I_i(x,t)\,. \tag{37}$$

We consider Eqs. (36) and (37) as the fundamental equations for the TEM modes in the MTL system with emission and absorption. As we have seen the inclusion of the retardation terms with the use of the properties of transmission-lines of thick wires with resistance is a natural extension of the standard multiconductor transmission-line theory. We comment here that similar equations without the retardation terms for the case of one transmission line was derived by Kirchhoff (Kirchhoff (1857)). The development later of the Kirchhoff work is described in a book of Ohta (Ohta (2005)). We shall see that these two retardation terms provide naturally the emission and absorption of electromagnetic waves through the multiconductor transmission-line system. We emphasize here that electromagnetic waves go through a multiconductor transmission-line system in the TEM mode while making electromagnetic radiation.

4. TEM mode of one line antenna

Since we have worked out the MTL equation including radiation, we would like to discuss an isolated system of one-conductor transmission-line so that we write explicitly how the electromagnetic energy is converted into Joule energy and radiation energy. In principle, we may have to consider the influence of the circumstance even for one-line antenna. However, for simplicity and also for the sake of understanding the antenna mode, we study the one-line antenna system using the new theory. Here, we have in mind the case of one transmission-line antenna and deal with the case that the current changes from its full value to the vanishing value. We write a set of the antenna mode equation using Eqs. (36) and (37) as

$$\frac{\partial V(x,t)}{\partial t} = -P\frac{\partial I(x,t)}{\partial x} + jM_e\frac{\partial Q^I(l,x,t)}{\partial t}\,. \tag{38}$$

and

$$\frac{\partial V(x,t)}{\partial x} = -L\frac{\partial I(x,t)}{\partial t} - jM_m\frac{\partial I^I(l,x,t)}{\partial t} - RI(x,t) \,. \tag{39}$$

The quantities Q^I and I^I are those defined in Eqs. (29) by dropping the suffix t because here we treat one-line antenna. These expressions for the integrated charge and current together with those of the coefficients of potential P and inductance L remind us the functions of sine and cosine integrals S_i and C_i in the antenna theory (Stratton (1941)).
Hence, we write the one antenna equation as

$$\begin{aligned}\frac{\partial V(x,t)}{\partial t} &= -cZ\frac{\partial I(x,t)}{\partial x} + jMc\frac{\partial Q^I(l,x,t)}{\partial t} \,, \\ \frac{\partial V(x,t)}{\partial x} &= -\frac{Z}{c}\frac{\partial I(x,t)}{\partial t} - j\frac{M}{c}\frac{\partial I^I(l,x,t)}{\partial t} - RI(x,t) \,.\end{aligned} \tag{40}$$

Here, we would like to write explicitly the characteristic impedance Z, the resistance R and the characteristic antenna mode coefficient M for an one-conductor transmission-line system. Since we are dealing with one line, the characteristic impedance is written as

$$Z = \frac{1}{2\pi}\sqrt{\frac{\mu}{\varepsilon}}\left(\ln\frac{2\tilde{l}}{\tilde{a}_{11}} - 1\right) \,. \tag{41}$$

This impedance is featured to include the length of the line-antenna explicitly. The resistance is simply the one of the transmission-line $R = R_1$ and the characteristic antenna mode coefficient is $M = \frac{1}{4\pi}\sqrt{\mu/\varepsilon}$. The characteristic impedance Z resembles the coefficient of the antenna theory (Stratton (1941)).
The integrated quantities Q^I and I^I are those related with the charge and current integrated over the length, and we can fix the time dependence of the potential $V(x,t) = V(x)e^{-j\omega t}$ and correspondingly for the current $I(x,t) = I(x)e^{-j\omega t}$. We can write then a coupled differential equation for a certain ω.

$$\begin{aligned}\frac{dV(x)}{dx} &= j\frac{Z\omega}{c}I(x) - RI(x) - \frac{M\omega}{c}I^I(l,x) \,, \\ -j\omega V(x) &= -Zc\frac{dI(x)}{dx} + Mc\omega Q^I(l,x) \,.\end{aligned} \tag{42}$$

In order to proceed from here, we consider the case of long wave length. This approximation corresponds to the long wave length approximation in the antenna theory.

$$\begin{aligned}Q^I(l) &\sim \frac{\omega}{c}\int_0^l dx' Q(x') \\ I^I(l) &\sim \frac{\omega}{c}\int_0^l dx' I(x') \,.\end{aligned} \tag{43}$$

We insert the second equation to the first one of Eq. (42) and obtain a second order integro-differential equation for the current.

$$\frac{d^2 I(x)}{dx^2} = -\frac{\omega^2}{c^2}I(x) - j\frac{R\omega}{Zc}I(x) - j\frac{M\omega^2}{Zc^2}I^I(l) \,. \tag{44}$$

We note that this equation is a second order linear differential equation with a constant, if we consider the last term is known. In this case, we can write a general solution as

$$I(x) = ie^{jkx} + i'e^{-jkx} - \frac{j}{1 + j\frac{Rc}{Z\omega}} \frac{M}{Z} I^I(l) \,. \tag{45}$$

Here $k = k_R + jk_I = \frac{\omega}{c}\sqrt{1 + j\frac{Rc}{Z\omega}}$. By inserting this solution to the second equation (42) with the long wave length approximation, we get

$$V(x) = \frac{Zkc}{\omega}(ie^{jkx} - i'e^{-jkx}) + jMcQ^I(l) \,. \tag{46}$$

We should keep in mind that these solutions for $I(x)$ and $V(x)$ are implicit solutions. Namely, $I^I(l)$ and $Q^I(l)$ are obtained by knowing the current $I(x)$ and $Q(x)$. Of course, when the boundary conditions at the center and its ends of the transmission line are given, we are able to use the above solutions for any case of interest.

As the most interesting case, we consider the standard linear antenna which could operate for radiation-emission as a transmitter or for radiation-absorption as a receiver. For this purpose, a power supply or a passive lumped circuit element is placed in the middle of a transmission-line, respectively, and both ends are open. Hence, the boundary conditions in this case are

$$\begin{aligned} V(x = +\epsilon) &= V(0) \,, \\ V(x = -\epsilon) &= -V(0) \,, \\ I(x = l) &= 0 \,, \\ I(x = -l) &= 0 \,. \end{aligned} \tag{47}$$

These boundary conditions fix a relation of i and i' of Eq. (45) and hence the current $I(x)$ and the potential $V(x)$ in terms of i. We write the equation to fix i' in terms of i by using the condition that the current vanishes at the end of the antenna.

$$I(l) = ie^{jkl} + i'e^{-jkl} - \frac{j}{1 + j\frac{Rc}{Z\omega}} \frac{M}{Z} I^I(l) = 0 \tag{48}$$

With this relation and Eq. (45), we are able to write $I(0)$ in a compact form.

$$I(0) = (1 - e^{jkl})i + (1 - e^{-jkl})i' \tag{49}$$

In the mean time, we get an implicit expression of $I^I(l)$ by integrating $I(x)$ of Eq. (45) from $x = 0$ to $x = l$ and find a more compact expression for $I^I(l)$ in terms of i and i'.

$$I^I(l) = \frac{(1 + j\frac{Rc}{Z\omega})\frac{j\omega}{kc}[(1 - e^{jkl})i - (1 - e^{-jkl})i']}{1 + j\frac{Rc}{Z\omega} + j\frac{M\omega l}{Zc}} \tag{50}$$

We can then solve for i' in terms of i by using Eq. (48).

$$i' = -\frac{(1 + j\frac{Rc}{Z\omega} + j\frac{M\omega l}{Zc})e^{jkl} + \frac{M\omega}{Zkc}(1 - e^{jkl})}{(1 + j\frac{Rc}{Z\omega} + j\frac{M\omega l}{Zc})e^{-jkl} - \frac{M\omega}{Zkc}(1 - e^{-jkl})} i \tag{51}$$

Substituting this expression to Eq. (49), we get $I(0)$ in terms of i.

$$I(0) = \frac{(1 + j\frac{Rc}{Z\omega} + j\frac{M\omega l}{Zc})(e^{-jkl} - e^{jkl}) - 2\frac{M\omega}{Zkc}(2 - e^{jkl} - e^{-jkl})}{(1 + j\frac{Rc}{Z\omega} + j\frac{M\omega l}{Zc})e^{-jkl} - \frac{M\omega}{Zkc}(1 - e^{-jkl})} i\,. \tag{52}$$

In order to obtain $V(0)$, we have to know $Q^I(l)$, which is obtained as $Q^I(l) = -\frac{j}{c}(I(l) - I(0))$. Since we take the boundary condition $I(l) = 0$, we find

$$\begin{aligned} Q^I(l) &= \frac{j}{c} I(0) \\ &= \frac{j}{c} \frac{(1 + j\frac{Rc}{Z\omega} + j\frac{M\omega l}{Zc})(e^{-jkl} - e^{jkl}) - 2\frac{M\omega}{Zkc}(2 - e^{jkl} - e^{-jkl})}{(1 + j\frac{Rc}{Z\omega} + j\frac{M\omega l}{Zc})e^{-jkl} - \frac{M\omega}{Zkc}(1 - e^{-jkl})} i\,. \end{aligned} \tag{53}$$

We can obtain also $I^I(l)$ by using the above expressions.

$$\begin{aligned} I^I(l) &= \frac{(1 + j\frac{Rc}{Z\omega})\frac{j\omega}{kc}(e^{-jkl} + e^{jkl} - 2)}{(1 + j\frac{Rc}{Z\omega} + j\frac{M\omega l}{Zc})e^{-jkl} - \frac{M\omega}{Zkc}(1 - e^{-jkl})} i \\ &= \frac{(1 + j\frac{Rc}{Z\omega})\frac{j\omega}{kc}(e^{-jkl} + e^{jkl} - 2)}{(1 + j\frac{Rc}{Z\omega} + j\frac{M\omega l}{Zc})(e^{-jkl} - e^{jkl}) - 2\frac{M\omega}{Zkc}(2 - e^{jkl} - e^{-jkl})} I(0) \end{aligned} \tag{54}$$

We can obtain $V(0)$ by using Eq. (46) with Eqs. (51), (52) and (53).

$$\begin{aligned} V(0) &= \frac{Zkc}{\omega}(i - i') + jMcQ^I(l) \\ &= \left(\frac{Zkc}{\omega} \frac{(1 + j\frac{Rc}{Z\omega} + j\frac{M\omega l}{Zc})(e^{-jkl} + e^{jkl}) + \frac{M\omega}{Zkc}(e^{-jkl} - e^{jkl})}{(1 + j\frac{Rc}{Z\omega} + j\frac{M\omega l}{Zc})(e^{jkl} - e^{-jkl}) - 2\frac{M\omega}{Zkc}(2 - e^{jkl} - e^{-jkl})} - M \right) I(0) \end{aligned} \tag{55}$$

We can obtain now the input impedance by taking the ratio of $V(0)$ and $I(0)$.

$$\begin{aligned} Z_s &= \frac{2V(x=0)}{I(x=0)} \\ &= 2\frac{Zkc}{\omega} \frac{(1 + j\frac{Rc}{Z\omega} + j\frac{M\omega l}{Zc})(e^{-jkl} + e^{jkl}) + \frac{M\omega}{Zkc}(e^{-jkl} - e^{jkl})}{(1 + j\frac{Rc}{Z\omega} + j\frac{M\omega l}{Zc})(e^{jkl} - e^{-jkl}) - 2\frac{M\omega}{Zkc}(2 - e^{jkl} - e^{-jkl})} - 2M \end{aligned} \tag{56}$$

Although lengthy, this expression does not depend on the initial input energy and is written in terms of Z, R, M and l for a given ω, which are the properties of the transmission-line. It is the first time to obtain the input impedance of one resistive conductor antenna. This expression should be contrasted with the EMF method for the input impedance, which is obtained by assuming the expression for the current in a line antenna. Here, the current is obtained by solving the TEM mode wave equation with the boundary condition at the center and its ends of a line antenna.

We try to understand the meaning of the input impedance by setting $M = 0$. In this case, the input impedance is written as

$$Z_s = 2\frac{Zkc}{\omega}\frac{e^{-jkl}+e^{jkl}}{e^{jkl}-e^{-jkl}} \tag{57}$$
$$= 2\frac{Z(k_R+jk_I)c}{\omega}\frac{(e^{-jk_Rl}+e^{jk_Rl})(e^{k_Il}+e^{-k_Il})+(e^{-jk_Rl}-e^{jk_Rl})(e^{k_Il}-e^{-k_Il})}{(e^{-jk_Rl}-e^{jk_Rl})(e^{k_Il}+e^{-k_Il})+(e^{-jk_Rl}+e^{jk_Rl})(e^{k_Il}-e^{-k_Il})}$$

We consider the case that R is small and write $k = \frac{\omega}{c} + j\frac{R}{2Z} = k_R + jk_I$. We set $k_I l \ll 1$ and expand the exponent up to the first order.

$$Z_s = 2\frac{Z(k_R+jk_I)c}{\omega}\frac{k_Il + jsin(k_Rl)cos(k_Rl)(1-k_Il)}{sin^2(k_Rl)+cos^2(k_Rl)(k_Il)^2} \tag{58}$$
$$\sim Rl\frac{1}{sin^2(k_Rl)} + j2Z\frac{cos(k_Rl)}{sin(k_Rl)}$$

In the last step, we take the dominant terms for the case that $sin(k_Rl)$ is not close to 0. The above expression indicates that the real part corresponds to the resistance and the imaginary part corresponds to the characteristic impedance for the TEM mode. We stress here that the TEM mode can exist even for one-line antenna in contrast to the standard understanding that the TEM mode is associated with at least two conductors. At the same time, the input impedance has a resonance structure around $k_Rl = n\pi$ with n being an integer due to the sine-function in the denominator. The real part has a peak structure at this point, while the imaginary part changes sign and the small additional term makes the imaginary part to go through zero around this point.
With these expressions for the current and potential and the input impedance, we are able to calculate the electromagnetic power

$$P(x) = \frac{1}{4}(V(x)I^*(x) + V^*(x)I(x)) \tag{59}$$

and the input power $P(0)$. We can then write all the power consumed by this one-line antenna system.

$$P_{total}(x=0) = 2P(x=0) = \frac{1}{2}\left(V^*(0)I(0) + V(0)I^*(0)\right) \tag{60}$$
$$= \frac{1}{4}(Z_s^* + Z_s)|I(0)|^2 = \frac{1}{2}ReZ_s|I(0)|^2$$

We note here that the power consumed by the one line antenna is not only used by the radiation but also by the resistance to heat up the one line antenna.
We can express the change rate of the EM power using the coupled differential equation.

$$\frac{dP(x)}{dx} = \frac{1}{4}\left(\frac{dV(x)}{dx}I^*(x) + \frac{dV^*(x)}{dx}I(x) + V^*(x)\frac{dI(x)}{dx} + V(x)\frac{dI^*(x)}{dx}\right) \tag{61}$$

Substituting Eq. (42) to this expression, we can calculate the change rate as a sum of the resistance and radiation terms.

$$\frac{dP(x)}{dx} = -\frac{1}{2}R|I|^2 - \frac{1}{4}\frac{M\omega}{c}(I^{I*}(l)I + I^I(l)I^*) - j\frac{1}{4}Mc\left(Q^{I*}(l)\frac{dI}{dx} - Q^I(l)\frac{dI^*}{dx}\right) \tag{62}$$

We can write the integrated change rate in a compact form.

$$P_t^{antenna} = \int_0^l \frac{dP^{antenna}(x)}{dx} dx = -\frac{1}{2} M |I^I(l)|^2 + \frac{1}{2} M c^2 |Q^I(l)|^2 \tag{63}$$

It is clear that the total change rate due to the antenna mode consists of the emission and absorption terms, which are indicated by the minus sign term and the plus sign term.
We would like to comment here which process as emission or absorption occurs. When a power supply is connected at the middle of one-conductor transmission-line, this antenna operates for radiation-emission as a transmitter because the emission term is larger than the absorption term. When a passive lumped circuit element is connected at the middle of one-conductor transmission line, this antenna operates for radiation-absorption as a receiver because the absorption term is larger than the emission term.

5. Three-conductor transmission-line system

We consider now the three-conductor transmission-line theory with emission and absorption through the antenna mode. This is a very interesting case where the two-conductor transmission-lines include the effect of the circumstance. In our previous publication (Toki & Sato (2009)), we have discussed the case where the total current is zero and hence the case without the antenna mode. The present situation with the antenna mode corresponds to the realistic case. In this case we introduce the normal, common and antenna modes. They are written with the currents and potentials of the three lines. Here, we consider that the lines 1 and 2 are the main lines and the line 3 denotes the circumstance.

$$\begin{aligned} I_n &= \frac{1}{2}(I_1 - I_2) \\ I_c &= \frac{1}{2}(I_1 + I_2 - I_3) \\ I_a &= \frac{1}{2}(I_1 + I_2 + I_3) \\ V_n &= V_1 - V_2 \\ V_c &= \frac{1}{2}(V_1 + V_2) - V_3 \\ V_a &= \frac{1}{2}(V_1 + V_2) + V_3 \end{aligned} \tag{64}$$

We work out the coupled integro-differential equations for the TEM mode with the retardation term treated explicitly. There is a factor two difference between the antenna mode current and the total current $I_t = 2I_a$. We write the results here for the normal, common and antenna modes. We use first the integro-differential equations for $N = 3$ in Eq. (37) and express the equations in terms of various modes,

$$\begin{aligned} \frac{\partial V_n(x,t)}{\partial x} &= -L_n \frac{\partial I_n(x,t)}{\partial t} - L_{nc} \frac{\partial I_c(x,t)}{\partial t} - L_{na} \frac{\partial I_a(x,t)}{\partial t} - R_n I_n - R_{nc} I_c - R_{na} I_a \\ \frac{\partial V_c(x,t)}{\partial x} &= -L_{cn} \frac{\partial I_n(x,t)}{\partial t} - L_c \frac{\partial I_c(x,t)}{\partial t} - L_{ca} \frac{\partial I_a(x,t)}{\partial t} - R_{cn} I_n - R_c I_c - R_{ca} I_a \\ \frac{\partial V_a(x,t)}{\partial x} &= -L_{an} \frac{\partial I_n(x,t)}{\partial t} - L_{ac} \frac{\partial I_c(x,t)}{\partial t} - L_a \frac{\partial I_a(x,t)}{\partial t} - j2M_m \frac{\partial I_t^I(l,x,t)}{\partial t} \\ &\quad - R_{an} I_n - R_{ac} I_c - R_a I_a \end{aligned} \tag{65}$$

In the above equations all the coefficients are written as follows,

$$\begin{aligned} L_n &= L_{11} - 2L_{21} + L_{22} \\ L_c &= \frac{1}{4}(L_{11} + 2L_{12} + L_{22}) - (L_{13} + L_{23}) + L_{33} \\ L_a &= \frac{1}{4}(L_{11} + 2L_{12} + L_{22}) + L_{13} + L_{23} + L_{33}\,, \end{aligned} \tag{66}$$

for the diagonal coefficients and

$$\begin{aligned} L_{nc} &= \frac{1}{2}(L_{11} - L_{22}) - (L_{13} - L_{23}) \\ L_{na} &= \frac{1}{2}(L_{11} - L_{22}) + (L_{13} - L_{23}) \\ L_{ca} &= \frac{1}{4}(L_{11} + 2L_{12} + L_{22}) - L_{33}\,, \end{aligned} \tag{67}$$

for the non-diagonal coefficients. We get the resistance terms as

$$\begin{aligned} R_n &= R_1 + R_2 \\ R_c &= \frac{1}{4}(R_1 + R_2) + R_3 \\ R_a &= \frac{1}{4}(R_1 + R_2) + R_3 \\ R_{nc} &= \frac{1}{2}(R_1 - R_2) \\ R_{na} &= \frac{1}{2}(R_1 - R_2) \\ R_{ca} &= \frac{1}{4}(R_1 + R_2) - R_3 \end{aligned} \tag{68}$$

We obtain similar relations for transmission-line equations (36) including P_{ij} as written below.

$$\begin{aligned} \frac{\partial V_n(x,t)}{\partial t} &= -P_n\frac{\partial I_n(x,t)}{\partial x} - P_{nc}\frac{\partial I_c(x,t)}{\partial x} - P_{na}\frac{\partial I_a(x,t)}{\partial x} \\ \frac{\partial V_c(x,t)}{\partial t} &= -P_{cn}\frac{\partial I_n(x,t)}{\partial x} - P_c\frac{\partial I_c(x,t)}{\partial x} - P_{ca}\frac{\partial I_a(x,t)}{\partial x} \\ \frac{\partial V_a(x,t)}{\partial t} &= -P_{an}\frac{\partial I_n(x,t)}{\partial x} - P_{ac}\frac{\partial I_c(x,t)}{\partial x} - P_a\frac{\partial I_a(x,t)}{\partial x} + j2M_e\frac{\partial Q_t^I(l,x,t)}{\partial t} \end{aligned} \tag{69}$$

In the above equations all the coefficients are written as follows,

$$\begin{aligned} P_n &= P_{11} - 2P_{12} + P_{22} \\ P_c &= \frac{1}{4}(P_{11} + 2P_{12} + P_{22}) - (P_{13} + P_{23}) + P_{33} \\ P_a &= \frac{1}{4}(P_{11} + 2P_{12} + P_{22}) + P_{13} + P_{23} + P_{33}\,, \end{aligned} \tag{70}$$

for the diagonal coefficients and

$$P_{nc} = \frac{1}{2}(P_{11} - P_{22}) - (P_{13} - P_{23}) \tag{71}$$
$$P_{na} = \frac{1}{2}(P_{11} - P_{22}) + (P_{13} - P_{23})$$
$$P_{ca} = \frac{1}{4}(P_{11} + 2P_{12} + P_{22}) - P_{33} ,$$

for the non-diagonal coefficients. All the coefficients of potential P_{ij} are written in a compact form as the coefficients of inductance L_{ij}.

$$L_{ij} = \frac{\mu}{2\pi}(\ln \frac{2\tilde{l}}{\tilde{a}_{ij}} - 1) \tag{72}$$
$$P_{ij} = \frac{1}{2\pi\varepsilon}(\ln \frac{2\tilde{l}}{\tilde{a}_{ij}} - 1)$$

Using these coefficients, we can write all the coefficients associated with the normal, common and antenna modes. They are written as

$$P_n = \frac{1}{2\pi\varepsilon} \ln \frac{\tilde{a}_{12}^2}{\tilde{a}_{11}\tilde{a}_{22}} \tag{73}$$
$$P_c = \frac{1}{8\pi\varepsilon} \ln \frac{\tilde{a}_{13}^4\tilde{a}_{23}^4}{\tilde{a}_{11}\tilde{a}_{22}\tilde{a}_{12}^2\tilde{a}_{33}^4}$$
$$P_a = \frac{1}{8\pi\varepsilon}(\ln \frac{(2\tilde{l})^{16}}{\tilde{a}_{11}\tilde{a}_{22}\tilde{a}_{12}^2\tilde{a}_{33}^4\tilde{a}_{13}^4\tilde{a}_{23}^4} - 16)$$

Only the antenna mode coefficient P_a contains the length of the transmission-lines explicitly and is appropriate for the antenna mode. The coupling terms are written as

$$P_{nc} = \frac{1}{2\pi\varepsilon} \ln \frac{\tilde{a}_{22}a_{13}}{\tilde{a}_{11}\tilde{a}_{23}} \tag{74}$$
$$P_{na} = \frac{1}{2\pi\varepsilon} \ln \frac{\tilde{a}_{22}a_{23}}{\tilde{a}_{11}\tilde{a}_{13}}$$
$$P_{ca} = \frac{1}{8\pi\varepsilon} \ln \frac{\tilde{a}_{33}^4}{\tilde{a}_{11}\tilde{a}_{22}\tilde{a}_{12}^2}$$

We get similar expressions for L_i. They are related with P_i as $L_i = P_i c^2$. It is very interesting to note that the coefficient of capacity for the normal mode is written as $C_n = 1/P_n = 2\pi\varepsilon/ln\frac{\tilde{a}_{12}^2}{\tilde{a}_{11}\tilde{a}_{22}}$.

The coupled differential equations tell many interesting facts. When there is a symmetry between the lines 1 and 2 in their relations to the third line due to the symmetric arrangement, the coupling terms of the normal mode to both the common and antenna modes can be made zero. The normal mode decouples from the common and antenna modes. On the other hand, when the symmetry is lost between the lines 1 and 2, the normal mode couples not only with the common mode but also with the antenna mode. This coupling of three wave-type modes is considered to be the origin of EM noise, which can not be understood due to the reflection and interference. It is therefore very important to take care of the symmetry of the lines 1 and 2. We repeat that if the third line represents the circumstance, it is impossible to make the coupling terms zero. Hence, the ordinary two-conductor transmission-line system is influenced by the circumstance and the electromagnetic emission and absorption take place. Therefore, we cannot avoid the noise problem. In addition, we comment that the common mode always couples with the antenna mode and the emission and absorption take place simultaneously with the generation of the EM noise in the circuit.

It is interesting to write the differential equation for the normal mode for the case of symmetrization, where the coupling terms of the normal mode to the common and antenna modes are zero. The TEM mode differential equations for the normal mode are written as

$$\begin{aligned} \frac{\partial V_n(x,t)}{\partial t} &= -P_n \frac{\partial I_n(x,t)}{\partial x} \\ \frac{\partial V_n(x,t)}{\partial x} &= -L_n \frac{\partial I_n(x,t)}{\partial t} - R_n I_n \,. \end{aligned} \tag{75}$$

These differential equations for the normal mode together with the coefficients L_n, P_n and R_n agree with the two-conductor transmission-line equations (Paul (2008)). Usually the upper equation in Eq. (75) is written in terms of $1/C_n$ in the place of P_n. The expression for C_n calculated by $C_n = 1/P_n = 2\pi\varepsilon / ln \frac{\tilde{a}_{12}^2}{\tilde{a}_{11}\tilde{a}_{22}}$ agrees with the capacitance per unit length of the usual two line expression. We stress again the use of the coefficient of potential is essential for the formulation of the three-line system. Hence, the ordinary TEM mode propagation of the EM wave is achieved only when the symmetrization is introduced for the electric circuit in the circumstance.

We shall calculate the electromagnetic power of the three-conductor transmission-line system.

$$\begin{aligned} P(x) &= \frac{1}{4}(V_1^* I_1 + V_2^* I_2 + V_3^* I_3 + V_1 I_1^* + V_2 I_2^* + V_3 I_3^*) \\ &= \frac{1}{4}(V_n^* I_n + V_c^* I_c + V_a^* I_a + V_n I_n^* + V_c I_c^* + V_a I_a^*) \end{aligned} \tag{76}$$

It is interesting to calculate the change of the power with distance so that we can pick up only the terms which change with distance. We take the time dependence of all the modes as $\exp(-j\omega t)$. We can work out the change rate $\frac{dP(x)}{dx}$ on the basis of Eqs. (65) and (69) in exactly the same way as the case of the one-conductor transmission line. We write only the final result.

$$\begin{aligned} \frac{dP(x)}{dx} &= -\frac{1}{2}(R_n|I_n(x)|^2 + R_c|I_c(x)|^2 + R_a|I_a(x)|^2 \\ &+R_{nc}(I_n(x)I_c^*(x) + I_n^*(x)I_c(x)) + R_{na}(I_n(x)I_a^*(x) \\ &+I_n^*(x)I_a(x)) + R_{ca}(I_c(x)I_a^*(x) + I_c^*(x)I_a(x))) \\ &-\frac{1}{2}\frac{M\omega}{c}(I_t^{I*}(l,x)I_a(x) + I_t^{I,x}(l)I_a^*(x)) \\ &-j\frac{1}{2}Mc\left(Q_t^{I*}(l,x)\frac{dI_a(x)}{dx} - Q_t^I(l.x)\frac{dI_a^*(x)}{dx}\right) \end{aligned} \tag{77}$$

This expression agrees with the one of the line-antenna (62), when the change rate is expressed with the total current by using the relation $I_a(x) = \frac{1}{2}I_t(x)$. It is interesting to point out that the change of the electric power is made by the resistance terms and the antenna mode terms.

There are two effects in the antenna mode terms. One is a term associated with emission and the other is a term associated with absorption.
Here, we would like to comment the strict symmetrization of the power supply. In a standard two-stage power supply in which two identical power supplies are connected in series but switchings are controlled alternately, a common mode current is unavoidable. Even when there is a symmetry between lines 1 and 2 in relation to the third line, coupling terms between the common and antenna modes do not vanish unavoidably. This implies that the radiation of EM wave occurs unless the common mode current is eliminated. The most effective method of eliminating the common and antenna modes is a strict symmetrization in which switchings of two-stage power supply should be synchronized in a symmetrized three-conductor transmission-line system.

6. Symmetrized electric circuit

In the previous section, we have discussed the performance of two transmission-lines in the circumstance and hence a three-conductor transmission-line system. The electromagnetic noise in the circumstance goes through the third line in the form of electromagnetic wave. In the standard two-stage power supply as mentioned above, the noise in the circumstance influences the main two-lines through the common mode. Since we are not able to control the circumstance, it is impossible to remove the electromagnetic noise in the case of the two line electric circuit in the circumstance. At the same time, the modern power supply and also both the inverter and converter use the chopping method and generate electromagnetic noise with high frequency. This noise goes through the standard two line circuit and at the same time goes out from the circuit in the form of electromagnetic wave.
The way out is to introduce a new third line to the main two line system in order to minimize the effect of the circumstance by asking the new third line to take care of the common mode effect. We are then able to control the whole electric circuit by arranging all the elements (conductor-lines, powers, loads etc.) so as to minimize the effect of noise. One very important thing is to introduce two identical power supplies and connect the third line to the middle point of the two power supplies. In this way, the common mode noise produced in the power supply system finds a way to go through the third line and does not go out from the electric circuit.
It is then important to decouple the normal mode from the common mode. This is achieved by using the same size and same quality transmission-lines for the main two lines and by arranging the geometrical distances of the two lines to the third line equal. At the same time, we have to arrange lumped-loads symmetrically around the third line. We have discussed why the normal mode decouples from the common mode in a symmetrical arrangement by calculating three-line lumped-circuit in our first publication (Sato & Toki (2007)), which reviewed the design principle of HIMAC synchrotron (Kumada (1994)) and provided a guide of alteration of magnet wiring of J-PARC MR (Kobayashi (2009)). The present new MTL theory with the antenna mode tells that the conditions of the normal mode to decouple from the common and antenna modes are to impose the symmetrization among the three-conductor transmission-lines in addition to the symmetrization of the lumped elements.
We show one example of electric circuit to use the normal mode current with largely reduced noise by the symmetric arrangement around the third line as shown in Fig.1. The present day power supply uses a chopping device and produces high frequency noise. We use a standard two-stage power supply and connect the third line at the middle point of the two-stage power supply denoted by P to confine high frequency noise produced by the power supply. The filtering device F should cut down high frequency noise and allows only low frequency noise to pass through the filter. It should be noted that the filtering device F consists of the common

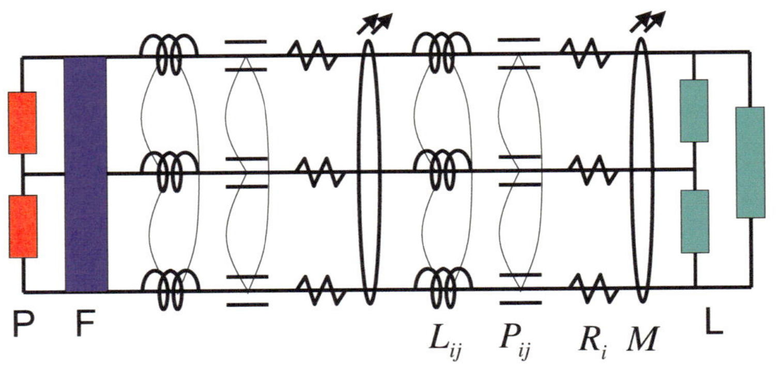

Fig. 1. A symmetrized electric circuit for the use of the normal mode decoupled from the common and antenna modes. Around the third line (middle line) two power supplies P, the filtering element F are placed in the left end and the electric loads L are placed in the right end. The connecting three-conductor transmission lines have the properties of self and mutual inductances L_{ij} denoted by coil, self and mutual coefficients of potential P_{ij} denoted by short parallel lines, resistance R_i and antenna mode coefficient M. The three-conductor transmission lines are coupled each other and their performance follows a coupled integro-differential equation with these coefficients.

mode filter in addition to the normal mode filter in order to cut down not only the normal mode noise but also the common mode noise. In the right end of the three-lines placed are electric loads L symmetrically. The arrangement of these lumped devices symmetrically is the requirement of the decoupling of the normal mode from the common mode.

Very important fact is now that these power-filter element $P - F$ and the electric loads L are connected by transmission-lines, which are not just structureless lines, but contain several functions as inductance L_{ij}, coefficient of potential P_{ij}, resistance R_i and antenna coefficient M. There are self- and mutual-inductances L_{ij} and they are denoted by coils on lines and connections of coils as usual in Fig.1. We denote the coefficients of potential P_{ij}, which have both self- and mutual-coefficients, by two short parallel lines and connections of short parallel lines. We abandon here the concept of capacitor C_{ij} and use a similar but rotated symbol for P_{ij}. The resistances R_i are denoted by the standard symbol and are attached to each line. In addition, we have the antenna coefficient M for radiation, which is associated with the whole transmission-lines and use the connection symbol with two arrows indicating radiation. The performance of the transmission-lines is controlled then by a set of coupled integro-differential equations with these coefficients L_{ij}, P_{ij}, R_i and M. At a glance of Fig.1, readers might picture that the symmetric arrangement provides decoupling of normal mode from common and antenna modes because of $L_{nc} = L_{na} = 0$ in Eq. (67), $R_{nc} = R_{na} = 0$ in Eq. (68) and $P_{nc} = P_{na} = 0$ in Eq. (71). Consequently, the symmetric arrangement of the transmission-lines and the lumped elements are the necessary step for the good performance of an electric circuit. We should on top consider that the noise is EM wave and goes through the transmission-lines in the TEM mode with loss of Joule and radiation energies.

7. Conclusion

We have constructed a new multi-conductor transmission-line (MTL) theory with the antenna mode. To this end, we have started from the electrodynamics field theory, which denotes that sources of electric and magnetic fields outside the conductors are true charge and conduction current inside the conductors. Based on the definite statement of the field theory, it is allowed to consider the dynamics of charge and current in resistive conductors of transmission-lines of thick wires and their coupling to the electromagnetic fields surrounding the transmission-lines. We have used the continuity equation of the charge and current and the combined equation of the boundary condition at the surface and the Ohm's law with the resistance, which controls the movement of the charge and current. The Maxwell equation then relates the dynamics of the charge and current to the scalar and vector potentials surrounding the transmission-lines. Since we are interested in the performance of the electromagnetic fields outside of the MTL system, we solve the wave equations for the scalar and vector potentials in the Lorenz gauge with the retardation charge and current. The scalar and vector potentials are now expressed in the integral forms and they are called retarded potentials. These four equations are the fundamental equations for the MTL system with the antenna mode. The coupled integro-differential equations are to be solved for the propagation of the electromagnetic wave and the energy loss due to the Joule and radiation processes.

To proceed, we have analyzed the retardation potentials for each frequency mode. The retardation charge and current introduces the real part with a cosine function and the imaginary part with a sine function. We are then able to make the TEM mode approximation for the real part, but should keep the imaginary part in the integral form. This TEM mode approximation can relate the scalar and vector potentials at the surface of each transmission-line with the charge and current for the introduction of the coefficients of potential P_{ij} and inductance L_{ij}. In this process, we consider the retardation charge and current effect explicitly. Hence, we modify the coefficients of inductance L_{ij} and potential P_{ij} by including the ω dependent term $cos(\omega\sqrt{(x-x')^2+d_{ij}^2}/c)$ in the integrand. As for the imaginary terms, we have now the omega dependent term $sin(\omega|x-x'|/c)$ in the numerator and should keep the integral form. We call the newly added integral terms coming from the imaginary parts of the retardation charge and current as the antenna mode terms with antenna mode coefficients M_e and M_m. We are then able to express MTL integro-differential equations for the scalar potential and the current by eliminating the charge and the vector potentials by using the continuity equation which is equivalent to the current conservation equation of the field theory and the combined equation of the boundary condition and the Ohm's law equation.

We have worked out an one-conductor transmission-line system to discuss the standard line-antenna with the propagation of a TEM mode through the transmission-line. In this case, we use the long wavelength approximation for the antenna mode terms originating from the retardation terms. Due to the fact that we are able to calculate the coefficients of inductance and potential, we can write down coupled integro-differential equations for potential V and current I with L, P and M_e, M_m and R. We solve the coupled equations formally and work out the input impedance, which is now a function of the size, the length and the resistance of the transmission-line. We have explicitly worked out the case for one linear transmission-line antenna. We can provide the solution of the differential equation and give an expression of the input impedance Z_s for the first time with the long wave length approximation after the MTL equations are fixed for the TEM mode. We work out the power of the system at the origin, which is eventually consumed by the Joule energy and the radiation energy. We have provided the input impedance for a typical case of a line antenna of thick wire with resistance.

We have studied also a three-conductor transmission-line system with emission and absorption. In addition to mathematical expressions, we propose a new circuit diagram of multi-conductors on the basis of coefficient of potential, coefficient of inductance, coefficient of antenna mode, and resistance. There appear three kinds of waves of normal, common, and antenna modes. All these modes propagate around the transmission-lines in the TEM mode waves. It is very interesting to point out if there is a symmetry between the lines 1 and 2 due to a symmetric arrangement, then the normal mode decouples from the common and antenna modes simultaneously. On the other hand, when the symmetry between the lines 1 and 2 is lost, the normal mode couples with both the common and antenna modes. This is a realistic situation of the ordinary two-conductor transmission-line system with the inclusion of the circumstance. We have to introduce the third line to the main two line system, instead of circumstance of which electrical performance is unclear, and symmetrize the system in order to confine the electromagnetic fields within the three-line system with the help of the common mode filter. In the near future we shall work out the skin effect by taking into account the motion of the current to the radial direction of each transmission-line in the MTL theory with the antenna mode (Sato & Toki (2011)).

Finally we would like to comment on a distinction between the present MTL theory and the former standard two-conductor transmission-line theory from the view point of electromagnetism. We consider resistive conductors for transmission-lines and abandon the concept of perfect conductor. The TEM mode wave could exist even in the case of one-conductor transmission line so that the TEM mode approximation is useful in the present study. The coefficients of potential are important to determine not only the coupling impedance between mutual transmission lines but also the characteristic impedance of a single transmission line itself. Consequently, it is unnecessary for a transmission line theory to introduce capacitance per unit length between two transmission lines and the displacement current flowing through the capacitance for a transmission-line theory any more. The boundary conditions for the electromagnetic fields at the surface of the resistive conductor provide the propagation of the TEM mode and replaces the concept of the Kirchhoff's current law between two lines due to displacement current.

8. Acknowledgment

The authors are grateful to Prof. H. Kobayashi and Prof. H. Horiuchi for fruitful discussions and encouragements.

9. References

Toki, H. and Sato, K., Journ. Phys. Soc. Jap. 78 No.9 (2009) 094201.

Paul, C.R., 'Analysis of Multiconductor Transmission Lines', (Wiley-Internscience (IEEE), New Jersey, 2008) 1.

Sato, K. and Toki, H., Nucl. Instr. Meth. (NIM) A565 (2007) 351.

Toki, H. and Sato, K., to be published in Journ. Phys. Soc. Jap. (2011).

Kumada, M. et al., Proc. 4th Euro. Part. Acc. Conf. (EPAC) (1994) 2338.

Kobayashi, H., Proc. of Particle Accelerator Conf. (2009) PAC09-WE1GRI02.

Takeyama, S., "Phenomenological Electromagnetic Theory" (Japanese), Maruzen pub. (1983) 1.

Maxwell, J. C., A Treatise on Electricity & Magnetism (Dover Publication Inc., New York, 1876) Vols. 1 and 2.

Bjorken, J.D. and Drell, S.D., 'Relativistic quantum mechanics' McGraw Hill (New-York) (1970) 1.

Lorenz, L., Dansk. Vid. Selsk. Forch. (1867) 26.

Rieman, B., Ann. Phys. 131 (1867) 237.

Jackson,J.D., 'Classical Electrodynamics', 3rd ed. (John Wiley & Sons, 1998).

Sato, K. and Toki, H., to be published (2011).

Kirchhoff, G., Ann. Phys. 100 (1857) 193.

Ohta, K. (2005). *Vortex of Maxwell and Watch of Einstein in Japanese*, Tokyo University Press.

Stratton, J.A., 'Electromagnetic Theory', McGraw-Hill Book Company, New York and London (1941).

10

Propagation in Lossy Rectangular Waveguides

Kim Ho Yeap[1], Choy Yoong Tham[2],
Ghassan Yassin[3] and Kee Choon Yeong[1]
[1]*Tunku Abdul Rahman University*
[2]*Wawasan Open University*
[3]*University of Oxford*
[1,2]*Malaysia*
[3]*United Kingdom*

1. Introduction

In millimeter and submillimeter radio astronomy, waveguide heterodyne receivers are often used in signal mixing. Wave guiding structures such as circular and rectangular waveguides are widely used in such receiver systems to direct and couple extraterrestrial signals at millimeter and submillimeter wavelengths to a mixer circuit (Carter et al., 2004; Boifot et al., 1990; Withington et al., 2003).

To illustrate in detail the applications of waveguides in receiver systems, a functional block diagram of a typical heterodyne receiver in radio telescopes is shown in Fig. 1 (Chattopadhyay et al., 2002). The electromagnetic signal (RF signal) from the antenna is directed down to the front end of the receiver system via mirrors and beam waveguides (Paine et al., 1994). At the front end of the receiver system, such as the sideband separating receiver designed for the ALMA band 7 cartridge (Vassilev and Belitsky, 2001a; Vassilev and Belitsky, 2001b; Vassilev et al., 2004), the RF signal is channelled from the aperture of the horn through a circular and subsequently a rectangular waveguide, before being coupled to the mixer. In the mixer circuit, a local oscillator (LO) signal which is generally of lower frequency is then mixed with the RF signal, to down convert the RF signal to a lower intermediate frequency (IF) signal. Here, a superconductor-insulator-superconductor (SIS) heterodyne mixer is commonly implemented for the process of down conversion. At the back end of the system, the IF signal goes through multiple stages of amplification and is, eventually, fed to a data analysis system such as an acousto-optic spectrometer. The data analysis system will then be able to perform Fourier transformation and record spectral information about the input signal.

The front-end receiver noise temperature T_R is determined by a number of factors. These include the mixer noise temperature T_M, the conversion loss C_{Loss}, the noise temperature of the first IF amplifier T_{IF}, and the coupling efficiency between the IF port of the junction and the input port of the first IF amplifier η_{IF} . A comparison of the performance of different SIS waveguide receivers is listed in Table 1 (Walker et al., 1992). It can be seen that the value of T_R for the 230 GHz system is a factor of 3 to 4 less than that achieved with the 492 GHz system. The decrease in system performance at 492 GHz is due to the increase of C_{Loss} and T_M by a factor of approximately 3.

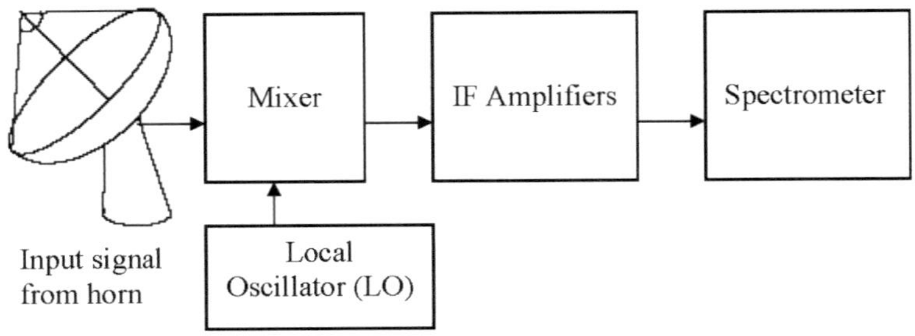

Fig. 1. Block diagram of a heterodyne receiver.

SIS Junction	Nb	Pb	Nb
Center Frequency (GHz)	230	345	492
T_R (K)	48	159	176
T_M (K)	34	129	123
C_{Loss} (dB)	3.1	8.1	8.9
T_{IF} (K)	7.0	4.2	6.8

Table 1. Comparison of SIS receiver performance.

Since the input power level of the weak millimeter and submillimeter signals is quite small – i.e. of the order of 10^{-18} to 10^{-20} W (Shankar, 1986), it is therefore of primary importance to minimize the conversion loss C_{Loss} of the mixer circuit. One way of doing so, is to ensure that the energy of the LO and, in particular, the RF signals is channelled and coupled from the waveguides to the mixer circuit in a highly efficient manner. It is simply too time consuming and too expensive to develop wave guiding structures in a receiver system on a trial-and-error basis. To minimize the loss of the propagating signals, the availability of an accurate and easy-to-use mathematical model to compute the loss of such signals in wave guiding structures is, of course, central to the development of receiver circuits.

2. Related work

Analysis of the propagation of wave in circular cylindrical waveguides has already been widely performed (Glaser, 1969; Yassin et al., 2003; Claricoats, 1960a; Claricoats, 1960b; Chou and Lee, 1988). The analyses by these authors are all based on the rigorous method formulated by Stratton (1941). In Stratton's formulation, the fields at the wall surface are made continuous into the wall material. Assumption made on the field decaying inside the wall material yields relations which allow the propagation constant to be determined. Due to the difficulty in matching the boundary conditions in Cartesian coordinates, this approach, however fails to be implemented in the case of rectangular waveguides. A similar rigorous technique to study the attenuation of rectangular waveguides is not available hitherto.

The perturbation power-loss method has been commonly used in analyzing wave attenuation in lossy (Stratton, 1941; Seida, 2003; Collin 1991; Cheng, 1989) and superconducting (Winters and Rose, 1991; Ma, 1998; Wang et al., 1994; Yalamanchili et al., 1995) rectangular waveguides; respectively. This is partly due to its ability to produce simple analytical solution, and also partly because it gives reasonably accurate result at frequencies f well above its cutoff frequency f_c. In this method, the field expressions are derived by assuming that the walls to be of infinite conductivity. This allows the solution to be separated into pure Transverse Electric (TE) and Transverse Magnetic (TM) modes. For a waveguide with finite conductivity, however, a superposition of both TE and TM modes is necessary to satisfy the boundary conditions (Stratton, 1941; Yassin et al., 2003). This is because, unlike those of the lossless case, the modes in a lossy waveguide are no longer mutually orthogonal to each other (Collin, 1991). To calculate the attenuation, ohmic losses are assumed due to small field penetration into the conductor surface. Results however show that this method fails near cutoff, as the attenuation obtained diverges to infinity when the signal frequency f approaches the cutoff f_c. Clearly, it is more realistic to expect losses to be high but finite rather than diverging to infinity. The inaccuracy in the power-loss method at cutoff is due to the fact that the field equations are assumed to be identical to those of a lossless waveguide. Since a lossless waveguide behaves exactly like an ideal high pass filter, signals cease to propagate at f below f_c.

It can be seen that the assumption of lossless fields fail to give an insight or deeper understanding on the mechanism of the propagation of wave in practical lossy waveguides. Moreover, at very high frequency - especially that approaches the millimeter and submillimeter wavelengths - the loss tangent of the conducting wall decreases. Therefore, such assumption turns out to be inaccurate at very high frequency. Although Stratton (1941) has developed a truly fundamental approach to analyze waveguides, his approach is only restricted to the case of circular waveguides and could not be applied to rectangular waveguides. The workhorse of this chapter is, therefore, to develop a novel and accurate formulation - i.e. one that does not assume lossless boundary conditions - to investigate the loss of waves in rectangular waveguides. In particular, the new method shall be found more accurate and useful for waveguides operating at very high frequencies, such as those in the millimeter and submillimeter wavelengths. In Yeap et al. (2009), a simple method to compute the loss of waves in rectangular waveguides has been developed. However, the drawback of the method is that different sets of characteristic equation are required to solve for the propagation constants of different modes. Here, the method proposed in Yeap et al. (2009) shall be developed further so that the loss of different modes can be conveniently computed using only a single set of equation. For convenience purpose, the new method shall be referred to as the boundary-matching method in the subsequent sections.

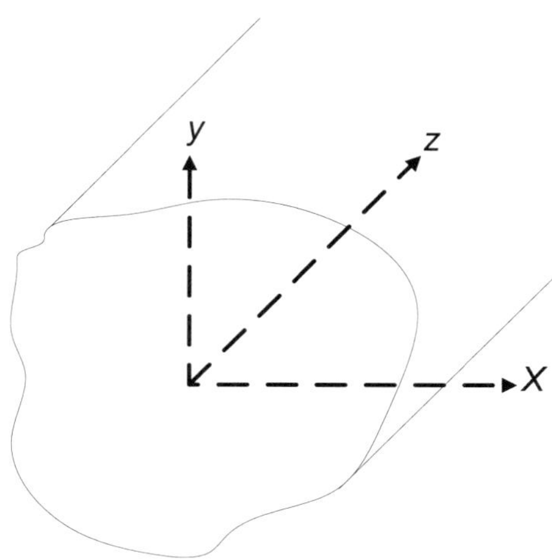

Fig. 2. A waveguide with arbitrary geometry.

3. General wave behaviours along uniform guiding structures

As depicted in Fig. 2, a time harmonic field propagating in the z direction of a uniform guiding structure with arbitrary geometry can be expressed as a combination of elementary waves having a general functional form (Cheng, 1989)

$$\psi = \psi^0(x,y)\exp[\mathrm{j}(\omega t + k_z z)] \tag{1}$$

where $\psi^0(x, y)$ is a two dimensional vector phasor that depends only on the cross-sectional coordinates, $\omega = 2\pi f$ the angular frequency, and k_z is the propagation constant.
Hence, in using phasor representation in equations relating field quantities, the partial derivatives with respect to t and z may be replaced by products with jω and jk_z, respectively; the common factor $\exp[\mathrm{j}(\omega t + k_z z)]$ can be dropped. Here, the propagation constant k_z is a complex variable, which consists of a phase constant β_z and an attenuation constant a_z

$$k_z = \beta_z - \mathrm{j}\alpha_z \tag{2}$$

The field intensities in a charge-free dielectric region (such as free-space), satisfy the following homogeneous vector Helmholtz's equation

$$\nabla^2\psi_z + (k^2 - k_z^{\ 2})\psi_z = 0 \tag{3}$$

where ψ_z is the longitudinal component of $\boldsymbol{\psi}$, ∇^2 is the Laplacian operator for the transverse coordinates, and k is the wavenumber in the material. For waves propagating in a hollow waveguide, $k = k_0$, the wavenumber in free-space.
It is convenient to classify propagating waves into three types, in correspond to the existence of the longitudinal electric field E_z or longitudinal magnetic H_z field:

1. Transverse electromagnetic (TEM) waves. A TEM wave consists of neither electric fields nor magnetic fields in the longitudinal direction.
2. Transverse magnetic (TM) waves. A TM wave consists of a nonzero electric field but zero magnetic field in the longitudinal direction.
3. Transverse electric (TE) waves. A TE wave consists of a zero electric field but nonzero magnetic field in the longitudinal direction.

Single-conductor waveguides, such as a hollow (or dielectric-filled) circular and rectangular waveguide, cannot support TEM waves. This is because according to Ampere's circuital law, the line integral of a magnetic field around any closed loop in a transverse plane must equal the sum of the longitudinal conduction and displacement currents through the loop. However, since a single-conductor waveguide does not have an inner conductor and that the longitudinal electric field is zero, there are no longitudinal conduction and displacement current. Hence, transverse magnetic field of a TEM mode cannot propagate in the waveguide (Cheng, 1989).

4. Fields in cartesian coordinates

For waves propagating in a rectangular waveguide, such as that shown in Fig. 3, Helmholtz's equation in (3) can be expanded in Cartesian coordinates to give

$$\frac{\partial^2 \psi_z}{\partial x^2} + \frac{\partial^2 \psi_z}{\partial y^2} + \left(k^2 - k_z^2\right)\psi_z = 0 \tag{4}$$

By applying the method of separation of variables, ψ_z can be expressed as

$$\psi_z = X(x)Y(y) \tag{5}$$

Equation (4) can thus be separated into two sets of second order differential equations, as shown below (Cheng, 1989)

$$\frac{d^2X(x)}{dx^2} + k_x^2 X(x) = 0 \tag{6}$$

$$\frac{d^2Y(y)}{dy^2} + k_y^2 Y(y) = 0 \tag{7}$$

where k_x and k_y are the transverse wavenumbers in the x and y directions, respectively. The longitudinal fields can be obtained by solving (6) and (7) based on a set of boundary conditions and substituting the solutions into (5).

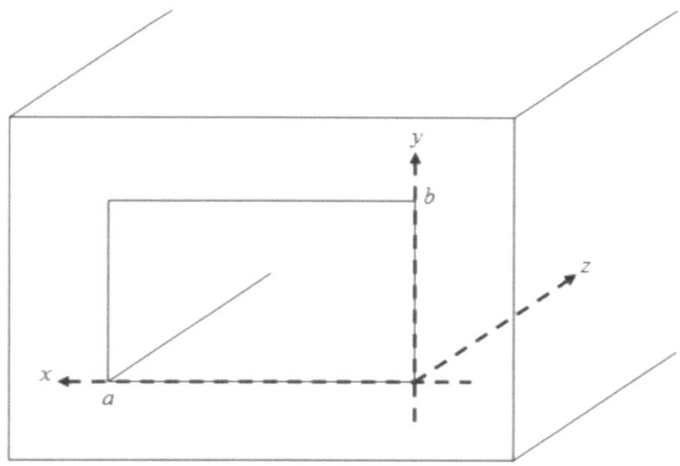

Fig. 3. The cross section of a rectangular waveguide.

The transverse field components can be derived by substituting the longitudinal field components into Maxwell's source free curl equations

$$\nabla \times E = -j\omega\mu H \tag{8}$$

$$\nabla \times H = j\omega\varepsilon E \tag{9}$$

where ε and μ are the permittivity and permeability of the material, respectively. Expressing the transverse field components in term of the longitudinal field components E_z and H_z, the following equations can be obtained (Cheng, 1989)

$$H_x = \frac{-j}{k_x^{\,2} + k_y^{\,2}} \left(k_z \frac{dH_z}{dx} - \omega\varepsilon \frac{dE_z}{dy} \right) \tag{10}$$

$$H_y = \frac{-j}{k_x^{\,2} + k_y^{\,2}} \left(k_z \frac{dH_z}{dy} + \omega\varepsilon \frac{dE_z}{dx} \right) \tag{11}$$

$$E_x = \frac{-j}{k_x^{\,2} + k_y^{\,2}} \left(k_z \frac{dE_z}{dx} + \omega\mu \frac{dH_z}{dy} \right) \tag{12}$$

$$E_y = \frac{-j}{k_x^{\,2} + k_y^{\,2}} \left(k_z \frac{dE_z}{dy} - \omega\mu \frac{dH_z}{dx} \right) \tag{13}$$

5. Review of the power-loss method

In the subsequent sections, analysis and comparison between the perturbation power-loss method and the new boundary matching method shall be performed. Hence, in order to present a complete scheme, the derivation of the conventional power-loss method is briefly outlined in this section.

The attenuation of electromagnetic waves in waveguides can be caused by two factors – i.e. the attenuation due to the lossy dielectric material $a_{z(d)}$, and that due to the ohmic losses in imperfectly conducting walls $a_{z(c)}$ (Cheng, 1989)

$$a_z = a_{z(d)} + a_{z(c)}\ , \tag{14}$$

For a conducting waveguide, the inner core is usually filled with low-loss dielectric material, such as air. Hence, $a_{z(d)}$ in (14) shall be assumed zero in the power-loss method and the loss in a waveguide is assumed to be caused solely by the conduction loss. It could be seen later that such assumption is not necessary in the new boundary-matching method. Indeed, the new method inherently accounts for both kinds of losses in its formulation.

The approximate power-loss method assumes that the fields' expression in a highly but imperfectly conducting waveguide, to be the same as those of a lossless waveguide. Hence, k_x, k_y, and k_z are given as (Cheng, 1989)

$$k_x = \frac{m\pi}{a} \tag{15}$$

$$k_y = \frac{n\pi}{b} \tag{16}$$

$$k_z = \beta_z \tag{17}$$

where a and b are the width and height, respectively, of the rectangular waveguide; whereas m and n denote the number of half cycle variations in the x and y directions, respectively. Every combination of m and n defines a possible mode for TE_{mn} and TM_{mn} waves.

Conduction loss is assumed to occur due to small fields' penetration into the conductor surfaces. According to the law of conservation of energy, the attenuation constant due to conduction loss can be derived as (Cheng, 1989)

$$\alpha_z = \frac{P_L}{2P_z} \tag{18}$$

where P_z is the time-average power flowing through the cross-section and P_L the time-average power lost per unit length of the waveguide.
Solving for P_L and P_z based on Poynting's theorem, the attenuation constant a_z for TM and TE modes – i.e. $a_{z(\mathrm{TM})}$ and $a_{z(\mathrm{TE})}$, respectively, can thus be expressed as (Collin, 1991)

$$\alpha_{z(TM)} = \frac{2R_s(m^2b^3 + n^2a^3)}{\eta ab\sqrt{1-\left(\frac{f_c}{f}\right)^2}\left(m^2b^2 + n^2a^2\right)} \tag{19}$$

$$\alpha_{z(TE)} = \frac{2R_s}{\eta b\sqrt{1-\left(\frac{f_c}{f}\right)^2}}\left\{\left(1+\frac{b}{a}\right)\left(\frac{f_c}{f}\right)^2 + \frac{b}{a}\left[1-\left(\frac{f_c}{f}\right)^2\right]\left[\frac{m^2ab + n^2a^2}{(mb)^2 + (na)^2}\right]\right\} \tag{20}$$

where R_s is the surface resistance, f_c the cutoff frequency, and η the intrinsic impedance of free space.

6. The new boundary-matching method

It is apparent that, in order to derive the approximate power-loss equations illustrated in section 5, the field equations must be assumed to be lossless. In a lossless waveguide, the boundary condition requires that the resultant tangential electric field E_t and the normal derivative of the tangential magnetic field $\partial H_t/\partial a_n$ to vanish at the waveguide wall, where a_n is the normal direction to the waveguide wall. In reality, however, this is not exactly the case. The conductivity of a practical waveguide is finite. Hence, both E_t and $\partial H_t/\partial a_n$ are not exactly zero at the boundary of the waveguide. Besides, the loss tangent of a material decreases in direct proportion with the increase of frequency. Hence, a highly conducting wall at low frequency may exhibit the properties of a lossy dielectric at high frequency, resulting in inaccuracy using the assumption at millimeter and submillimeter wavelengths.
In order to model the field expressions closer to those in a lossy waveguide and to account for the presence of fields inside the walls, two phase parameters have been introduced in the new method. The phase parameters – i.e ϕ_x and ϕ_y, are referred to as the field's penetration factors in the x and y directions, respectively. It is worthwhile noting that, with the introduction of the penetration factors, E_t and $\partial H_t/\partial a_n$ do not necessarily decay to zero at the boundary, therefore allowing the effect of not being a perfect conductor at the waveguide wall.

6.1 Fields in a lossy rectangular waveguide

For waves propagating in a lossy hollow rectangular waveguide, as shown in Figure 3, a superposition of TM and TE waves is necessary to satisfy the boundary condition at the wall (Stratton, 1941; Yassin et al., 2003). The longitudinal electric and magnetic field components E_z and H_z, respectively, can be derived by solving Helmholtz's homogeneous equation in

Cartesian coordinate. Using the method of separation of variables (Cheng, 1989), the following set of field equations is obtained

$$E_z = E_0 \sin\left(k_x x + \phi_x\right)\sin\left(k_y y + \phi_y\right) \tag{21}$$

$$H_z = H_0 \cos\left(k_x x + \phi_x\right)\cos\left(k_y y + \phi_y\right) \tag{22}$$

where E_0 and H_0 are constant amplitudes of the fields.
The propagation constant k_z for each mode will be found by solving for k_x and k_y and substituting the results into the dispersion relation

$$k_z = \sqrt{k_0^{\,2} - k_x^{\,2} - k_y^{\,2}} \tag{23}$$

Equations (21) and (22) must also apply to a perfectly conducting waveguide. In that case E_z and $\partial H_z / \partial a_n$ are either at their maximum magnitude or zero at both $x = a/2$ and $y = b/2$ – i.e. the centre of the waveguide, therefore

$$\sin\left(\frac{k_x a}{2} + \phi_x\right) = \sin\left(\frac{k_y b}{2} + \phi_y\right) = \pm 1 \text{ or } 0 \tag{24}$$

Solving (24), the penetration factors are obtained as,

$$\phi_x = \frac{\left(m\pi - k_x a\right)}{2} \tag{25}$$

$$\phi_y = \frac{\left(n\pi - k_y b\right)}{2} \tag{26}$$

For waveguides with perfectly conducting wall, $k_x = m\pi/a$ and $k_y = n\pi/b$, (25) and (26) result in zero penetration and E_z and H_z in (21) and (22) are reduced to the fields of a lossless waveguide. To take the finite conductivity into account, k_x and k_y are allowed to take complex values yielding non-zero penetration of the fields into the waveguide material

$$k_x = \beta_x - j\alpha_x \tag{27}$$

$$k_y = \beta_y - j\alpha_y \tag{28}$$

where β_x and β_y are the phase constants and α_x and α_y are the attenuation constants in the x and y directions, respectively. Substituting the transverse wavenumbers in (27) and (28) into (23), the propagation constant of the waveguide k_z results in a complex value, therefore, yielding loss in wave propagation.
Substituting (21) and (22) into (10) to (13), the fields are obtained as

$$H_x = \frac{j\left[k_z k_x H_0 + \omega\varepsilon_0 k_y E_0\right]\sin(k_x x + \varphi_x)\cos(k_y y + \phi_y)}{k_x^{\,2} + k_y^{\,2}} \tag{29}$$

$$H_y = \frac{j\left[k_z k_y H_0 - \omega\varepsilon_0 k_x E_0\right]\cos(k_x x + \phi_x)\sin(k_y y + \phi_y)}{k_x^{\ 2} + k_y^{\ 2}} \tag{30}$$

$$E_x = -\frac{j\left[k_z k_x E_0 - \omega\mu_0 k_y H_0\right]\cos(k_x x + \phi_x)\sin(k_y y + \phi_y)}{k_x^{\ 2} + k_y^{\ 2}} \tag{31}$$

$$E_y = -\frac{j\left[k_z k_y E_0 + \omega\mu_0 k_x H_0\right]\sin(k_x x + \phi_x)\cos(k_y y + \phi_y)}{k_x^{\ 2} + k_y^{\ 2}} \tag{32}$$

where μ_0 and ε_0 are the permeability and permittivity of free space, respectively.

6.2 Formulation

At the wall, the tangential fields must satisfy the relationship defined by the constitutive properties μ_c and ε_c of the material. The ratio of the tangential component of the electric field to the surface current density at the conductor surface is represented by (Yeap et al., 2009b; Yeap et al., 2010)

$$\frac{E_t}{a_n \times H_t} = \sqrt{\frac{\mu_c}{\varepsilon_c}} \tag{33}$$

where μ_c and ε_c are the permeability and permittivity of the wall material, respectively, and $\sqrt{\frac{\mu_c}{\varepsilon_c}}$ is the intrinsic impedance of the wall material. The dielectric constant is complex and ε_c may be written as

$$\varepsilon_c = \varepsilon_0 - j\frac{\sigma_c}{\omega} \tag{34}$$

where σ_c is the conductivity of the wall.

In order to estimate the loss of waves in millimeter and submillimeter wavelengths more accurately, a more evolved model than the conventional constant conductivity model used at microwave frequencies is necessary. Here, Drude's model is applied for the frequency dependent conductivity σ_c (Booker, 1982)

$$\sigma_c = \frac{\sigma}{(1 + j\omega\tau)} \tag{35}$$

where σ is the conventional constant conductivity of the wall material and τ the mean free time. For most conductors, such as Copper, the mean free time τ is in the range of 10^{-13} to 10^{-14} s (Kittel, 1986).

At the width surface of the waveguide, $y = b$, $E_z/H_x = -E_x/H_z = \sqrt{\frac{\mu_c}{\varepsilon_c}}$. Substituting (21), (22), (29), and (31) into (33), the following relationships are obtained

$$\frac{-E_x}{H_z}=\frac{j}{k_x^{\ 2}+k_y^{\ 2}}\left(\frac{E_0}{H_0}k_z k_x-\omega\mu_0 k_y\right)\tan\left(k_y b+\phi_y\right)=\sqrt{\frac{\mu_c}{\varepsilon_c}} \tag{36a}$$

$$\frac{H_x}{E_z}=\frac{j}{k_x^{\ 2}+k_y^{\ 2}}\left(\frac{H_0}{E_0}k_z k_x+\omega\varepsilon_0 k_y\right)\cot\left(k_y b+\phi_y\right)=\sqrt{\frac{\varepsilon_c}{\mu_c}} \tag{36b}$$

Similarly, at the height surface where $x = a$, we obtain $E_y/H_z = -E_z/H_y = \sqrt{\frac{\mu_c}{\varepsilon_c}}$. Substituting (21), (22), (30), and (32) into (33), the following relationships are obtained

$$\frac{E_y}{H_z}=\frac{-j}{k_x^{\ 2}+k_y^{\ 2}}\left(\frac{E_0}{H_0}k_z k_y+\omega\mu_0 k_x\right)\tan\left(k_x a+\phi_x\right)=\sqrt{\frac{\mu_c}{\varepsilon_c}} \tag{37a}$$

$$\frac{-H_y}{E_z}=\frac{-j}{k_x^{\ 2}+k_y^{\ 2}}\left(\frac{H_0}{E_0}k_z k_y-\omega\varepsilon_0 k_x\right)\cot\left(k_x a+\phi_x\right)=\sqrt{\frac{\varepsilon_c}{\mu_c}} \tag{37b}$$

In order to obtain nontrivial solutions for (36) and (37), the determinant of the equations must be zero (Yeap et al., 2009a). By letting the determinant of the coefficients of E_0 and H_0 in (36) and (37) vanish the following transcendental equations are obtained

$$\left[\frac{j\omega\mu_0 k_y\tan\left(k_y b+\phi_y\right)}{k_x^{\ 2}+k_y^{\ 2}}+\sqrt{\frac{\mu_c}{\varepsilon_c}}\right]\left[\frac{j\omega\varepsilon_0 k_y\cot\left(k_y b+\phi_y\right)}{k_x^{\ 2}+k_y^{\ 2}}-\sqrt{\frac{\varepsilon_c}{\mu_c}}\right]=\left[\frac{k_z k_x}{k_x^{\ 2}+k_y^{\ 2}}\right]^2 \tag{38a}$$

$$\left[\frac{j\omega\mu_0 k_x\tan\left(k_x a+\phi_x\right)}{k_x^{\ 2}+k_y^{\ 2}}+\sqrt{\frac{\mu_c}{\varepsilon_c}}\right]\left[\frac{j\omega\varepsilon_0 k_x\cot\left(k_x a+\phi_x\right)}{k_x^{\ 2}+k_y^{\ 2}}-\sqrt{\frac{\varepsilon_c}{\mu_c}}\right]=\left[\frac{k_z k_y}{k_x^{\ 2}+k_y^{\ 2}}\right]^2 \tag{38b}$$

Since the dominant TE_{10} mode has the lowest cutoff frequency among all modes and is the only possible mode propagating alone, it is of engineering importance. In the subsequent sections, comparison and detail analysis of the TE_{10} mode shall be performed. For the TE_{10} mode, m and n are set to 1 and 0, respectively. Substituting $m = 1$ and $n = 0$ into the penetration factors in (25) and (26), the transcendental equations in (38) for TE_{10} mode can be simplified to

$$\left[\sqrt{\frac{\mu_c}{\varepsilon_c}}+\frac{j\omega\mu_0 k_y\tan\left(k_y\frac{b}{2}\right)}{k_x^{\ 2}+k_y^{\ 2}}\right]\left[\sqrt{\frac{\varepsilon_c}{\mu_c}}-\frac{j\omega\varepsilon_0 k_y\cot\left(k_y\frac{b}{2}\right)}{k_x^{\ 2}+k_y^{\ 2}}\right]=\left[\frac{jk_z k_x}{k_x^{\ 2}+k_y^{\ 2}}\right]^2 \tag{39a}$$

$$\left[\sqrt{\frac{\mu_c}{\varepsilon_c}}-\frac{j\omega\mu_0 k_x\cot\left(k_x\frac{a}{2}\right)}{k_x^{\ 2}+k_y^{\ 2}}\right]\left[\sqrt{\frac{\varepsilon_c}{\mu_c}}+\frac{j\omega\varepsilon_0 k_x\tan\left(k_x\frac{a}{2}\right)}{k_x^{\ 2}+k_y^{\ 2}}\right]=\left[\frac{jk_z k_y}{k_x^{\ 2}+k_y^{\ 2}}\right]^2 \tag{39b}$$

In the above equations, k_x and k_y are the unknowns and k_z can then be obtained from the dispersion relation in (23). A multi root searching algorithm, such as the Powell Hybrid root searching algorithm in a NAG routine, can be used to find the roots of k_x and k_y. The routine requires initial guesses of k_x and k_y for the search. For good conductors, suitable guess values are clearly those close to the perfect conductor values i.e. $\beta_x = m\pi/a$, $\beta_y = n\pi/b$, $\alpha_x = \alpha_y = 0$. Hence, for the TE_{10} mode, the initial guesses for k_x and k_y are π/a and 0 respectively.)

6.3 Experimental setup

To validate the results, experimental measurements had been carried out. The loss as a function of frequency for a rectangular waveguide was measured using a Vector Network Analyzer (VNA). A 20 cm copper rectangular waveguide with dimensions of a = 1.30 cm and b = 0.64 cm such as that shown in Fig. 4 were used in the measurement.
To minimize noise in the waveguide, a pair of chokes had also been designed and fabricated as shown in Fig. 5. A detail design of the choke drawn using AutoCAD are shown in Fig. 6(a) and Fig. 6(b).
In order to allow the waveguide to be connected to the adapters which are of different sizes, a pair of taper transitions had also been used as shown in Fig. 7. Fig. 8 depicts the complete setup of the experiment where the rectangular waveguide was connected to the VNA via tapers, chokes, coaxial cables, and adapters. Before measurement was carried out, the coaxial cables and waveguide adapters were calibrated to eliminate noise from the two devices. The loss in the waveguide was then observed from the S_{21} or S_{12} parameter of the scattering matrix. The measurement was performed in the frequencies at the vicinity of cutoff.

Fig. 4. Rectangular waveguides with width a = 1.30 cm and height b = 0.64 cm.

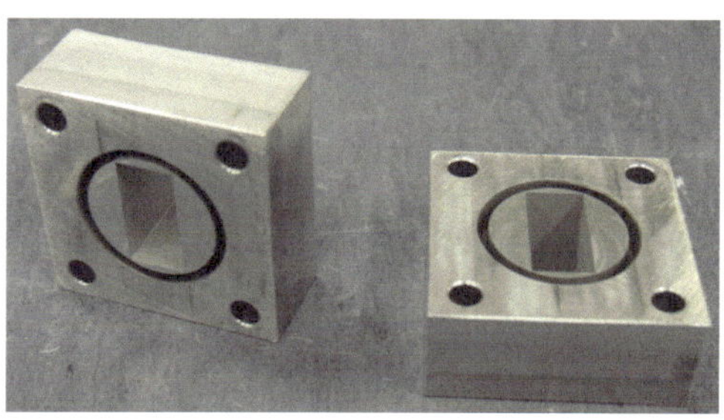

Fig. 5. A pair of chokes made of aluminum.

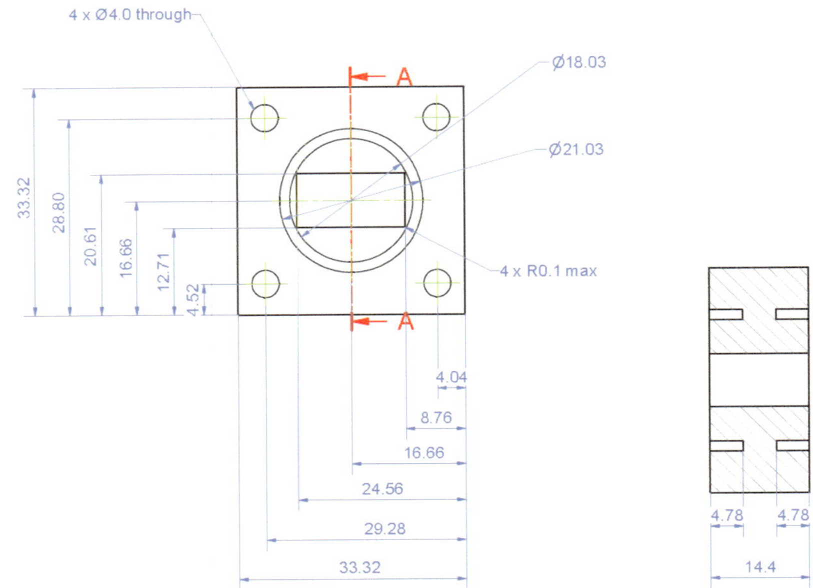

(a) (b)

Fig. 6. Parameters of the (a) cross section and (b) side view of the choke.

Fig. 7. Taper transitions.

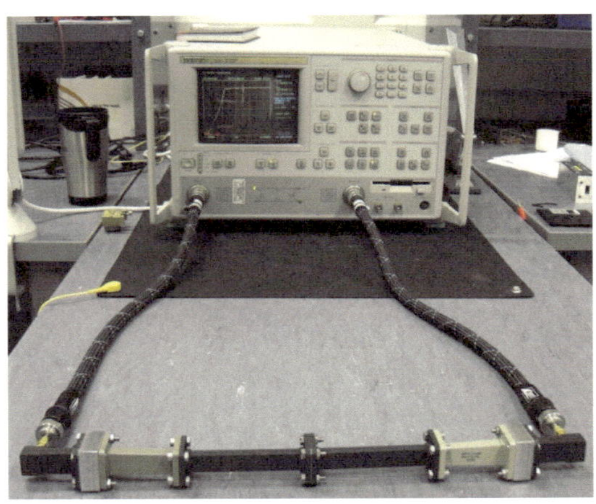

Fig. 8. A 20 cm rectangular waveguide connected to the VNA, via tapers, chokes, adapters, and coaxial cables.

6.4 Results and discussion

As shown in Fig. 9, a comparison among the attenuation of the TE_{10} mode near cutoff as computed by the new method, the conventional power-loss method, and the measured S_{21} result was performed. Clearly, the attenuation constant a_z computed from the power-loss method diverges sharply to infinity, as the frequency approaches f_c, and is very different to the simulated results, which show clearly that the loss at frequencies below f_c is high but finite. The attenuation computed using the new boundary-matching method, on the other hand, matches very closely with the S_{21} curve, measured using from the VNA. As shown in Table 2, the loss between 11.47025 GHz and 11.49950 GHz computed by the boundary-matching method agrees with measurement to within 5% which is comparable to the error in the measurement. The inaccuracy in the power-loss method is due to the fact that the fields expressions are assumed to be lossless – i.e. k_x and k_y are taken as real variables. Analyzing the dispersion relation in (23), it could be seen that, in order to obtain a_z, k_x and/or k_y must be complex, given that the wavenumber in free space is purely real. Although the initial guesses for k_x and k_y applied in the new boundary-matching method are assumed to be identical with the lossless case, the final results actually converge to complex values when the characteristic equations are solved numerically.

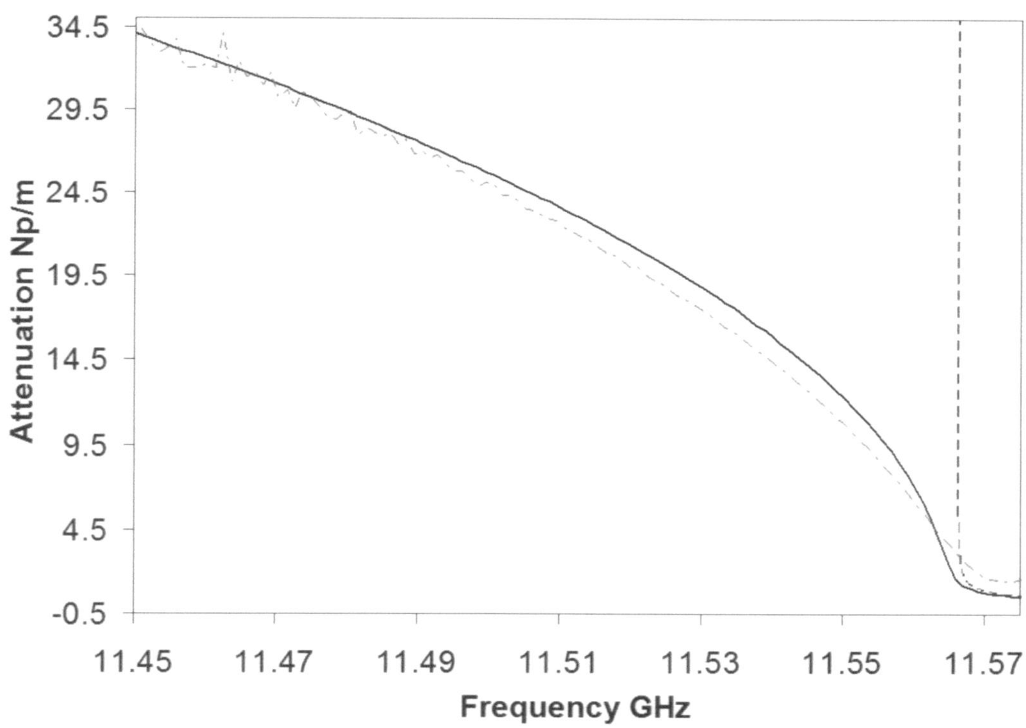

Fig. 9. Attenuation of TE_{10} mode at the vicinity of cutoff. ———————— the new boundary matching method. - - - - - - - - - - - - power loss method. - · - · - · - · - · - · - S21 measurement.

Fig. 10 shows the attenuation curve when the frequency is extended to higher values. Here, the loss due to TE_{10} alone could no longer be measured alone, since higher-order modes, such as TE_{11} and TM_{11}, etc., start to propagate. Close inspection shows that the loss predicted by the two methods at higher frequencies is in very close agreement. It is, therefore, sufficed to say that, although the power-loss method fails to predict the attenuation near f_c accurately, it is still considered adequate in computing the attenuation of TE_{10} in lossy waveguides, provided that the frequency f is reasonably above the cutoff f_c.

As depicted in Fig. 11, at frequencies beyond millimeter wavelengths, however, the loss computed by the boundary-matching method appears to be much higher than those by the power-loss method. The differences can be attributed to the fact that at extremely high frequencies, the loss tangent of the wall material decreases and the field in a lossy waveguide can no longer be approximated to those derived from a perfectly conducting waveguide. At such high frequencies, the wave propagating in the waveguide is a hybrid mode and the presence of the longitudinal electric field E_z can no longer be neglected.

Frequency GHz	**Experiment**	**Boundary-matching method**	%Δ
11.47025	30.17693	30.95782	2.59
11.47138	30.68101	30.77417	0.30
11.47250	29.53345	30.5894	3.58
11.47363	30.51672	30.40349	0.37
11.47475	30.16449	30.21642	0.17
11.47588	29.68032	30.02816	1.17
11.47700	29.09721	29.8387	2.55
11.47813	28.85077	29.648	2.76
11.47925	29.25528	29.45606	0.69
11.48038	29.20923	29.26283	0.18
11.48150	27.99881	29.06831	3.82
11.48263	28.38341	28.87245	1.72
11.48375	28.18551	28.67524	1.74
11.48488	27.91169	28.47664	2.02
11.48600	28.08407	28.27663	0.69
11.48713	27.44495	28.07517	2.30
11.48825	27.67956	27.87224	0.70
11.48938	26.84192	27.66779	3.08
11.49050	26.95767	27.46181	1.87
11.49163	26.60108	27.25425	2.46
11.49275	26.78715	27.04508	0.96
11.49388	26.14928	26.83426	2.62
11.49500	25.83003	26.62174	3.07
11.49613	25.82691	26.4075	2.25
11.49725	25.26994	26.19148	3.65
11.49838	24.82685	25.97365	4.62
11.49950	25.1100	25.75395	2.56

Table 1. Attenuation of TE_{10} at the vicinity of the cutoff frequency.

Unlike the power-loss method which only gives the value of the attenuation constant, one other advantage of the boundary-matching method is that it is able to account for the phase

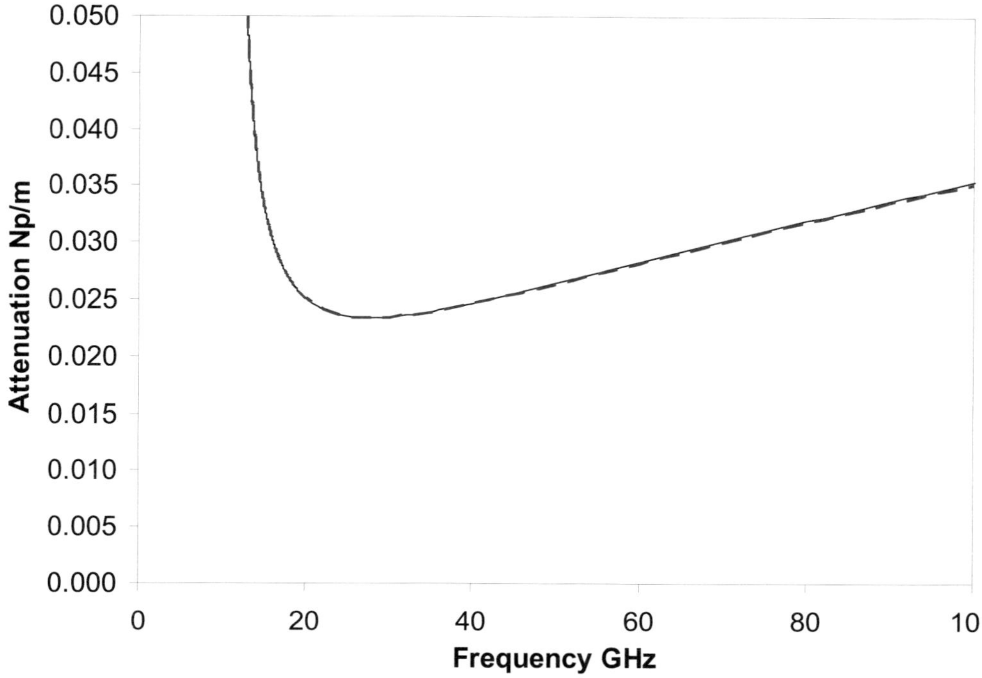

Fig. 10. Attenuation of TE_{10} mode from 0 to 100 GHz. ———— the new boundary matching method. - - - - - - - - - - - power loss method.

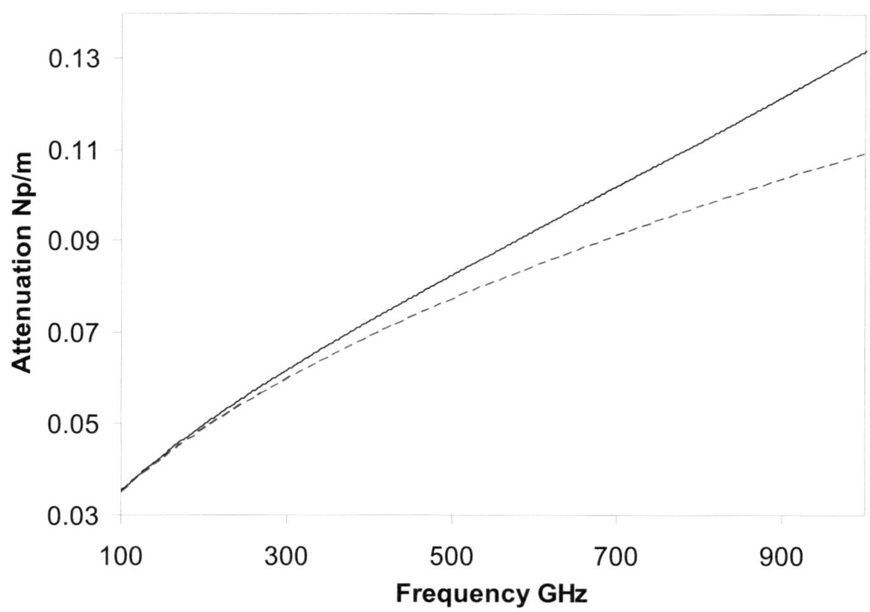

Fig. 11. Attenuation of TE_{10} mode from 100 GHz to 1 THz. ———— the new boundary matching method. - - - - - - - - - - - power loss method.

constant of the wave as well. A comparison between the attenuation constant and phase constant of a TE_{10} mode is shown in Fig. 12. As can be observed, as the attenuation in the waveguide gradually decreases, the phase constant increases. Fig. 12 illustrates the change in the mode – i.e. from evanescent below cutoff to propagating mode above cutoff.

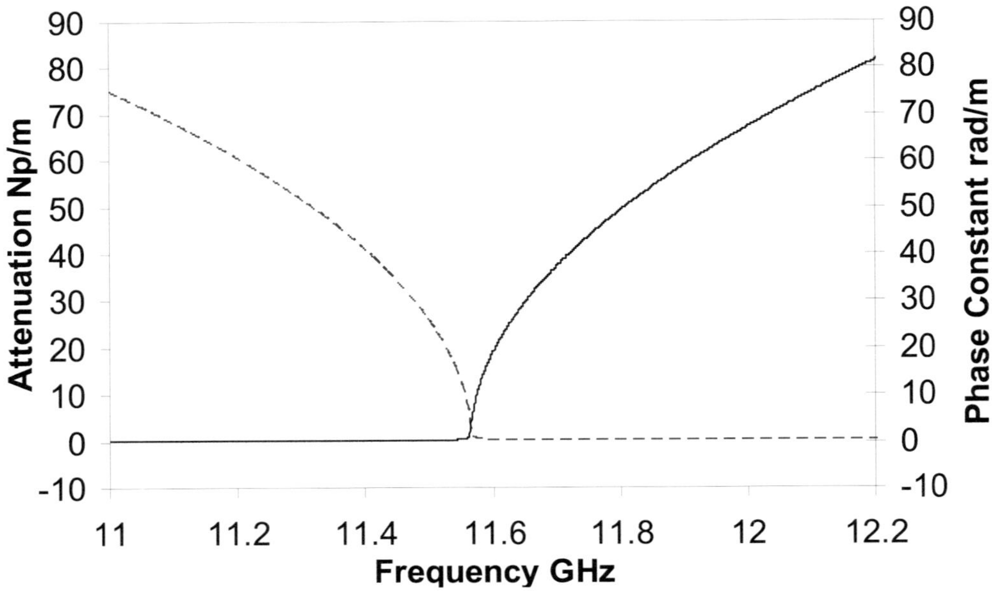

Fig. 12. Propagation constant (phase constant and attenuation constant) of TE10 mode in a lossy rectangular waveguide. ———— phase constant.- - - - - - - - - - - attenuation constant.

7. Summary

A fundamental and accurate technique to compute the propagation constant of waves in a lossy rectangular waveguide is proposed. The formulation is based on matching the fields to the constitutive properties of the material at the boundary. The electromagnetic fields are used in conjunction of the concept of surface impedance to derive transcendental equations, whose roots give values for the wavenumbers in the x and y directions for different TE or TM modes. The wave propagation constant k_z could then be obtained from k_x, k_y, and k_0 using the dispersion relation.

The new boundary-matching method has been validated by comparing the attenuation of the dominant mode with the S21 measurement, as well as, that obtained from the power-loss method. The attenuation curve plotted using the new method matches with the power-loss method at a reasonable range of frequencies above the cutoff. There are however two regions where both curves are found to differ significantly. At frequencies below the cutoff f_c, the power-loss method diverges to infinity with a singularity at frequency $f = f_c$. The new method, however, shows that the signal increases to a highly attenuating mode as the frequencies drop below f_c. Indeed, such result agrees very closely with the measurement result, therefore, verifying the validity of the new method. At frequencies above 100 GHz, the attenuation obtained using the new method increases beyond that predicted by the power-loss method. At f above the millimeter wavelengths, the field in a lossy waveguide can no longer be approximated to those of the lossless case. The additional loss predicted by the new boundary-matching method is attributed to the presence of the longitudinal E_z component in hybrid modes.

8. Acknowledgment

K. H. Yeap acknowledges Boon Kok, Paul Grimes, and Jamie Leech for their advise and discussion.

9. References

Boifot, A. M.; Lier, E. & Schaug-Petersen, T. (1990). Simple and broadband orthomode transducer, *Proceedings of IEE*, 137, pp. 396 - 400

Booker, H. (1982). *Energy in Electromagnetism*. 1st Edition. Peter Peregrinus.

Carter, M. C.; Baryshev, A.; Harman, M.; Lazareff, B.; Lamb, J.; Navarro, S.; John, D.; Fontana, A. -L.; Ediss, G.; Tham, C. Y.; Withington, S.; Tercero, F.; Nesti, R.; Tan, G. -H.; Sekimoto, Y.; Matsunaga, M.; Ogawa, H. & Claude, S. (2004). ALMA front-end optics. *Proceedings of the Society of Photo Optical Instrumentation Engineers*, 5489, pp. 1074 - 1084.

Chattopadhyay, G.; Schlecht, E.; Maiwald, F.; Dengler, R. J.; Pearson, J. C. & Mehdi, I. (2002). Frequency multiplier response to spurious signals and its effects on local oscillator systems in millimeter and submillimeter wavelengths. *Proceedings of the Society of Photo-Optical Instrumentation Engineers*, 4855, pp. 480 - 488.

Cheng, D. K. (1989). *Field and Wave Electromagnetics*, Addison Wesley, ISBN 0201528207, US.

Chou, R. C. & Lee, S. W. (1988). Modal attenuation in multilayered coated waveguides. *IEEE Transactions on Microwave Theory and Techniques*, 36, pp. 1167 - 1176.

Claricoats, P. J. B. (1960a). Propagation along unbounded and bounded dielectric rods: Part 1. Propagation along an unbounded dielectric rod. *IEE Monograph*, 409E, pp. 170 - 176.

Claricoats, P. J. B. (1960b). Propagation along unbounded and bounded dielectric rods: Part 2. Propagation along a dielectric rod contained in a circular waveguide. *IEE Monograph*, 409E, pp. 177 - 185.

Collin, R. E. (1991). *Field Theory of Guided Waves*, John Wiley & Sons, ISBN 0879422378, New York.

Glaser, J. I. (1969). Attenuation and guidance of modes on hollow dielectric waveguides. *IEEE Transactions on Microwave Theory and Techniques (Correspondence)*, 17, pp. 173 - 176.

Kittel, C. (1986). *Introduction to Solid State Physics*, John Wiley & Sons, New York.

Ma, J. (1998). TM-properties of HTS's rectangular waveguides with Meissner boundary condition. *International Journal of Infrared and Millimeter Waves*, 19, pp. 399 - 408.

Paine, S.; Papa, D. C.; Leombruno, R. L.; Zhang, X. & Blundell, R. (1994). Beam waveguide and receiver optics for the SMA. *Proceedings of the 5th International Symposium on Space Terahertz Technology*, University of Michigan, Ann Arbor, Michigan.

Seida, O. M. A. (2003). Propagation of electromagnetic waves in a rectangular tunnel. *Applied Mathematics and Computation*, 136, pp. 405 - 413.

Shankar, N. U. (1986). *Application of digital techniques to radio astronomy measurements*, Ph.D. Thesis. Raman Research Institute. Bangalore University.

Stratton, J. A. (1941). *Electromagnetic Theory*, McGraw-Hill, ISBN 070621500, New York.

Vassilev, V. & Belitsky, V. (2001a). A new 3-dB power divider for millimetre-wavelengths. *IEEE Microwave and Wireless Components Letters*, 11, pp. 30 – 32.

Vassilev, V. & Belitsky, V. (2001b). Design of sideband separation SIS mixer for 3 mm band. *Proceedings of the 12th International Symposium on Space Terahertz Technology*, Shelter Island, San Diego, California.

Vassilev, V.; Belitsky, V.; Risacher, C.; Lapkin, I.; Pavolotsky, A. & Sundin, E. (2004). Design and characterization of a sideband separating SIS mixer for 85 - 115 GHz. *Proceedings of the 15th International Symposium on Space Terahertz Technology*, Hotel Northampton, Northampton, Masachusetts.

Walker, C. K.; Kooi, J. W.; Chan, M.; Leduc, H. G.; Schaffer, P. L.; Carlstrom, J. E. & Phillips, T. G. (1992). A low-noise 492 GHz SIS waveguide receiver. *International Journal of Infrared and Millimeter Waves*, 13, pp. 785 – 798.

Wang, Y.; Qiu, Z. A. & Yalamanchili, R. (1994). Meissner model of superconducting rectangular waveguides. *International Journal of Electronics*, 76, pp. 1151 – 1171.

Winters, J. H. & Rose, C. (1991). High-T_c superconductors waveguides: Theory and applications. *IEEE Transactions on Microwave Theory and Techniques*, 39, pp. 617 – 623.

Withington, S. (2003). Terahertz astronomical telescopes and instrumentation. *Philosophical Transactions of the Royal Society of London*, 362, pp. 395 – 402.

Yalamanchili, R., Qiu, Z. A., Wang, Y. (1995). Rectangular waveguides with two conventional and two superconducting walls. *International Journal of Electronics*, 78, pp. 715 – 727.

Yassin, G., Tham, C. Y. & Withington, S. (2003). Propagation in lossy and superconducting cylindrical waveguides. *Proceedings of* the *14th International Symposium on Space Terahertz Technology*, Tucson, Az.

Yeap, K. H., Tham, C. Y., and Yeong, K. C. (2009a). Attenuation of the dominant mode in a lossy rectangular waveguide. *Proceedings of the IEEE 9th Malaysia International Conference on Communications*, KL., Malaysia.

Yeap, K. H., Tham, C. Y., Yeong, K. C. & Lim, E. H. (2010). Full wave analysis of normal and superconducting microstrip transmission lines. *Frequenz Journal of RF-Engineering and Telecommunications*, 64, pp. 59 – 66.

Yeap, K. H., Tham, C. Y., Yeong, K. C. & Yeap, K. H. (2009b). A simple method for calculating attenuation in waveguides. *Frequenz Journal of RF-Engineering and Telecommunications*, 63, pp. 236 – 240.

Part 5

Numerical Solutions based on Parallel Computations

11

Optimization of Parallel FDTD Computations Based on Program Macro Data Flow Graph Transformations

Adam Smyk[1] and Marek Tudruj[2]
[1]*Polish-Japanese Institute of Information Technology*
[2]*Institute of Computer Science Polish Academy of Science*
Poland

1. Introduction

This chapter concerns numerical problems that are solved by parallel regular computations performed in rectangular meshes that span over irregular computational areas. Such parallel problems are more difficult to be optimized than problems concerning regular areas since the problem cannot be solved by a simple geometrical decomposition of the computational area. Usually, a kind of step-by-step algorithm has to be designed to balance parallel computations and communication in and between executive processors. The Finite Difference Time Domain (FDTD) simulation of electromagnetic wave propagation in irregular computational area, numerical linear algebra or VLSI layout design belong to this class of computational problems solved by unstructured computational algorithms (Lin, 1996) with irregular data patterns. Some heuristic methods are known that enable graphs partitioning necessary to solve such problems (NP-complete problem (Garey et al., 1976)), but generally two kinds of such methods are used: direct methods (Khan et al., 1995) and iterative methods (Khan et al., 1995; Kerighan & Lin, 1970; Kirkpatrick et al., 1983; Karypis & Kumar, 1995; Dutt & Deng, 1997). Direct methods are usually based on the min-cut optimization (Stone & Bokhari, 1978). The iterative methods are mainly based on extensions of the algorithms of Kernighan-Lin (Kerighan & Lin, 1970), next improved by Fidducia-Mattheyses methods (FM)(Fiduccia & Mattheyses, 1982). There are also many kinds of various program graph partitioning packages like JOSTLE (Walshaw et al., 1995), SCHOTCH (Scotch, 2010) and METIS (Metis, 2008) etc. All of them enable performing efficient graph partitioning but there are two unresolved problems that have been found out. In the case of very irregular graphs, partitioning algorithms used in these packages can produce a partition that can be divided into two or more graph parts placed in various disjointed locations of the computofational area. As it follows from observed practice, there are no prerequisites to create such disjoint partitions, because in almost all cases it increases a total communication volume during execution in distributed systems. The second disadvantage is that the partitioning methods mentioned above do not take into account any architectural requirements of a target computational system. It is very important especially in heterogeneous systems, where proper load balancing allows efficient exploiting all computational resources and simultaneously, it allows reducing the total time of computations.

In (Smyk & Tudruj, 2006) we have presented a comparison of two algorithms: redeployment algorithm and CDC (Connectivity-based Distributed Node Clustering) algorithm (Ramaswamy et al., 2005). The first one is an extension of the FM algorithm and it is divided into three main phases. In the first phase, a partitioning of the FDTD computational area is performed. It provides an initial macro data flow graph to be used in further optimizations. The number of created initial macro nodes is usually much larger than the number of processors in the parallel system. Therefore, usually a merging algorithm phase is next executed. Several merging criteria are used to balance processor computational loads and to minimize total inter-processor communication. The obtained macro data flow graphs are usually adjusted to current architectural requirements in the last algorithm phase. A simple architectural model can be used for this in a computational cells redeployment. The second algorithm is a modification of the CDC algorithm known in the literature (Ramaswamy et al., 2005). It is decentralized and is based on information exchange on the whole computational area executed between neighboring nodes. In this chapter we present a hierarchical approach for program macro data flow graph partitioning for the optimized parallel execution of the FDTD method. In the proposed algorithm, we try to exploit the advantages of two mentioned above algorithms. In general, the redeployment algorithm is used to reduce the execution time of the optimization process, while the main idea of the CDC algorithm enables obtaining an efficient partitioning.

The chapter is composed of five parts. In the first part, the main idea of the FDTD problem and its execution according to macro data flow paradigm is described. In the next three parts, the redeployment and the CDC algorithms are described. We present experimental results which compare both of these algorithms. We also present a special memory infrastructure (RB RDMA) used for efficient communication in distributed systems. In the last part of this chapter we present an implementation of the hierarchical algorithm of FDTD program graph partitioning.

2. FDTD implementation with the macro data flow paradigm

Finite Difference Time Domain (FDTD) method is used in simulation of high frequency electromagnetic wave propagation. In general, the simulated area (two or three-dimensional irregular shape) can contain different characteristic sub-areas like excitation points, dielectrics etc. (see Fig. 1). The whole simulation is divided into two phases. In the first phase, whole computational area must be transformed into a discrete mesh (a set of Yee cells).

Each discrete point, obtained in this process, contains alternately (for two dimensional problem) electric component Ez of electromagnetic field and one from two magnetic components Hx or Hy (Smyk & Tudruj, 2006). In the second phase of the FDTD method, we perform wave propagation simulation (see Fig. 2). In each step of simulation, the values of all electric vectors (Ez) or the magnetic components (Hx, Hy) are alternately computed. Electromagnetic wave propagation in an isotropic environment is described by time-dependent Maxwell equations (1):

$$\nabla \times H = \gamma E + \varepsilon \frac{\partial E}{\partial t}, \qquad \nabla \times E = -\mu \frac{\partial H}{\partial t} \tag{1}$$

and can be easily transformed into their differential forms (2)

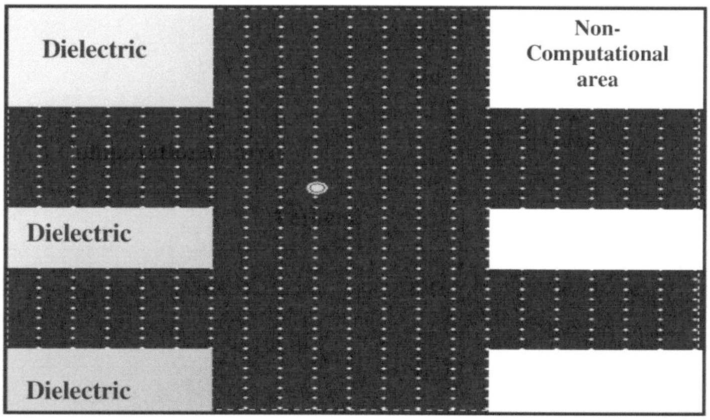

Fig. 1. Model of the 2 Dimensional FDTD Simulation.

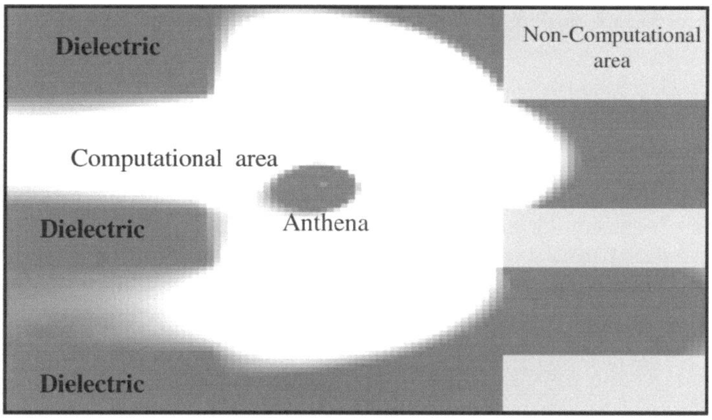

Fig. 2. FDTD Simulation in Action.

$$\begin{cases} \bar{E}_z^n(i,j) = CA_z(i,j)\bar{E}_z^{n-1}(i,j) + CB_z(i,j)\cdot \\ \qquad [\bar{H}_y^{n-0.5}(i+1,j) - \bar{H}_y^{n-0.5}(i+1,j) + \\ \qquad + \bar{H}_x^{n-0.5}(i,j-1) - \bar{H}_x^{n-0.5}(i,j+1)] \\ \bar{H}_x^n(i,j) = \bar{H}_x^{n-1}(i,j) + RC\cdot[\bar{E}_z^{n-0.5}(i-1,j) - \bar{E}_z^{n-0.5}(i+1,j)] \\ \bar{H}_y^n(i,j) = \bar{H}_y^{n-1}(i,j) + RC\cdot[\bar{E}_z^{n-0.5}(i,j-1) - \bar{E}_z^{n-0.5}(i,j+1)] \end{cases} \quad (2)$$

This computational process can be executed in parallel way. In this case, FDTD computations in the mesh are divided into fragments assigned to computational partitions. Each partition contains a number of computations that are mapped onto separate processing elements of a parallel machine. For regular shapes of the computational area (e.g. rectangular), it can be done by a stripe or block partitioning that allows obtaining almost ideal balance of computations on all available processors with minimal communication volume of data transmissions. For computational areas with irregular shapes, such an approach will not provide satisfactory partitioning. It needs a more advanced analysis of data dependencies. In this case, FDTD computation is represented by a data flow graph (Fig. 3) which is iteratively

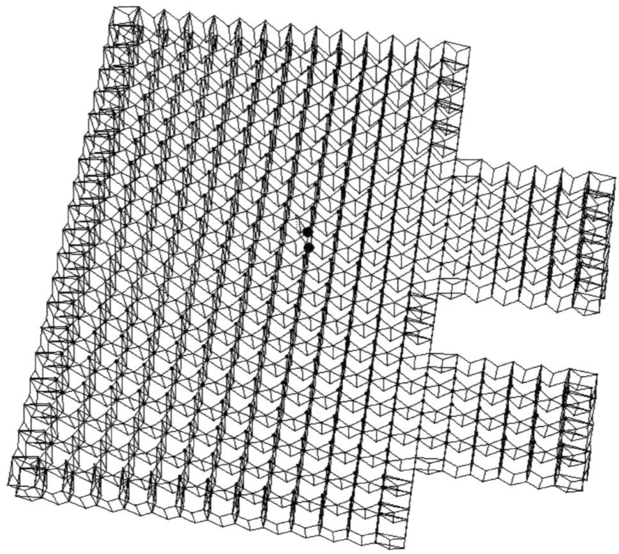

Fig. 3. Data Flow Graph for Two Iterations of the FDTD Problem for Irregular Computational Area.

transformed into a macro data flow graph. The number of macro nodes corresponds to given number of processing elements in an executive computer system. Such transformation from a data flow graph to a macro data flow graph takes into consideration both proper load of all processing nodes and the minimal number of data transmissions. To solve this problem, we have decided to design the FDTD method programs based on the macro data flow paradigm using an algorithm which is a combination of two algorithms described above: the redeployment (R) and the CDC algorithms. We call this algorithm RCDC (Redeployment with Connectivity-based Distributed Node Clustering). It is described in the last section.

3. Macro data flow graph optimization with cell redeployment algorithm

In this section, an outline of the redeployment algorithm will be presented. In the first step, we create a data flow graph of computations which represents basic data dependencies (see Fig. 3 and 4). This data flow graph will be used by a redeployment algorithm to define and optimize a macro data flow graph with n macro data nodes, where n is the number of processors. At the start, we create M macro nodes (where M is number of computational cells). Each computational cell is assigned to a separate macro node. Computations in each simulation sub-area are represented by one macro node. According to macro data flow paradigm, a macro node can be executed only if all external input data have arrived to the physical processor on which this macro node has been mapped. In the redeployment algorithm we do not perform any geometrical analysis of computational area. The whole optimization process is based on analysing data dependencies in the computational FDTD mesh.

The optimization algorithm is composed of three steps: simulation area partitioning, macro nodes merging and redeployment of cells. During the partitioning step we define macro data flow nodes for a given computational area. First, we determine computational "leader" nodes in the FDTD mesh. We have implemented two methods for choosing leaders. In the first method, we create a coarse regular mesh (LM) of cells spanned over the computational

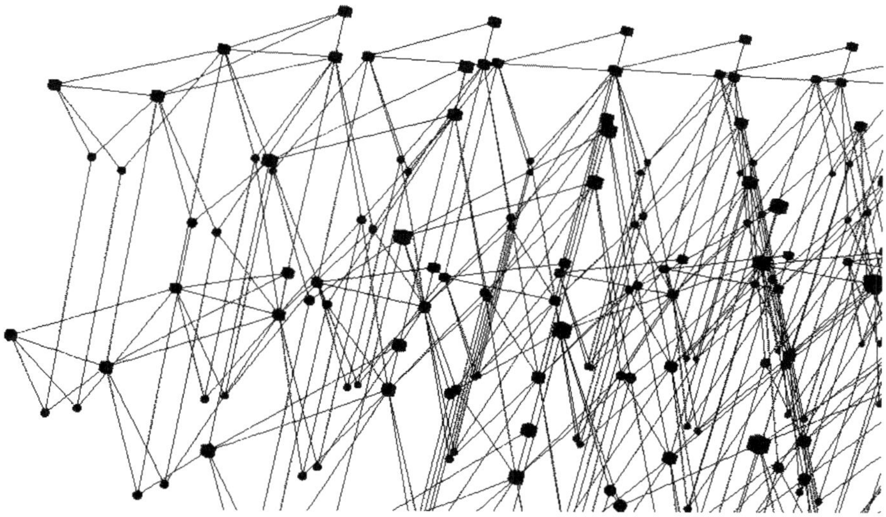

Fig. 4. Data Flow Graph for Two Iterations of the FDTD Problem (zoomed left bottom corner of Fig. 3).

mesh (CM). If cell A from LM is covered by cell B from CM, the cell B is qualified to be a leader. In an alternative method (P1), leaders are chosen only from circumference of the simulated area. Next, computational cells (that have not been earlier qualified as leaders) are gradually included into macro nodes assigned to the nearest leader. Additionally, all data-dependent macro nodes are connected by weighted edges. The weight represents the communication volume. The number of macro data flow nodes significantly exceeds the assumed number of processors. Additionally, already created macro data flow nodes can vary in their computational load. Such a macro data flow graph will be further optimized in the next phase, which is a merging phase. It enables reducing a total number of macro nodes to be equal to the number of processors with simultaneous coarse load balancing. Two macro nodes A and B are merged if: node A has the smallest execution time in the whole macro data flow graph and node B fulfills one chosen criterion from those shown in Table 1.

In the final phase, the program macro flow graph obtained after merging must be transformed so as to reduce program execution time by load balancing in executive processors for a given system configuration. Standard MDFG (macro data flow graph generated for all iterations of the FDTD problem) will be transformed into a Macro Node Communication Graph (MNCG - macro data flow graph generated for one iteration of the FDTD problem). The MDFG can be "compressed" into a MNCG because in each iteration, the pattern of data dependency is the same. So, in our algorithm, we can analyze only data dependencies in one iteration. After that, we identify the set of all cliques existing in the MNCG. A clique is a set of all macro nodes that are directly connected by edges with one, "central" macro node.

The main idea of this phase of the algorithm is based on equalizing execution time among all cliques. To achieve it, we will redeploy chosen subsets of computational cells between two selected cliques. The number of cells that can be redeployed is determined by the difference between average execution time of one clique and average execution time of one step in the MDFG. The whole redeployment phase is constantly monitored by execution time checking, so our optimization method in a step-by-step way tries to find the best data distribution among the executive processors, in which a given simulation problem will be performed.

Id	Rule priority	Description
MR0	Computational load balancing	Two least loaded adjacent nodes will be merged
MR1	Computational load balancing	The most loaded node will be merged with the least loaded adjacent node
MR2	Computational load balancing	The least loaded node will be merged with the most loaded adjacent node
MR3	Communication optimisation - edge cut reduction	Two most loaded adjacent nodes will be merged
MR4	Communication optimisation - edge cut reduction	The node with the biggest communication volume will be merged with its neighbour with the biggest communication volume
MR5	Communication optimisation - edge cut reduction	The node with the smallest communication volume will be merged with its neighbour with the biggest communication volume
MR6	Communication optimisation - edge cut reduction	The node with the smallest communication volume will be merged with its neighbour with the smallest communication volume
MR7	Communication optimisation - edge cut reduction	The node with the biggest communication volume will be merged with its neighbour with the biggest communication volume
MR8	Computational load balancing with edge cut reduction	The least loaded node will be merged with the adjacent node with the biggest communication volume
MR9	Computational load balancing with edge cut reduction	The least loaded node will be merged with the adjacent node with the lowest communication volume

Table 1. Chosen Merging Rules for Redeployment Algorithm.

The optimization algorithm requires precising three input parameters which describe problem configuration: a macro data flow graph, the speed of computational node and the throughput of available communication system. A simplified scheme of the redeployment algorithm is presented in Fig. 5. After all redeployment steps, we must re-compute an execution time for a modified MNCG graph. If new execution time is bigger than the old one, the redeployment is not profitable, and two cliques chosen for the redeployment operation are marked as examined. Because moving the computational cells between these two cliques is not profitable, they will not be chosen for next redeployment operation. If the execution time is better, the last redeployment operation can be validated. The optimization algorithm will be finished in the three following cases:

1. All cliques are marked as examined – it is not possible to perform any new redeployment operation.

2. The assumed number of consecutive redeployment steps, does not produce any increase of speedup – it is not possible to perform any redeployment with success (it is useful for large number of cliques).
3. The *CliqueB* parameter is close to 1, where *CliqueB* is a value, which determines clique balance in the MNCG graph, computed from the following equation:

$$CliqueB = \frac{absolute(L - S)}{total\ number\ of\ cliques}$$

where: L (S) is the number of cliques whose maximal execution time is larger (smaller) than the average execution time for the MNCG. If none of these conditions are met, then next two cliques are chosen and the redeployment step is repeated.

4. CDC - partitioning algorithm

The CDC (Connectivity-based Distributed Node Clustering) algorithm is a graph partitioning algorithm which is used to divide peer to peer networks into a given number of clusters. Unlike the redeployment method, the CDC is a decentralized algorithm. In this algorithm, only the nearest vicinity of nodes is needed to be analyzed to perform efficient partitioning operation whereas in the redeployment algorithm we must know the shape of whole computational area.
The CDC algorithm consists of two phases:

1. Phase 1 - choosing leader nodes (called originators (Ramaswamy et al., 2005)) – it can be done similarly as in the redeployment algorithm, but here the number of originators must be exactly equal to the number of computational nodes. In order to fulfill this condition, we have implemented another method of choosing originators. In our new method, we sort all computational cells by their coordinates (first by Y co-ordinate and after by X coordinate). After that we set every P cells to be originators, where

$$P = \frac{total\ number\ of\ computational\ cells}{number\ of\ processors}$$

2. Phase 2 - It is an iterative phase. It begins, when each node, which was previously chosen as an originator, sends a messages to its neighbor nodes. In all next iterations, each node that has received any messages in the previous iteration, re-sends the messages to their neighbors. Each message consists of following attributes:
   - source originator id (OID) – it indicates also unique identity of a cluster. This attribute is set by originator node, and it cannot be changed by any other nodes.
   - weight (W) – it describes a distance from originator node with a given OID number. The originator node sets W to 1.
   - time to live (TTL) – it describes, how many times this message will be resent.

The whole optimization process is parameterized using several factors. The most important are: MinWeight and MaxTTL. Both of them are used to reduce the total number of messages during the execution of the CDC algorithm. Parameter MaxTTL is an initial value for TTL attribute of each message. As value of MaxTTL increases, the number of generated messages increases as well. MinWeight parameter determines the minimal value for a message. If attribute W in a message M is smaller than MinWeight parameter, the message M will be

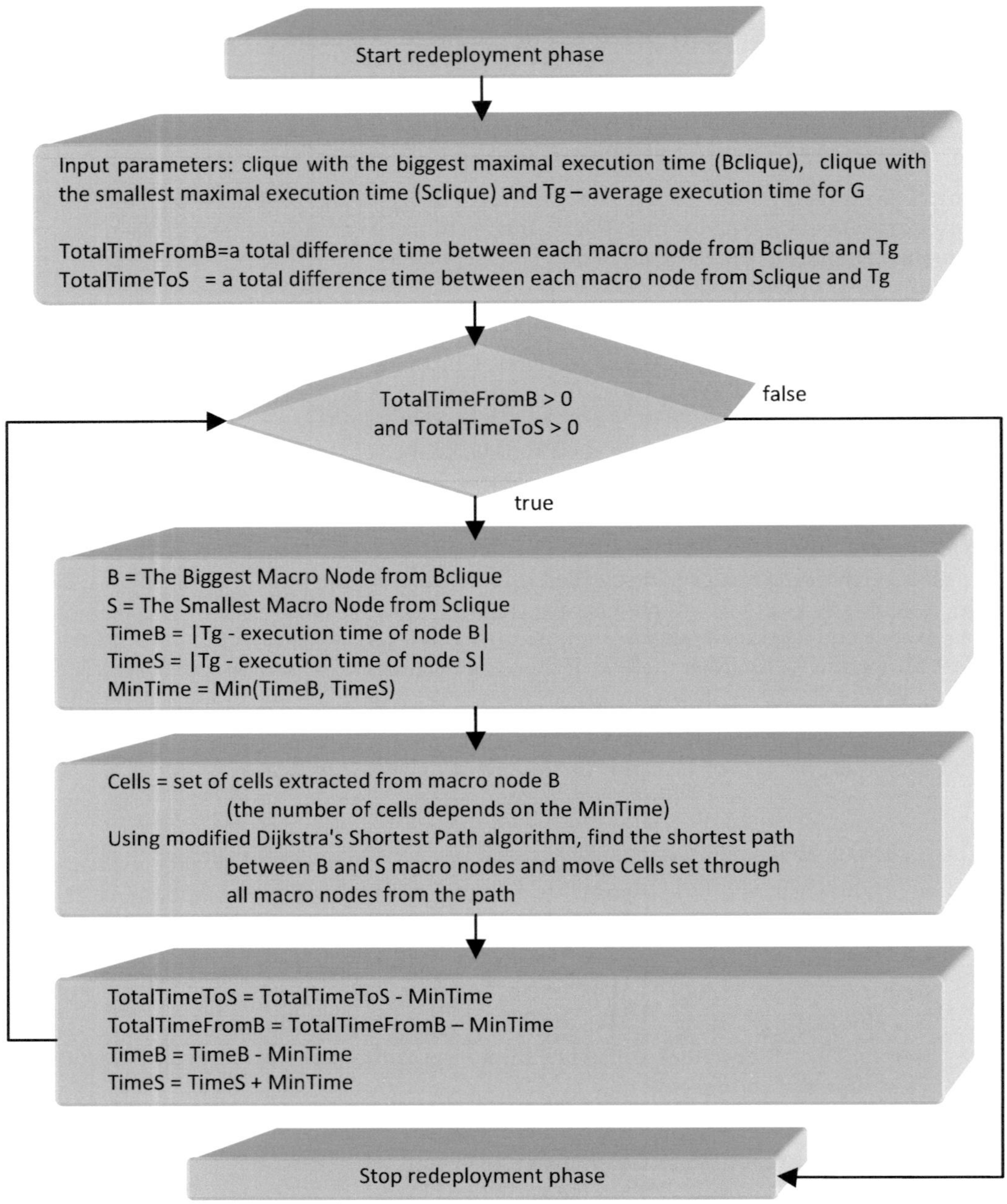

Fig. 5. Scheme of the Redeployment Phase of the Algorithm.

```
Cell C receives a message M
Table T on cell C is updated T[M.OID]= T[M.OID]+ M.W
A new message (Mnew) on the cell C is created
Mnew.OID = M.OID
Mnew.TTL = M.TTL-1
Mnew.W = Mnew.W/(Degree of cell C)
if Mnew.TTL>0 and Mnew.W>= MinWeight then
    the message Mnew will be re-sent to all neighbors of C cell
if Mnew.TTL<0 or Mnew.W<MinWeight then
    the message Mnew will be destroyed
if any messages exist then
    goto 1
```

Fig. 6. Second Phase of the CDC Algorithm.

destroyed. Each node contains a table T with total sum of weights from all received messages sourced by originators. The CDC algorithm in phase 2 is presented in Fig. 6. Phase 2 is repeated until all messages are destroyed. After that, each cell is attached to a cluster C, where C is equal to the index of the element from T with maximal value.

5. Experiment results

Both of these methods have been implemented, executed and tested for several shapes of simulation area and for several system configurations. To perform it we have introduced a simple architectural model of the executive system (Bharadwaj et al., 1996). It includes homogenous multiprocessor systems and is described by two values: processor computational speed, communication performance, see Table 2. The RB RDMA (Rotating Buffers-based Remote Direct Memory Access) is a special kind of communication facility. It enables data transmissions with a very small engagement of processor time (Smyk & Tudruj, 2003; 2004; Hitachi, 1997), which is usually done in the background of computations. The transmissions proceed without any data buffering by the operating system, so the overheads of this kind of communication is very small.

5.1 Rotating buffers infrastructure RB RDMA

The logical structure of the memory used in the rotating buffers facility is presented in Fig. 7. The RB RDMA infrastructure was designed for Hitachi SR2201 supercomputer, but it can be easily adapted to various distributed parallel computational systems like Cell/BE PS3 for example or even to cluster systems with network cards supporting DMA communication (or other compatible).

Local memory of a computational node is logically divided into two parts *LAM* (*Locally Accessed Memory*) with data used only for local computations with direct access from application program level and *GAM* (*Globally Accessed Memory*) with data used for data exchange. Access to *GAM* is available (from local and remote sites) only through the rotating buffers memory infrastructure.

GAM area is divided into *N* separate sub-area pairs: *RCA* (*Remote Confirmation Area*) and *RDM* (*Remote Data Memory*), where *N* is the number of remote processors. Each pair is used to perform communication between two given remote processors. The numbers of rotating buffers in the send and receive memory parts are fixed and denoted by *NSB* and *NRB*,

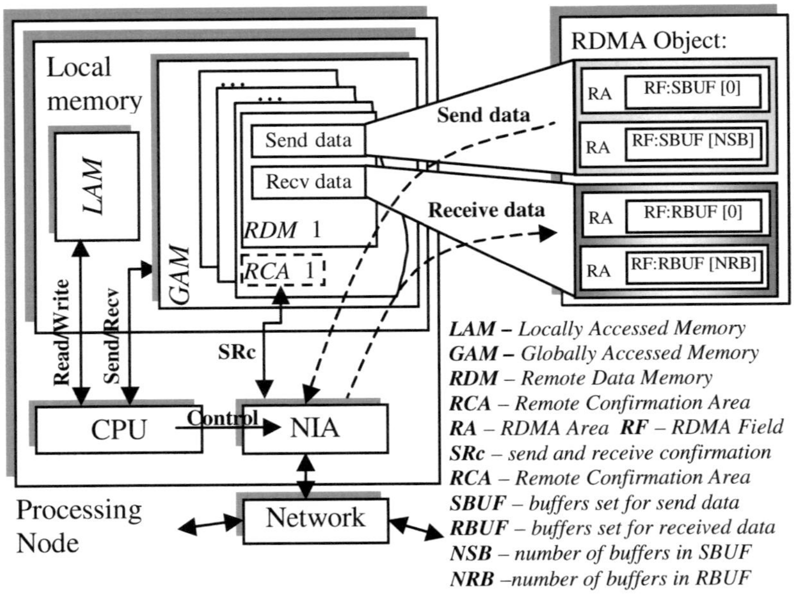

Fig. 7. Memory Structure in Rotating-Buffers Method (For One processing Node on the HITACHI SR2201 supercomputer).

respectively. All buffers which are defined in a *RDM* area are used only for data transmission. To avoid possible data overwriting, an additional control has to be introduced. This control is based on the *RCA* areas which are assigned independently to each *RDM* area. Each *RCA* area is intended to send and receive additional control messages which determine if the buffers from a *RDM* are ready to receive new data. A *RCA* consists of only two sets of one buffer each (by analogy to a *RDM* where the number of buffers can change, here the *NSB* and *NRB* numbers are always equal 1). They are used to exchange synchronized messages between two processors.

The control flow in the rotating buffers method for one processing node is presented in Fig. 8. Data are exchanged between a local processor and a remote processor *K*. On both these nodes, the described above control and communication infrastructure was created. Additionally, for each processor two pointers *K.SPTR* and *K.RPTR* are created. They are used to indicate a next free buffer in which new data (to be sent to processor *K*) will be placed (*K.SPTR*) or new data just received from *K* will be written (*K.RPTR*). These two pointers determine control for a rotating access to available buffers and introduce periodical synchronization between two communicating nodes, which assures that no data which are transferred from one node to another, will be lost (overwritten).

5.2 Redeployment algorithm efficiency

First, we have tested the speedup of macro data flow graph execution after program profiling without and with the use of cell redeployment optimization algorithm, see Figure 9.

We can observe that the best speedup was obtained for computational systems with shared memory (FF, MF, SF) and in two cases with RB RDMA communication (MM, SM). It is independent of the number of processors, with or without redeployment of computational nodes, and even of the efficiency of a single computation node. We can see the significant

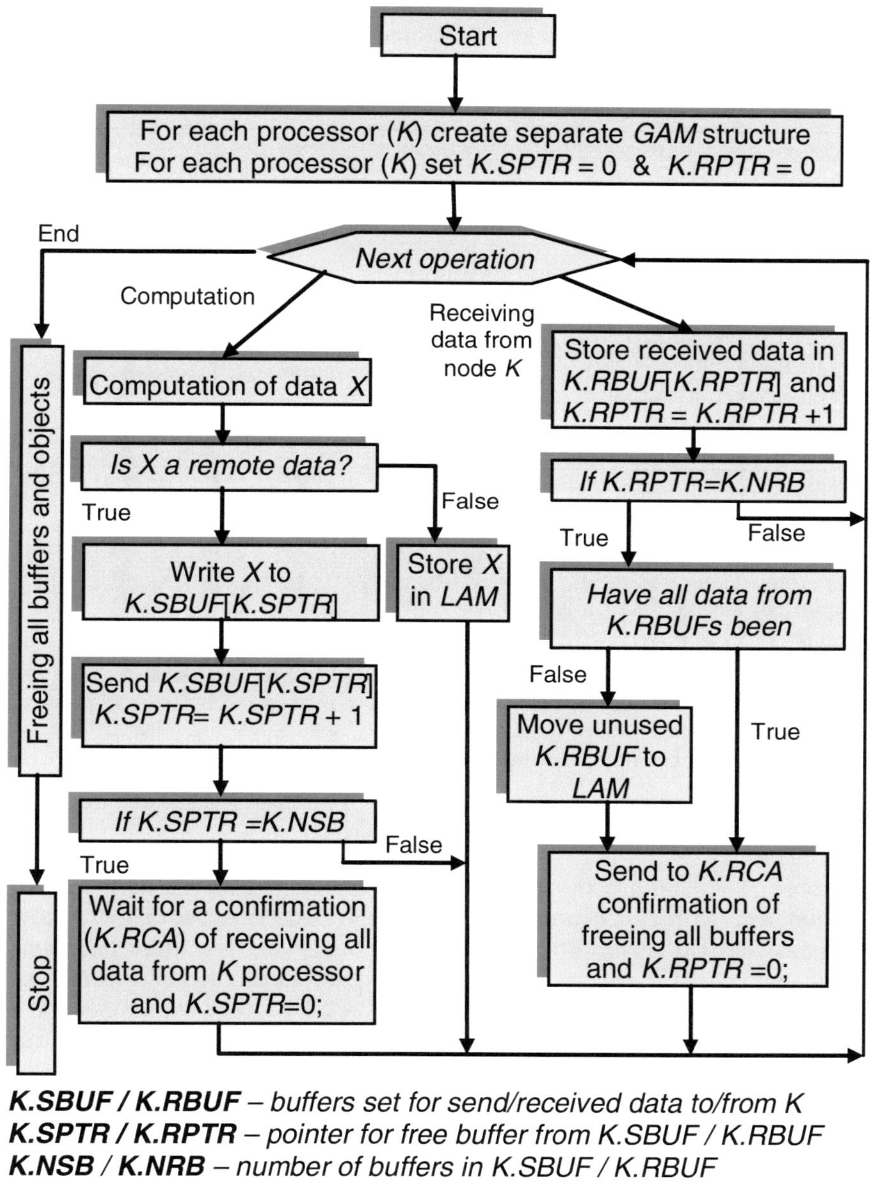

Fig. 8. Control Flow of the Rotating-Buffers Method for One Processing Node.

Architecture symbol	Computational speed of a single processor	Communication speed between two processors
FF	Fast – 1GFlops	Fast – Shared memory
FM	Fast – 1GFlops	Medium – RB RDMA
FS	Fast – 1GFlops	Slow – MPI
MF	Medium – 0.3 GFlops	Fast – Shared memory
MM	Medium – 0.3 GFlops	Medium – RB RDMA
MS	Medium – 0.3 GFlops	Slow – MPI
SF	Slow – 0.02 GFlops	Fast – Shared memory
SM	Slow – 0.02 GFlops	Medium – RB RDMA
SS	Slow – 0.02 GFlops	Slow – MPI

Table 2. Symbols and Architectural Model Parameters.

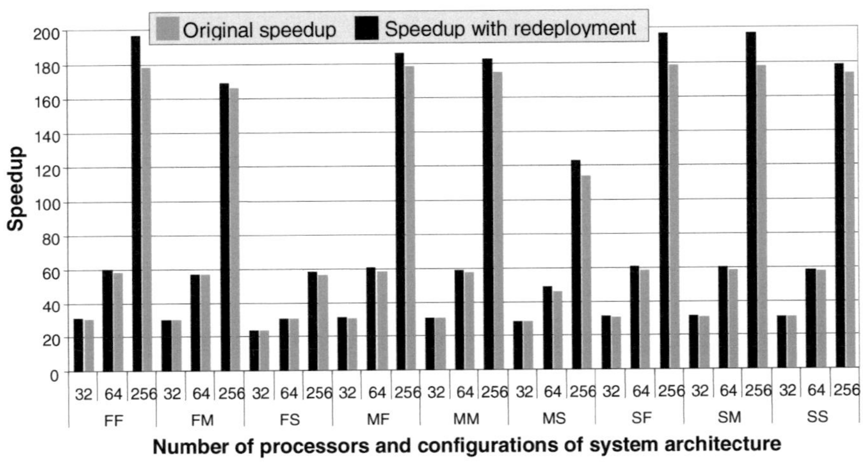

Fig. 9. MDFG Execution Speedup Versus Single Node Execution for Various Computational System Configurations.

increase of speedup for 256 processors. We can also notice that for systems with RB RDMA communication and with fast processors (FM) slightly worse speedup was obtained. It can be explained by bad relation between computational and communication parameters of the system, which could not match program requirements. It is especially noticeably in the case of systems with slow communication (FS, MS) where sometimes the speedup decreased dramatically. We can observe also that a noticeable increase of speedup (~10% for 256 processors) appears when a redeployment operation has been applied. It means, that the balancing of execution time is possible only for configurations with large number of processors and with relatively fast communication systems.

5.3 Redeployment algorithm versus CDC

In next experiments, we have compared the efficiency of the redeployment algorithm with our implementation of the CDC algorithm. The results are presented in Figure 10.

We can see that the speedup of CDC in almost all cases has been much better than that of the redeployment optimization what would indicate the superiority of the CDC algorithm.

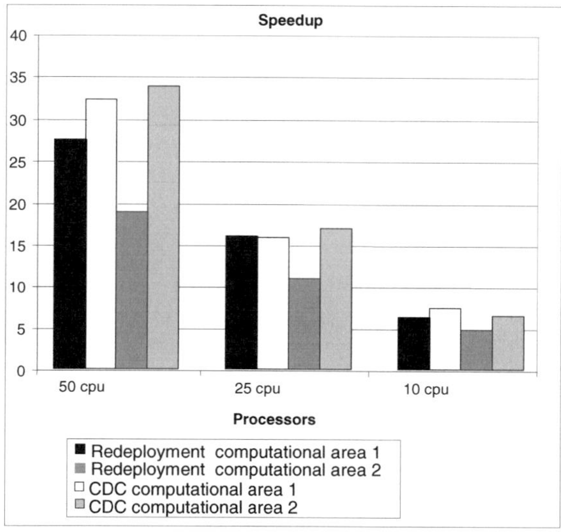

Fig. 10. Comparison of Speedup for the Redeployment Method with the CDC Algorithm.

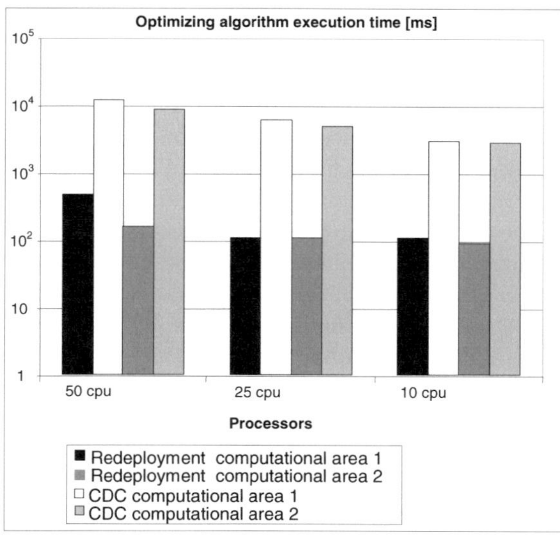

Fig. 11. Comparison of the Total Optimization Execution Time for Redeployment and CDC Algorithm.

Unfortunately, we have observed some unfavorable features of this algorithm. The first one is that the execution time of this algorithm (Figure 11) was from 25 to 50 times longer in comparison to redeployment algorithm. It can completely eliminate the use of CDC algorithm for computational areas with large number of cells. In our experiments, we could not simulate areas larger than 1000 cells. It is because large number of messages are generated. We observed also that CDC algorithm is very sensitive to two simulation parameters: MinWeight and initial value of TTL time. These parameters are decisive for messages lifespan and unfortunately, their values considerably depend on the shape of computational area.

6. Hierarchical algorithm

Based on the results presented in the previous section, we can observe that the redeployment algorithm is much faster than the CDC algorithm. The convergence time of the redeployment algorithm is almost two times shorter in comparison to CDC. However, for system configurations with a big number of processors, the parallel speedup of the FDTD programs obtained using the redeployment method is lower than that obtained with the use of the CDC algorithm. A difficult problem in CDC is setting the initial values of the parameters MinWeight and TTL time. It is especially true for a large number of executive processors. Taking into considerations all pros and cons of the two methods we propose a hierarchical method of FDTD program optimization (Redeployment with Connectivity-based Distributed Node Clustering - RCDC). This method consists of two main steps. In the first step, we apply the standard redeployment algorithm. In the second step, we switch to a modified CDC method. The standard redeployment algorithm step is used here to reduce the number of nodes in the input data flow graph (in fact to reduce the optimization time), while the CDC algorithm step will be exploited to obtain the best possible parallel simulation speedup.
Efficiency of the RCDC algorithm strongly depends on the choice of the time point in which switching between these two algorithms should take place. Because the parallel simulation speedup provided with the CDC algorithm is usually significantly better than that obtained by the redeployment, we decided to modify the standard redeployment algorithm so as to execute it in two phases: generation of an initial MDFG based on wave propagation area partitioning and the MDFG nodes merging with load balancing to obtain the given number of macro nodes. In fact, the number of macro nodes obtained in the second phase is from 5 to 20 times bigger than the assumed number of executive processors. The final reduction of the number of macro nodes together with communication optimization (to minimize and balance internodes data transmissions.) will be performed during the CDC step. In our implementation, the CDC algorithm does not take into account load balancing in the executive processors, so after this step, some load imbalance is possible. To avoid it, we have introduced the last phase with redeployment of computational cells.
The final scheme of the RCDC algorithm is as follows (see Fig. 12):

1. Generation of initial MDFG based on wave propagation area partitioning;
2. MDFG nodes merging (with initial load balancing) and optimization with re-deployment of computational cells;
3. CDC optimization (communication optimization).

A crucial problem in the RCDC algorithm is a decision when we should switch from phase 2 to 3. To answer this question, we can consider several independent methods. In the first method, we can compute the diameter of our MDFG and if it corresponds (it means that the most distant macro nodes can send and receive messages to/from each others in the

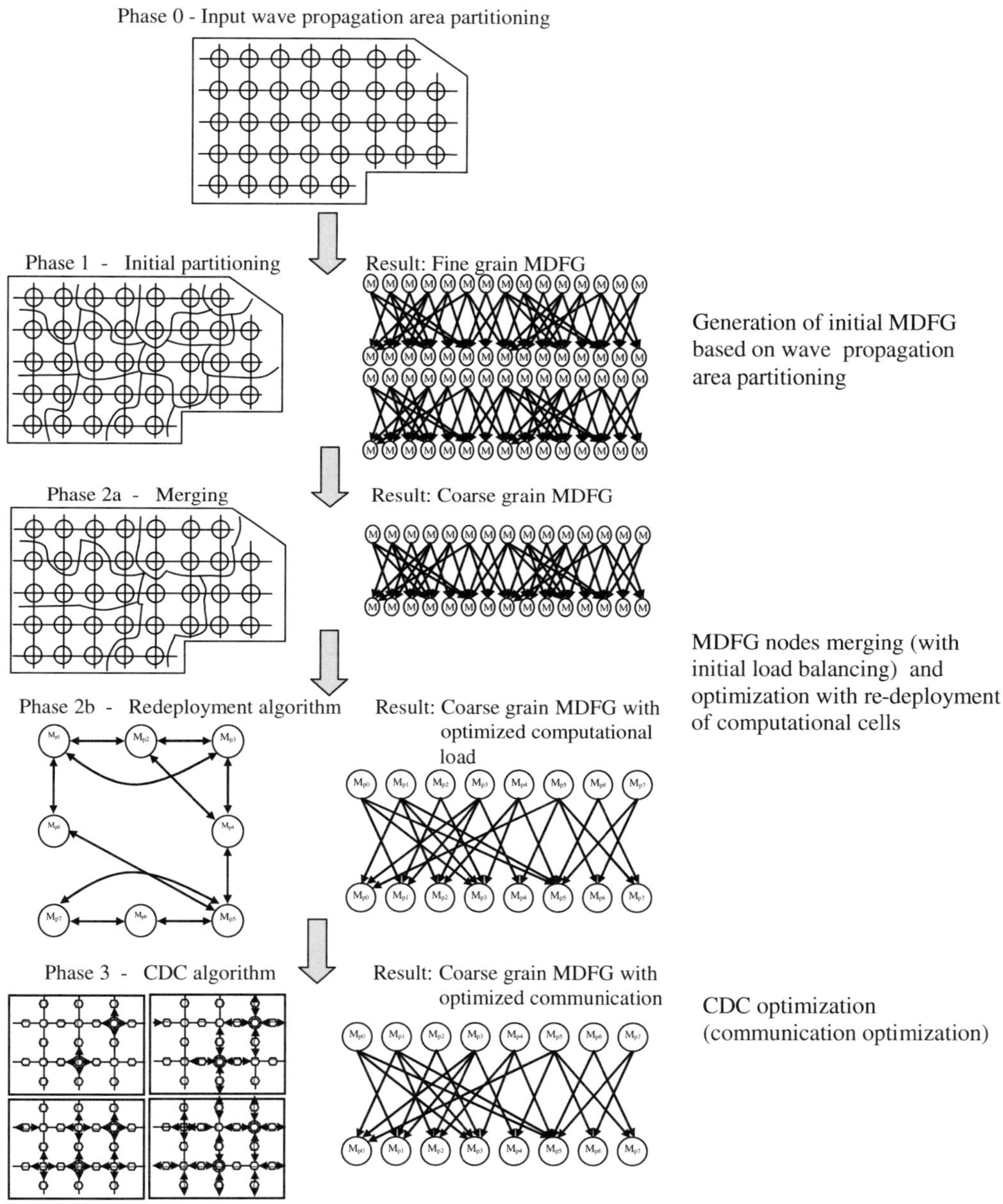

Fig. 12. RCDC Hierarchical Algorithm Scheme.

Problem configuration	Computational area size	Computational area type	Number of processors
1	Small	Regular	50
2	Large	Irregular	50
3	Small	Regular	25
4	Large	Irregular	25
5	Small	Regular	10
6	Large	Irregular	10

Table 3. Problem Configurations Used in the Experiments.

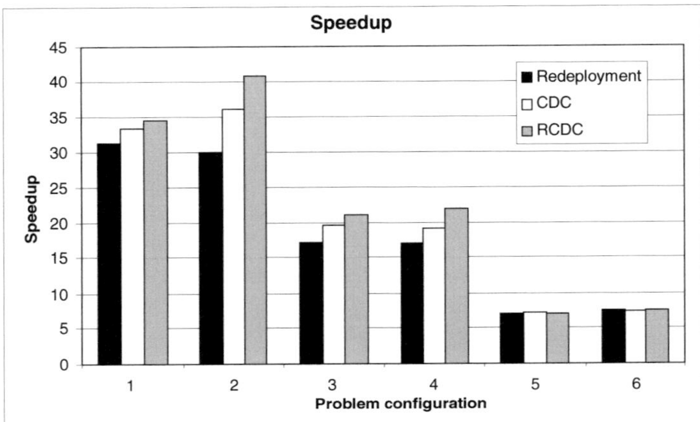

Fig. 13. Comparison of the Parallel Simulation Speedup for Redeployment, CDC and RCDC Algorithms.

sense of the CDC algorithm) with MinWeight and/or TTL time parameters, we can start phase 3. This method needs some additional computations but its estimation can be precise only for regular graphs. In case of very irregular graphs, it can distort information about the graph structure, which can cause that the optimization time for phase 3 will be much bigger. In the second method, we can assume a time of the optimization after which phase 3 is entered. In the third method, we can assume a static number of macro nodes created in our MDFG after which the phase 2 will be terminated and a passage to phase 3 will take place. In our implementation, we have assumed the use of this method. It does not introduce any additional costly computations in phase 2 and it allows us to find a coarse estimation of the parameters (MinWeight, TTL time) needed for phase 3. The number of macro nodes obtained in phase 2 is very important for the rest of the algorithm. If it is too large, the total optimization time will take too much time, and if it is too small, the obtained speedup can be unsatisfactory. We have assumed that dependently on the structure of the MDFG (regular or irregular), switching from phase 2 to 3 is done when the number of macro nodes is from 5 to 20 times bigger than the assumed number of computational macro nodes.

As it can be expected, the RCDC hierarchical method produced better partitioning in comparison to redeployment and the CDC methods. It is especially visible for configurations with large number of processors (see Figure 13, configurations 2 and 4).

For small computational areas (configurations 5 and 6), no significant differences between all discussed algorithms are observed. Additionally, we can see that when the number

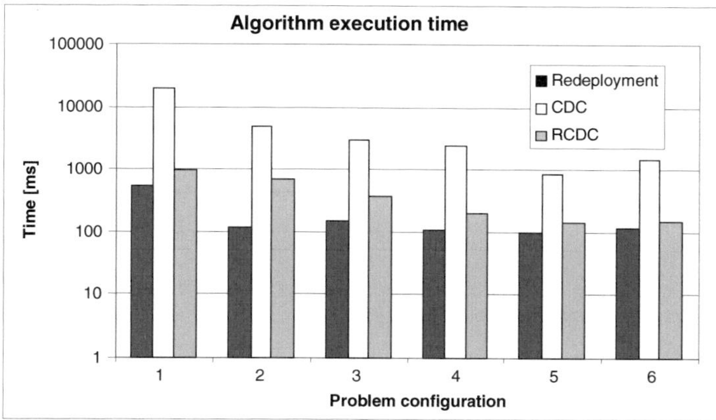

Fig. 14. Comparison of the Algorithm Execution Time for Redeployment, CDC and RCDC Algorithms.

of processors grows, the RCDC algorithm has much more degrees of freedom and it can produce better partitioning. Even though, in our implementation of the CDC method, we have not considered problems concerned with load balancing, the partitioning efficiency was satisfactory. It is because, the granularity of the input graph was very fine and computational load in each cell was equal to others. In case of the RCDC, we must carefully switch between the redeployment and the CDC algorithm phases. The redeployment algorithm will be stopped if the number of macro nodes in MDFG divided by k (where k is a positive integer) is equal to the number of processors. If it is fulfilled, we can switch to phase 3 of the RCDC algorithm. The execution time of the optimization by the RCDC algorithm is much shorter than CDC method and slightly bigger than the redeployment algorithm, see Figure 14. It can be dynamically adjusted to the k value. When the k value grows, the behavior of the RCDC algorithm is closer to the CDC method. Otherwise it works like the redeployment algorithm.

7. Conclusions

In this chapter, we presented a hierarchical approach to the optimized program macro data flow graph design for execution of FDTD simulations in parallel systems. The presented RCDC algorithm combines two independent methods for the FDTD data flow graph optimization: the cell redeployment and CDC algorithm. There are several differences between these two methods. The first method is fully centralized and the macro data flow graph is created in three phases: computational area partitioning, merging and redeployment. The CDC method is decentralized with only local current knowledge of the simulation area. In the RCDC algorithms we wanted to merge both of these methods in order to obtain more efficient parallel simulation speedup (comparable to the speedup obtained in the CDC) and to shorten the execution time of the optimization. It turned out that such a hierarchical combination of the two algorithms has improved partitioning of data flow graphs for the FDTD problem and additionally such a hierarchical optimization takes significantly less time than the CDC method.

8. References

Lin, H.X., van Gemund, A.J.C. & Meijdam, J. (1996). Scalability analysis and parallel execution of unstructured problems, *Eurosim'96 Conference.*

Garey, M., Johnson, D. & Stockmeyer L. (1976). Some simplified NP-complete graph problems, *Theoretical Computer Science*, 1, pp. 237-267.

Khan, M.S. & Li, K.F. (1995). Fast Graph Partitioning Algorithms, *Proceedings of IEEE Pacific Rim Conference on Communications, Computers, and Signal Processing*, Victoria, B.C., Canada, May 1995, pp. 337-342.

Kerighan, B.W. & Lin, S. (1970). An efficient heuristic procedure for partitioning graphs, *AT&T Bell Labs. Tech. J.*, 49:291-307, February 1970.

Kirkpatrick, S., Gelatt, C.D. & Vecchi, M.P. (1983). Optimization by simulated annealing, *Science*, 220(4598):671-680, May 1983.

Karypis, G. & Kumar, V. (1995). Unstructured Graph Partitioning and Sparse Matrix Ordering, *Technical Report*, Department of Computer Science, University of Minesota, 1995 (http://www.cs.umn.edu/~kumar).

Dutt, S. & Deng W. (1997). VLSI Circuit Partitioning by Cluster-Removal using Iterative Improvement Techniques, *Proc. IEEE International Conference on Computer-Aided Design*, pp.350-355.

Stone, H.S. & Bokhari, S.H. (1978). Control of distributed process, *IEEE Computer*, 11(7):97-106, July 1978.

Fiduccia, C.M. & Mattheyses, R.M. (1982). A Linear Time Heuristic for Improving Network Partitions., DAC, pp. 175–181.

Bharadwaj, V., Ghose, D., Mani, V., Robertazi T.G. (1996). Scheduling Divisible Loads in Parallel and Distributed Systems, *IEEE Computer Society Press*, Los Alamitos, California.

Smyk, A. & Tudruj, M. (2003). RDMA Communication Based on Rotating Buffers for Efficient Parallel Fine-Grain Computations, *PPAM 2003*, Czestochowa, Poland.

Smyk, A. & Tudruj, M. (2004). RDMA Control Support for Fine-Grain Parallel Computations, *PDP 2004*, La Coruna, Spain.

Hitachi Ltd. (1997). HI-UX/MPP - Remote DMA -C- User's Guide Manual Number: 6A20-3-021-10(E), Second Edition: January 1997.

Ramaswamy, L.; Gedik, B.; Liu, L. (2005). A distributed Approach to Node Clustering in Decentralized Peer-to-Peer Networks, *IEEE Transactions on Parallel and Distributed Systems*, Vol.16, No.9, September 2005.

Smyk, A. & Tudruj, M. (2006). Parallel FDTD Computations Optimized by Program Macro Data Flow Graph Redeployment, *PARELEC 2006*, Bialystok, Poland, September 2006.

Walshaw, C., Cross, M., Everett, M.G., Johnson, S. (1995). JOSTLE: Partitioning of Unstructured Meshes for Massively Parallel Machines, *Parallel Computational Fluid Dynamics: New Algorithms and Applications 1995*, Elsevier.

http://www.labri.fr/perso/pelegrin/scotch.

http://glaros.dtc.umn.edu/gkhome/views/metis.